MITCHELL'
CONSTRUCTION
1

Volume 1

# BUILDING
# CONSTRUCTION AND
# DRAWING 1906

# BUILDING CONSTRUCTION AND DRAWING 1906

A Textbook on the Principles and Details of Modern Construction

First Stage (Or Elementary Course)

CHARLES F. MITCHELL

LONDON AND NEW YORK

First published in 1906 by B. T. Batsford.

This edition first published in 2022
by Routledge
4 Park Square, Milton Park, Abingdon, Oxon OX14 4RN

and by Routledge
605 Third Avenue, New York, NY 10158

*Routledge is an imprint of the Taylor & Francis Group, an informa business*

*British Library Cataloguing in Publication Data*
A catalogue record for this book is available from the British Library

ISBN: 978-1-032-19904-7 (Set)
ISBN: 978-1-032-19906-1 (Volume 1) (hbk)
ISBN: 978-1-032-19920-7 (Volume 1) (pbk)
ISBN: 978-1-003-26147-6 (Volume 1) (ebk)

DOI: 10.1201/9781003261476

**Publisher's Note**
The publisher has gone to great lengths to ensure the quality of this reprint but points out that some imperfections in the original copies may be apparent.

**Disclaimer**
The publisher has made every effort to trace copyright holders and would welcome correspondence from those they have been unable to trace.

# BUILDING CONSTRUCTION AND DRAWING

## A TEXT BOOK ON THE PRINCIPLES AND DETAILS OF MODERN CONSTRUCTION

Compiled to assist Students preparing for the Examinations of the Board of Education, the Royal Institute of British Architects, the Surveyors' Institution, the City Guilds, the Civil Service, and other Technical Examinations.

BY

## CHARLES F. MITCHELL,

*Lecturer on Building Construction to the Regent Street Polytechnic, London, Head Master of the Polytechnic Technical School.*

(MEMBER OF THE SOCIETY OF ARCHITECTS, MEMBER OF THE ARCHITECTURAL ASSOCIATION, ASSOCIATE OF THE SANITARY INSTITUTE, AND MEMBER OF THE SOCIETY OF ARTS.)

ASSISTED BY

## GEORGE A. MITCHELL,

*Honours Medallist Building Construction, National Silver Medallist in Carpentry and Joinery, Bronze Medallist in Brickwork, Masonry and Plastering, &c.*

## FIRST STAGE (OR ELEMENTARY COURSE).

### Seventh Edition, thoroughly Revised & much Enlarged.

(SEVENTY-SEVENTH THOUSAND.)

With about 1,100 Illustrations.

LONDON:

B. T. BATSFORD, 94 HIGH HOLBORN.

1906.

# Mitchell's Building Construction and Drawing Series 1881 – present day

## The Preservation of Construction Knowledge

**Preface to the Reissue of the 1906 edition by Stephen J. Scaysbrook, FCIAT ACIOB.**

Ever since the initial issue of Charles Frederick Mitchell's *Building Construction and Drawing* books in 1881, no other series of books has had such a continued presence in the construction industry. For almost 140 years, students and practitioners have referred to both the *First Stage or Elementary Course* and the *Advanced and Honours Courses* as the go-to reference works for construction detailing and materials.

*Figure* Charles F Mitchell

Through a prestigious career as a lecturer at The Polytechnic Institute, Regent Street, London, which became known as the Regent Street Polytechnic and finally Westminster University, Charles F. Mitchell (1859–1916) was the senior lecturer.

Together with his younger brother George A. Mitchell they embarked on a 30-year project of writing, updating, and modifying what became the two separate construction books, published initially by Batsford, and now Routledge.

The series grew from the initial single volume in 1881 to a two-volume issue in 1893/4 and expanded very quickly from two books into a collection of support issues by specifically qualified authors, dealing with structure, environment & services, components & finishes, with a specific Issue of 40 plates of the detailed drawings used in many of the releases, as well as other more specific books on carpentry and bricklaying, offering a complete cross discipline guide to construction technology.

The book series was originally aimed at students attending the Regent Street college, but quickly spread with the current editions aimed at construction technology students, architects, architectural technologists, building surveys, & quantity surveyors, who are studying construction technology as part of a qualification route to a degree level attainment and above.

The above list of professions is by no mean the limit to the reader base, construction managers, property developers, planners and civil engineers all benefit from reading parts, if not all, of the current editions and adding it to their bibliography and personal knowledge base.

Mitchell continued the development and updating of his series until his death in 1916 when his younger brother George continued his legacy and continued to expand and correct the series until his death in 1953. It is important to note that the issue numbers of the two separate books titles do not correspond to each other with updates to the *First Stage or Elementary* editions being slower than that of the *Advanced*. Since 1953 several industry professionals took over the helm updating and adding new content.

The importance of these books to the construction industry as a description of construction practice at specific points in history is now beginning to be appreciated by modern-day architects, technologists and building surveyors as they prepare to update or repair existing buildings from the late 1890's.

The materials used during this period differ in both the way they were made, and installed, and they performed differently on site. One of the main purposes of this Reissue is to make current designers and technologists aware of these differences in terms of the internal as well as the external climate which affected buildings and their construction.

Since 1906, the weather, although not that dissimilar to 2021, should be considered as having waged over a century of continual warfare against the materials, changing them in so many ways: From the mortars used in the often-solid wall construction, and the mostly non-insulated environment. The weather, together with coal-fired heating on an almost room by room basis, released sooty smoke that smothered big cities and coated buildings in grime, further adding to the weathering.

The Mitchells series of books from this period give both an understanding of design and construction for this period, but also show the way to upgrade and repair historic structures, which demands an understanding of the manufacture of the materials used at the time, their installation and even more important, the ageing of the individual material over the lifetime of the building. This is individual to each project, and its position in the building, internal or external and its orientation to the sun.

Although this reissues the 1906 edition, each edition offers a similar insight to that particular period of construction history, showing materials and construction of that age, offering professionals a unique timeline of material development and construction technology. The highly regarded Mitchell's series, and this Reissue in particular, is an invaluable tool in helping the modern day architect to understand how a change of use to a historic building will impact on the materials used in its construction: The original gentleman's residence, (a somewhat draughty environment) with its open fires, later converted to gas central heating, with windows and doors sealed, inevitably requires the building and its construction materials to change according to the new inhabitants and the building's purpose.

The explosion of distinguished buildings for large corporations and businesses after the death of Queen Victoria, and the arrival of Edward VII left behind a legacy of architecture which Charles Mitchell would have been well aware of, working in the centre of London amidst the ongoing construction of so many notable buildings. His 1906 *Advanced and Honours* edition reflected this in the many revisions made from previous editions, both in the written text and the addition of so many well-drawn technical details which expand the detail described.

It is this area of technical drawing that so many modern-day technologists, surveyors and architects fail to understand, relying upon modern CAD to draw with and communicate the desired detail, little realising the constraints CAD puts upon free-thinking. It is of no surprise to see Mitchell's *First Stage or Elementary* editions starting with a detailed explanation of the instruments and methods of drawing with pencil ink and tracing paper, emphasising the need to learn basic drawing skills and the need to think about a detail and the materials used to create a detail capable of lasting well over 100 years or more. The simple act of making a scale from a drawing with only one dimension may be lost to modern-day students, but not to Charles Mitchell, who describes the method and its use.

1906 was a year of great upheaval, Edward VII had succeeded his mother Queen Victoria to the throne, and he was embarking upon and encouraging a huge building programme not just within the United Kingdom but throughout the British Empire whose industry was then at its zenith, pouring out goods to the world, many of which are still in existence and used as they were originally designed, all be it with modern updates.

Despite political turmoil in Europe prior to the First World War Edward VII ushered in a new style of construction and elegance, with buildings of immense importance, such as Admiralty Arch, London (1912), Australia House, London (1918) and the Central Criminal Court (Old Bailey), London, by Edward William Mountford (1902–07) being built, with similar properties being erected in this period around the UK and abroad in the British empire with classic examples such as the Central Railway Station, Sydney, Australia, the Birkbeck Building, Toronto, Canada and the Ripon Building, Chennai, India. Together with many local town halls, civic buildings and notable manufacturers like Alfred Bird who built his then modern factory in Digbeth, Birmingham.

Mitchell successfully includes many of the details, materials and concepts used during this global expansion of Classic Edwardian Architecture, in this 1906 edition. Later editions continued this legacy of updating and expanding the series to its current 9 volumes and taking into account new materials and methods of construction and technology advancement.

To preserve the ideals of the book and make it available to researchers, students and professionals, a sample year, 1906 has been chosen to represent the invaluable information contained in this unique series, with the preservation of the hardbacked print versions as facsimile editions and ebooks. They will be of use to professionals as a reference in research on specific topics of construction used in a specific time frame.

Many similar books have been created, all trying to fill in gaps left by the series, but never surpassing the continuity and depth of the Mitchells series that offers such a detailed insight into construction techniques and materials, from the original 1881 first edition to that of the present day. The complete series offers an unprecedented insight into construction covering over 140 years of construction development. This Reissue of Mitchell's 1906 *First Stage or Elementary* and *Advanced and Honours* editions will add a valuable aid to building pathology, allowing students and practitioners to research construction methods and materials pertinent to the period.

In addition to the Reissue of the 1906 edition, the current nine books, www.routledge.com/Mitchells-Building-Series/book-series/PEAMIBS are being updated and expanded to take account of new construction methods, and above all the transition to computer-aided design or CAD and its development in the communication of the design via BIM, and how data and sensors in the industrial 4.0 period, will aid the buildings use, and help owners run and maintain both their heritage and more modern buildings far more efficiently, against an ever-changing global climate.

To apply the lessons learned from the 1906 books, it is wise to try and visit some of the buildings that were designed and constructed during this period. There is so much for students and professionals to appreciate, from the elegance of the design to the detailing, that is still in current use, and has never been touched. Has it survived, and worked as originally intended, have alterations been made to replace or repair?

Observation is a skill that is so important to surveying. So much can be observed and noted, sketching trains the mind, and forces the observer to take time in looking at the detail. Photography is an easy route out, and offers clarity, but sketching offers so much more.

Stephen J. Scaysbrook, August 2021

# PREFACE

## TO THE SEVENTH EDITION.

———

The approval so universally accorded to this work by teachers and students has led to the rapid exhaustion of the Sixth Edition, a fact which is most gratifying to the Author.

The call for another edition of the work happily coincides with the same need in the case of the Author's "Advanced and Honours Course," and has thus enabled him to deal with the revision of both volumes as a coherent whole, thereby giving due regard to the requirements of each in relation to the course it covers.

The Author's endeavour has been to provide a concise Hand-book, sound in its enunciation of first principles and thoroughly abreast of current practice, which should be equally valuable as a guide for the student taking the examinations of the Board of Education and other bodies, and to the practical man engaged in building.

The thoroughness of the revision, with its consequent re-arrangement and enlargement, will be at once apparent to those familiar with the earlier issues. The more important alterations occur in the chapters dealing with Brickwork, Masonry, Girders, Partitions, Roofs, and Joinery, the majority of which have been wholly recast and greatly extended, while the remainder of the book has been subjected to a very careful revision and amplification where necessary. It should be added that new matter has been introduced with the sole object of helping the Student in his studies, and not merely of increasing the number of pages of the book.

It is the hope and wish of the Author that in its revised form the book may have an increased sphere of usefulness amongst those for whom it has been compiled.

In conclusion, the Author desires to acknowledge the valuable help he has received in the revision of the work from his colleagues at the Polytechnic, whose positions as specialists in the various sub-divisions of the building trades give a peculiar value to their services. In this connection he desires particularly to acknowledge the help rendered by Mr. T. Hobart Pritchard and Mr. A. E. Holbrow.

The Author's thanks are also due to the Publisher for numerous suggestions, and for the liberal way in which he has throughout seconded the efforts to improve the book.

CHARLES F. MITCHELL.

THE POLYTECHNIC INSTITUTE,
309, REGENT STREET, W.
*September*, 1906.

# CONTENTS.

CHAP.                                                                    PAGE
   I. INSTRUCTIONS FOR BEGINNERS...  ...  ...  ...  ...  1

  II. BRICKWORK  ...  ...  ...  ...  ...  ...  ...  17

 III. MASONRY  ...  ...  ...  ...  ...  ...  ...  .. 90

 IV. GIRDERS  ...  ...  ..  ...  ...  ..  ...  .. 152

  V. JOINTS IN CARPENTRY  ...  ...  ...  ...  ...  ... 173

 VI. FLOORS  ...  ...  ...  ...  ...  ...  ...  .. 200

 VII. PARTITIONS  ...  ...  ...  ...  ...  ...  ... 227

VIII. WOOD ROOFS  ...  ...  ...  ...  ...  ...  ... 245

 IX. COMPOSITE ROOFS  ...  ...  ...  ...  ...  .. 267

  X. IRON AND STEEL ROOFS..  ..  ...  ...  ..  ... 276

 XI. JOINERY  ...  ...  ...  ...  ...  ...  ...  ... 299

 XII. PLUMBING ...  ...  ...  ...  ...  ...  ...  ... 358

XIII. SLATING  ...  ...  ..  ...  ...  ...  ..  ... 382

XIV. BUILDING QUANTITIES AND MEMORANDA  ...  ...  ... 393

     EXERCISES ...  ...  ..  ...  ...  ...  ...  ... 412

## APPENDIX.

BOARD OF EDUCATION SYLLABUS  ...  ...  ...  ...  ... 414

EXAMINATION PAPERS OF THE BOARD OF EDUCATION  ...  ... 419

INDEX  ...  ...  ..  ...  ..  ...  ...  ...  ... 445

# LIST OF WORKS, ETC.

## REFERRED TO.

In preparing this treatise the Author is pleased to acknowledge his indebtedness to the Authors of the following works which he has consulted:—

| | | |
|---|---|---|
| "Chemistry of Building Materials" ... ... | Abney. |
| "Chemistry" (Matriculation) ... ... ... | Bailey. |
| "Gothic Architecture" ... ... ... ... | Bond. |
| "Graphic Statics" ... ... ... ... | G. S. Clarke. |
| "Plumbing Practice" ... ... ... ... | Wright Clarke. |
| "Modern Practical Joinery" ... ... ... | Ellis. |
| "History of Architecture" ... ... ... | Fletcher. |
| "Quantities" ... ... ... ... ... | Fletcher. |
| "Iron and Steel" ... ... ... ... ... | Greenwood. |
| "Vectors and Rotors" ... ... ... ... | Henrici. |
| "Architectural Surveyors' Handbook" ... ... | Hurst. |
| "Engineers' Year Book of Tables, Data, and Memoranda, 1906" ... ... .. ... | Kempe. |
| "Model Bye-Laws" ... ... ... ... | Knight. |
| "Quantity Surveying" ... ... ... ... | Leaning. |
| "Building Specifications ... ... ... ... | Leaning |
| "Reinforced Concrete" ... ... ... ... | Marsh. |
| "Plastering, Plain and Decorative" ... ... | Millar. |
| "Carpentry Workshop Practice" ... ... ... | Mitchell. |
| "Pocket Book of Engineering Formulæ" ... | Molesworth. |
| "Ferro-Concrete" (Hennebique) ... ... | Mouchel. |
| "Concrete, its Use in Building" ... ... | Potter. |
| "Applied Mechanics" ... ... ... ... | Rankine. |
| "Civil Engineering" ... ... ... ... | Rankine. |
| "How to Estimate" ... ... ... ... | Rea. |
| "Notes on Building Construction" ... ... | Rivington. |
| "Builders' Work" ... ... ... ... ... | Seddon, Col. H. |
| "Concrete" ... ... ... ... ... ... | Sutcliffe. |
| "Carpentry" ... ... ... ... ... | Tredgold and Hurst. |
| "Concrete Steel" ... ... ... ... ... | Twelvetrees. |
| "Testing of Materials of Construction" ... | Unwin. |
| "Instruction in Construction" ... ... | Wray, Col. H. |

Board of Education Examination Papers.
The "Builder."
The "Building News."
The "Builders' Journal."
Publications of the London County Council.
Building Laws of the New York City Council.
Publications of the R.I.B.A.
Publications of the Engineering Standards Committee

# CHAPTER I.

# INSTRUCTIONS FOR BEGINNERS.

---

*Instruments and Materials.*—The following instructions to beginners are among those given at the Building Classes held at the Polytechnic, Regent Street, London, W.

The student should bear in mind that it is better to have a few good instruments than a greater number that are inferior, and he would find it to his advantage to buy his instruments separately from some good maker or firm, rather than purchase a complete set of inferior quality, which will be sure ultimately to prove unsatisfactory. If the instruments are required to be carried about, a good way of keeping them in order is to have a roll made of chamois or wash-leather, with separate compartments for each instrument, such as are often made in leather for pocket sets of tools.

The following instruments should be obtained:—6-inch compasses with ink and pencil points, and with double-jointed legs and round dividing points (instruments with needle points for ordinary work should be avoided), dividers, ruling pen, with one blade hinged for facilitating cleaning, and two set-squares, 45° and 60° respectively. Set-squares are made of pear-wood, celluloid and vulcanite. Those of pear-wood are not to be recommended, as the angles are subject to alteration due to the shrinkage of the wood ; the celluloid are subject to warping, nevertheless are preferred by some as they are transparent, but for general work the vulcanite are undoubtedly the best, as they retain their shape, and not being transparent, there is less confusion of lines.

A set of spring bows (consisting of dividers, pen, and pencil bows) is also to be recommended for small work, though not absolutely necessary. It is better that these should not be provided with needle points, as, without abnormal care, they cannot be used for extremely fine work, as the holes get enlarged and the needles have a tendency to wobble, and the draughtsman cannot easily see if the point be placed in the exact position.

To prevent corrosion, great care should be taken to clean all instruments after they have been used.

Drawing board of yellow pine, the minimum dimensions of which should be 24 inches by 18 inches, $\frac{3}{4}$ inch thick, secured at the back with battens so arranged as to permit of the board shrinking without splitting, those with clamped ends are not to be recommended.

T-square, 24 inches long, with bevelled ebony edge, screwed on to the stock to allow the set-square to pass freely over the stock.

HB pencil for notes and sketches, H and HH pencils for drawings, india-rubber, four good drawing pins, sharp pen-knife, a box of colours, and glue. Cake colours are preferable for technical drawing, but the moist colours in pans are more readily mixed and are rapidly displacing the former, and the student will find it to be to his advantage to obtain these also from some good manufacturer.

A set of scales is required; as those of boxwood are expensive, their use is to a large extent replaced by sets of paper scales, such as the Polytechnic technical scales, sold by the publishers of the present work, which, having only one scale on each edge, prevent the student falling into mistakes common to the use of boxwood scales with two or more scales on each edge ; the cost of the paper scales is also considerably less.

*Pencil Sharpening.*—A sharp penknife should be used for this purpose. For ordinary use a round point should be made, as shown in figure 1, but for drawing lines against straight-edges, especially with the harder pencils, a chisel point, as shown in figure 2, may be advantageously used, as it will retain its edge longer. Pencils in compasses should always be chisel pointed.

*Drawing Paper.*—The following are the sizes in general use, although the exact dimensions vary slightly with different manufacturers :—

|  |  |  | Inches. |
|---|---|---|---|
| Demy . | Hot pressed and Not . | . | 20 by 15½ |
| Medium . | ,, ,, | . | 22 ,, 17½ |
| Royal . | Hot pressed, Not and Rough . | | 24 ,, 19½ |
| Imperial . | ,, ,, ,, | . | 30½ ,, 22 |
| Double Elephant ,, | ,, ,, | . | 40 ,, 26¾ |
| Antiquarian | ,, ,, ,, | . | 52½ ,, 30½ |

Whatman's (*not* hot pressed) medium thickness is a quality to be recommended for finished drawings ; it will stand wetting and stretching without injury, and when so treated receives

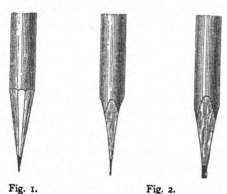

Fig. 1.          Fig. 2.

shading and colouring easily and freely. Whatman's *hot-pressed* paper is useful for exhibition pen-and-ink drawings, but great care must be exercised to prevent any mistakes, as it does not readily admit of alterations. The proper face of the paper upon which to draw is that surface on which, being held to the light, you can read the maker's name. For ordinary working drawings and details, cartridge paper, which is cheaper, or cartoon paper—of which there are many varieties, and which may be obtained up to 60 inches in width and in continuous rolls—are to be preferred.

*Damp Stretching.*—If the paper is to be used for a highly-finished drawing it must be very firmly attached to the board, and nothing is more suitable for the purpose than glue, or a strong paste made of flour and powdered resin. The stretched irregular edges of the paper are first cut off. The paper should then be thoroughly saturated on both faces, but if one face only is damped, it should be the upper one, as any dirt may be easily sponged off. A wide brush or clean sponge should be used, the edges of the paper being first thoroughly damped all round, and then the whole surface. The edges of the paper should not be quite as damp as the other parts of the surface. After the paper has been thoroughly damped, it must be left until the gloss entirely disappears and then turned over and placed in position on the board. Next, about three-eighths of an inch is turned up at the edge against a flat ruler or straight-edge, and a brush with glue or strong paste passed between this turned-up edge and the board; the ruler is then drawn over the prepared edge and pressed along. If the paper be found not to be thoroughly close when the ruler is removed, a paper-knife passed along the edge will remedy this. The next adjoining edge must be treated in a similar manner, then the next, and the next. When dry, the paper should present a perfectly plane surface for drawing and colouring upon. When the drawing is completed, the paper is cut with a cutting gauge or a penknife drawn along an iron straight-edge. On no account must the edge of a T-square be used for cutting against. The glued margin on the board should be carefully cleaned off with a sponge and cold water.

*Centre and Datum Lines.*—If the work to be produced offer any opportunity for using centre lines, draw them at the com· mencement as chain dotted lines. Especially should this be noted wherever window or door openings, columns, or any symmetrical details occur. These will be found of great convenience whenever any dimension is called into question, or whenever any reference or readjustment has to be made. If the drawing does not lend itself to be set out from centre lines, some important plane surfaces, such as floor levels, wall-plate levels, etc., should be drawn, and any line could be measured from the same. If, however, there be no such important surfaces or lines, which is of rare occurrence, the best thing to do is to make imaginary ones and let all measurements bear a relative position to them. These important plane surfaces or lines are known as datum surfaces or datum lines.

*Section Lines.*—Lines representing the traces of centre planes as shown in figure 171 should be inked in with a firm black line.

*Scale Construction.*—Where the dimensions of the object or detail to be drawn are too large to place conveniently on the paper, all its parts are represented reduced in some given ratio, and the drawing is said to be to scale, and the ratio of the reduction is termed the representative fraction. If the drawing is to be to an unusual scale, say 5 feet to 2 inches, construct one in the following manner:—

Draw an indefinite straight line, AB, and mark off 2 inches, AC. Draw line AD at any convenient angle, and from A on line AD, with dividers opened to any suitable distance, mark 12 equal divisions A to E; open dividers to distance AE, set out four equal distances, EFGHK. Join KC, and from divisions draw parallels to KC, dividing the line AC; then complete the scale as shown in figure 3. If the divisions on the line AD can be conveniently set off direct from a scale,

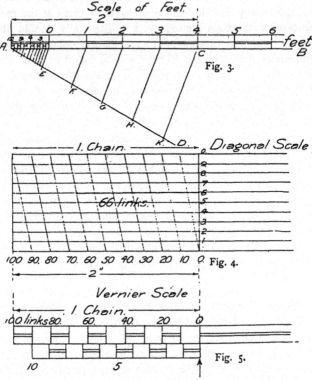

Fig. 3.

Fig. 4.

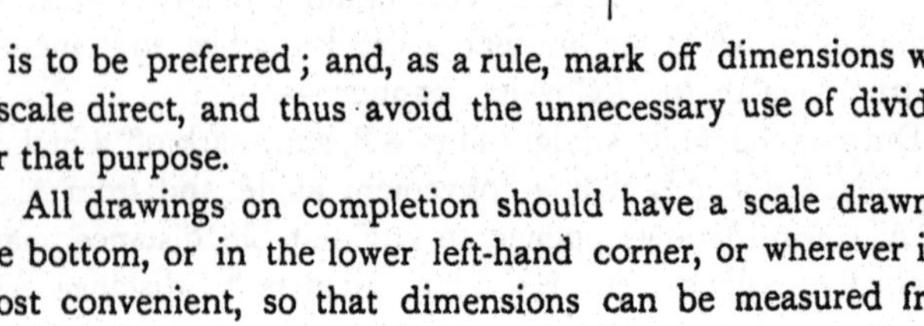

Fig. 5.

it is to be preferred; and, as a rule, mark off dimensions with a scale direct, and thus avoid the unnecessary use of dividers for that purpose.

All drawings on completion should have a scale drawn at the bottom, or in the lower left-hand corner, or wherever it is most convenient, so that dimensions can be measured from it, at any time, the scale and drawing being equally affected by the contraction and expansion of the paper, which is often a material quantity, especially in drawings made to a small scale.

*Diagonal Scale.*—Where the scale of a drawing is small, such as is the case in plans of land surveys, diagonal scales, as shown in figure 4, are constructed to obtain minute dimensions. Figure 4 shows a diagonal scale, 2 inches to the chain. The scale is constructed to measure links or one hundredths of the chain. It is necessary to take two numbers the product of which equals one hundred. Ten and ten have been taken in this case. The 2-inch line is then divided into ten equal parts and ten other lines parallel to the first are drawn at equal distances apart. Lines are drawn at right angles from the extremities of the 2-inch line forming a rectangle, from the right-hand lower corner of the rectangle shown in figure 4 draw a line to intersect the topmost line in the first division, from each of the other divisions draw parallels to this line. Number the points as shown in figure 4. In this case any measure required must be taken from the line, the number of which agrees with its last figure, thus 66 links would be measured on number 6 line.

*Vernier Scale.*—This method of measuring is employed where very minute dimensions are required. It is usually applied to measuring instruments, and is employed in lineal and angular measures. It consists of a small movable scale divided into a number of equal parts, each of which bears a definite ratio to the equal dimensions of the main or primary scale to which it is applied; thus, in the example shown in figure 5, a scale of 2 inches to the chain is drawn, divided into ten equal parts each representing ten links; if it is desired to measure to single links by this scale, the vernier or movable scale is made in length equal to 9 parts or 90 links of the main scale, and this length is then divided into ten equal parts, each of these divisions being therefore equal to $\frac{1}{10}$ of 90 links, that is, 9 links.

The difference therefore between the lengths of the first division of the vernier and the first division of the main scale will be one link, between the lengths of two divisions of the vernier and the two divisions of the main scale will be two links, and so on. Therefore in using the vernier the length to be measured is placed between zero on the main scale and the arrow head on the vernier, this will give the length up to the nearest ten links; then notice on the vernier the number of the division which coincides with any division on the main scale, and add this to the number previously determined, because this will indicate the difference in links; the vernier has to be adjusted to the nearest ten links below the exact number, to give the dimensions to the nearest link.

*Inking in.*—Always use a good Indian ink, one rubbed up from the stick in a smooth china saucer is to be preferred, especially if it is to be diluted to obtain gradations in brush work.

Ready-made Indian inks are now produced by many manufacturers, which give good results and save much time in preparation. The bottles should always be kept corked when the ink is not being used, otherwise it rapidly thickens. Waterproof inks are largely made, but for general work are inferior to the ordinary ink, as they tend to dry rapidly and clog the pen.

Inferior Indian inks, which run when drawings are coloured, must on no account be used, as many drawings when coloured are spoilt by the ink running. Should the ink run, the damage may be remedied by immediately washing the whole drawing with plenty of clean water, which will remove all free ink. The water must be carefully removed by means of a piece of clean blotting paper. When stick ink is employed, to ascertain if it is sufficiently mixed for use, blow in the saucer or palette.

or make a line on paper and smear it; if by either method the ink exhibits a brownish tint instead of a black, it is not mixed sufficiently. In using the ruling pen it should be drawn over the paper, being held in a nearly vertical position; and the pen should always be held in this position, otherwise a variation in the thickness of the line will ensue.

*Medium for Reproduction Work.*—For this class of work not only is great care necessary, but better results are obtainable by noting the following suggestions.

1st. For black-and-white work the black must be very intense; Indian ink is very good, but if required in large quantities is found somewhat expensive. Stephens's ebony-stain is in most cases a good substitute for Indian ink, but the stain cannot be used where there is danger of water coming into contact with the work.

2nd. For all line drawings, care must be taken that the lines are unbroken, and must be firm and of medium thickness.

*Erasure of Lines.*—Inked lines may be eliminated as follows:—Take a piece of clean paper and place it over the lines that are to be removed, cutting away a portion of the clean sheet just sufficient to enclose them; then with a clean sponge the lines may be washed out, the remaining portion of the drawing being protected by the covering paper, or they may be scratched out with a sharp penknife worked lightly over the surface, or by a piece of ink eraser. After employing either of the two latter methods, the injured surface of the paper should be burnished with a piece of polished bone or ivory before other lines or colour are passed over the spot.

Cleaning out pencil lines with india-rubber must be done in one direction, as rubbing backwards and forwards destroys the surface of the paper. Drawings, if dirty, should be cleaned with stale bread in preference to india-rubber.

*Hatching Sections.*—When the drawings are not to be

coloured, the materials most commonly used may be denoted in section by lines as shown in figures 6 to 16

*Colouring.*—If a drawing is to be coloured, no sectional nor hatching lines are to be drawn; the same colour as that used for elevations, but of a darker tint, is to be laid on for sections, unless otherwise stated.

*List of Conventional Colours to be used in Preparing Drawings :*—

Gamboge, Raw Sienna, Burnt Sienna, Burnt Umber, Vandyke Brown, Sepia, Venetian Red, Crimson Lake, Cobalt, Indigo, Prussian Blue, Neutral Tint.

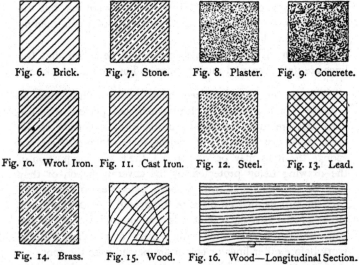

Fig. 6.  Brick.      Fig. 7.  Stone.      Fig. 8.  Plaster.      Fig. 9.  Concrete.

Fig. 10.  Wrot. Iron.  Fig. 11.  Cast Iron.  Fig. 12.  Steel.  Fig. 13.  Lead.

Fig. 14.  Brass.    Fig. 15.  Wood.    Fig. 16.  Wood—Longitudinal Section.

*Brickwork.*—Elevation, Venetian red or yellow ochre. Section, crimson lake.

*Stonework.*—Dressings, sepia; rough or hard stone, indigo.

*Plaster.*—Elevation, light Prussian blue. Section, neutral tint.

*Concrete.*—Neutral tint.

*Wrought Iron.*—Prussian blue.

*Cast Iron.*—Payne's grey.

*Steel.*—Purple. (Mixture of crimson lake and Prussian blue.)

*Lead.*—Indigo.

*Brass.*—Gamboge.

*Oak.*—Sepia and a tinge of yellow.

*Fir, unwrought and floors.*—Raw sienna. Sections, streaked with burnt sienna.

*Wrought Fir.*—Burnt sienna.

*Mahogany.*—Crimson lake, sepia, and burnt sienna.

*Tiles.*—Venetian red and a little yellow.

*Slates.*—*Green,* indigo and yellow; *Purple,* indigo and crimson lake.

*Walnut.*—Sepia and burnt sienna.

*Glass, interior.*—Flat wash of cobalt.

*Glass, exterior.*—Flat wash of Indian ink, or a wash of indigo.

*Generally.*—In elevation all washes should be light, while sections should have darker tints of the colours similar to those used for the elevation.

It is sometimes necessary to mix colours, but as a rule this method of obtaining tints should not be adopted, as it is extremely difficult to obtain the required tint, and then to match it again. When mixing colours care should be taken to have a quantity sufficient for the work in hand. Each time a mixed tint is taken from the palette, the whole should be stirred with a brush.

Any of the following tints may be mixed from the colours given in the list of conventional colours. Burnt umber: sepia and burnt sienna. Indian yellow: gamboge and pale burnt sienna. Hooker's green (glass in section): Prussian blue and gamboge. Payne's grey: indigo, Indian ink, and crimson lake.

*Colouring.*—It is usual before colouring to ink in the lines of architectural drawings, as alterations after colouring disfigure the work. Where the washes of colour are large it is advisable to damp the paper to obtain an evenness of tint. Care should be taken in colouring an inked drawing that the brush is not worked too much over the lines as there is a danger of the ink running, so, to avoid this, many colour first and ink after. This has the advantage of giving brilliant black lines. Light colours should precede dark ones. Colour straightway all parts to be of the same tint. Avoid the mistake of mixing colours of too dark a tint. If a dark tint is required, a second wash may be applied when the first wash has dried. Sable or camel-hair brushes should be used, and care must be taken to select such as have good sharp and not long straggling points.

*Dimensions.*—All the leading dimensions should be distinctly figured in. Inattention to this precaution may produce serious mistakes   The extent of the dimensions should be shown by neat arrow-heads connected by a continuous thin blue line broken only in the middle, the dimensions being inserted in this space in a vertical position, and all to read one way. Internal dimensions should be taken between the brick or stone walls, also between floor levels.

*Lettering.*—For titles and headings, Egyptian or block upright letters are effective. Roman letters are satisfactory and easier to form, while for general and descriptive writing an easily written form of small upright block italics is to be preferred.

SPECIMEN UPRIGHT BLOCK—CAPITALS.

A B C D E F G H I J K L M N O P O
R S T U V W X Y Z
1 2 3 4 5 6 7 8 9 0

SPECIMEN UPRIGHT BLOCK—SMALL.

a b c d e f g h i j k l m n o p q
r s t u v w x y z

SPECIMEN UPRIGHT ROMAN—CAPITALS.

A B C D E F G H I J K L
M N O P Q R S T U V W
X Y Z
1 2 .3 4 5 6 7 8 9 0

SPECIMEN UPRIGHT ROMAN—SMALL.

a b c d e f g h i j k l m n o p q
r s t u v w x y z

SPECIMEN—ITALICS.

*a b c d e f g h i j k l m n o p q r s t u v w x y z*
*1 2 3 4 5 6 7 8 9 0*

*Geometrical Explanations—Measurements.*—Throughout this treatise a single accent (') signifies feet, a double accent (") inches, and (°) degrees.

*Plan.*—If the boundaries of the surfaces of solids are imagined to be projected in paths, or by lines perpendicular to a horizontal plane, that portion of the horizontal plane enclosed by the projectors is called the plan, and thus there may be roof plans, floor plans, basement plans, etc.

*Elevation.*—In a similar manner the space covered on a vertical plane by projectors perpendicular to the vertical plane, from a solid, is the elevation; and therefore front elevations, back elevations, side elevations, etc., are so called after the sides seen.

*Section.*—Where a plane or imaginary plane divides a solid, the cut surfaces are called sections, and there may be cross (or transverse) sections, longitudinal sections, vertical, and horizontal sections.

*Sectional Plans* or *Elevations* are the projections of a cut surface, and any remaining uncut part of the solid.

The parts seen from the outside in plan or elevation (always being perpendicular to the plane that projectors are being taken from) are drawn in firm lines, those unseen in dotted lines, to indicate their position.

*Isometric Projection.*—It is often convenient, especially for details of solids with rectangular faces, to show the three dimensions in one view. This may be accomplished by drawing three lines meeting in a point containing equal angles of 120°, representing length, breadth, and depth, to scale, and from the free extremities of these lines draw lines parallel to the other two. This method is misleading for angular and curved work, as it gives distorted views, and all lines not parallel to any one of the three axes cannot be measured accurately with the scale used. It is usual to work with the ordinary scales.

*Perspective Drawing.*—This is the science of making representations of objects as they appear to the eye, and is very useful for showing how a proposed object will appear when made, and for illustrating details.

*Tracing.*—An original drawing is made of all proposed work ; from this standard drawing a number of copies are often required to be made. The copies are made by tracing or by photography.

For comparing the plotting of different stories during the process of designing, tracing paper is commonly used ; but copies of drawings that are to be much used are usually traced upon linen, which offers a greater resistance to tearing. The glazed side is usually inked and the coloured washes placed upon the back. There is often a difficulty in getting the ink to flow uniformly upon the glazed surface of the linen, but to overcome this obstacle a little oxgall is mixed with the ink ; or chalk may be rubbed upon the glazed surface of the linen for the same purpose.

*Black Line, Water-bath Reproduction.*—A tracing in ink is made from an original, which need not necessarily be in ink, upon white or bluish tinged tracing paper or tracing cloth. Yellow tracing paper must be avoided for this purpose, as the actinic power of the rays of light is reduced in passing through this coloured medium. The tracing is placed against the glass of a printing frame (such as is used in photography) in front of specially prepared sensitized paper in a dark room, and the frame is then closed and exposed to the direct and uninterrupted rays of light.

Sufficiency of exposure is indicated when the blank surface of the sensitized paper (that is, the part not covered by the ink lines on the tracing) turns from the yellow to the natural whiteness of the back of the sensitized paper. Then the print is held under running water till the black lines are developed, and the yellow coating which turns black on contact with the water is washed away.

The required exposure ranges from five minutes in bright sunshine to one hour in dull weather.

This method is now extensively used.

*White Lines on Blue Ground.*—This method, known as the ferro-prussiate process, is largely used for the reproduction of drawings that do not require colouring afterwards, and is accomplished in the following way:—A tracing is made from the original as in the previous method, it is placed in a printing frame with its face against the glass, the printing paper is then placed in the frame with its sensitized face in contact with the tracing, the frame is then closed and the exposure made, the time taken varying, according to the intensity of the light, from five minutes in bright sunshine to one hour in dull weather. The print is then washed in running water. This method gives white lines on a bright blue surface, and is much used for engineering drawings.

*Manifolding of Writing.*—Mechanical means should always be used to reproduce specifications, quantities, etc., as every copy will then be exactly similar. There are three methods largely employed : (1) typewriting with carbon paper, when not more than six copies are required. (2) By writing or typewriting on specially prepared stencil paper, which is placed in a printing frame, paper is placed on the bed of the machine, the printing frame, which is hinged, is lowered till the stencil is in contact with the sheet on which the copy is to be made— the upper side of the stencil having been previously covered with ink—a roller is run over the upper surface, this produces the impression, the frame is opened, the copy removed and the operation is repeated. Five hundred copies may be made by this method. There are many modifications of this stencil principle. (3) By lithography : a copy is made by hand from the original documents upon specially prepared paper, which is then pressed on to the face of a specially prepared stone ; this leaves an impression. The stone is then subjected to chemical action and the impression made permanent. From this thousands of copies may be printed.

# CHAPTER II.

# BRICKWORK.

———

*Definition.*—Brickwork is the art of arranging and bedding bricks in either lime or cement mortar, to unite them and form a homogeneous mass of a desired shape with sufficient stability to resist safely the stresses to which it is to be subjected.

*Bricks.*—These are blocks of argillaceous earth of convenient size for handling and bonding, made from plastic clays by moulding and burning.

*Moulding.*—The necessity of moulding gives an opportunity to prepare all the blocks of a uniform dimension, thus allowing of regular arrangements of bedding in a wall.

*Burning.*—This changes the peculiar physical characteristic of plasticity and renders the compound rigid; secondly, it vitrifies the material, endowing it with durability; and thirdly, renders it hard and capable of resisting great compressional stresses. The description and methods of manufacture of the best known bricks are given in the *Advanced Course.*

*Dimensions of Bricks.*—To admit of perfect bonding, the width and length of a brick, neglecting the thickness of a joint, should be in the ratio of one to two. The usual thickness of a mortar joint is three-eighths of an inch, and as there are twice the number of joints in the heading than there are in the stretching courses, it is usual to make the length of a brick equal to twice the width plus the thickness of one mortar joint. The thickness of bricks, providing there is uniformity throughout, is immaterial, but they are usually made between the limits of one fourth to one-third of the length.

The average size of a good stock brick is $8\frac{3}{4}''$ long, $4\frac{1}{4}''$ wide, and $2\frac{3}{4}''$ thick, and the dimensions, including one mortar joint, in this work, unless otherwise stated, will be taken as,

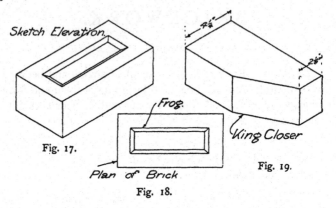

Sketch Elevation.

Fig. 17.

Frog.

Plan of Brick.

Fig. 18.

King Closer

Fig. 19.

length $9''$, width $4\frac{1}{2}''$, and thickness $3''$. Bricks are usually bedded on the $9'' \times 4\frac{1}{2}''$ surface in lime or cement mortar.

In the north of England, bricks are often made up to 4 inches in thickness. For partition walls, special bricks 6 inches in width are made.*

*Frog.*—In hand-made and machine-pressed bricks, indentations, known as frogs, are formed in the bedding surfaces of the bricks. In the former brick one frog only is formed, in the latter two. The object of the frog is twofold : first, to reduce the weight ; secondly, to form a key for the mortar. Frogs are not formed in wire-cut bricks, the method of manufacture rendering this impossible.

Instead of frogs, some bricks are perforated from bed surface to bed surface to attain the same advantages.

* Particulars of the R. I. B. A. standard size of brick are given in the *Advanced Course.*

Figures 17, 18, 19, illustrate the frog in the bed of a brick.

*Technical Terms.*—*Course* is the name given to the row of bricks between two bed joints, and the thickness is taken as one brick plus one mortar joint; in this work, unless otherwise stated, it will be considered as 3″ or, as technically described, four courses to the foot.

*Bed Joints.*—The mortar joints normal to the pressure.

- (*a*) In walls with vertical faces the bed joints would be horizontal.
- (*b*) In batter walls they would be at right angles to the batter.
- (*c*) In arches the joints would be normal to the curve of the arch.

*Quoins.*—The external corners of walls. The name is sometimes applied to the bricks or stones which form the quoins, *e.g.* quoin brick, quoin stone.

*Perpends.*—The vertical joints of the face of the wall. In plain walling it is necessary for good bond that these joints in alternate courses should be vertically one above the other.

*Stretchers.*—Bricks laid with their lengths of 9″ on the face of the wall or parallel to the face of the wall.

*Headers.*—Bricks laid with the width of $4\frac{1}{2}$″ on the face of the wall or parallel to the face of the wall.

*Bats.*—Pieces of bricks, usually known, according to their fraction of a whole brick, as $\frac{1}{2}$ or $\frac{3}{4}$ bats.

*Lap.*—The horizontal distance between the vertical joints in two successive courses. This should be one-fourth the length of a brick.

*Queen Closers.*—Bricks made the same length and thickness as ordinary bricks, but half the width, placed usually next to the quoin header to obtain the lap. Closers are not usually specially moulded; bricklayers form the closers economically by cutting their broken bricks into $\frac{1}{4}$ brick bats and placing two together to form closers, as shown in figures 46 to 56.

*King Closers.*—Bricks cut so that one end is half the width of the brick, as shown in figure 19. They are used in the construction of reveals to avoid having any face brick less than $4\frac{1}{2}''$ on the bed.

*Mortar.*—This is the matrix used in the beds and side joints of brickwork; its functions are as follows : first to distribute the pressure throughout the brickwork, secondly to adhere and bind the bricks together, and thirdly to act as a non-conductor and prevent the transmission of heat, sound, and moisture from one side of a wall to the other. Mortar is composed of an aggregate and a matrix. Clean, sharp pit sand forms the best aggregate. Old bricks, burnt ballast, or stones ground in a mortar mill are often used as substitutes for sand. The necessary characteristics of the aggregates are that they shall be clean and free from clayey or earthy matter, and that they shall be hard and gritty or angular. The usual matrices for constructional work are moderately hydraulic limes or Portland cement.

Limes and cements, their characteristics and properties, are fully dealt with in the *Advanced Course.*

*Lime Mortar.*—The ratio of lime to sand is usually one to two. It should be thoroughly mixed to ensure uniformity of composition, and left for at least a week before using.

*Cement Mortar.*—The ratio of Portland cement to sand is from one to two to one to four. The constituents should be thoroughly mixed while dry to ensure uniformity of composition, and should be used immediately after the addition of the water. The setting action commences soon after mixing, and if the mortar is disturbed after this has commenced, the strength is greatly reduced, consequently no cement mortar should be mixed up a second time.

*Wetting of Bricks.*—The presence of water in mortar is necessary for the setting action to take place; precautions should therefore be taken to prevent the work drying too quickly; as a means to this end all the bricks should be saturated before bedding, to prevent them absorbing the moisture from the mortar, and also to remove all loose dust from the surfaces that are to be in contact with the mortar.

*Building during Frosty Weather.*—All brickwork should be suspended during frosty weather, as the stability of the same is endangered by the disintegration of the mortar by the frost while it is wet. When the work is urgently required it should be carried up in cement mortar in the intervals between the frost; but all the freshly built portions should be carefully covered and protected on any recurrence of the frost, and always during the suspension of work for the night.

*Bedding of Bricks.*—Great care should be taken when bedding bricks that both the bed and side joints are thoroughly flushed, or filled up with mortar. This is done in three ways : (1) by the trowel; (2) by larrying; (3) by grouting. The first method is that usually adopted in all thin walls. The second, larrying, is largely adopted in thick walls. The face bricks are first laid; the mortar, in a semi-fluid condition, is then poured into the space between the face bricks; the bricks are then pushed rapidly horizontally for a short distance, and into their position; a certain amount of the mortar is thus displaced, this rises in the side joints, and completely fills all the interstices; should the mortar not rise to the top of the joints, the vacant spaces are filled up, when the next course is larried. (3) Grouting is an operation used in brickwork, generally for gauged arches and similar work, where fine joints are required; it consists in mixing the mortar to a fluid con. dition, of about the consistency of cream, this being poured into the joints of the work after the latter has been placed in

position.    Figure 174 shows a flat gauged arch voussoir prepared for grouting.

*Building Walls.* —Where new walls are erected the usual
method of procedure is to build what is technically termed a
corner, that is, the angles or the extremities of the walls, to a
height of two or three feet, the angle bricks being carefully
plumbed on both faces.   The base of the corner is extended
along the wall, and is racked back as the work is carried up,
as shown in figure 22.   The intermediate portion of the wall is
then built between the two corners, the bricks in the courses
being kept level and straight by building their upper edges to
a line strained between the two corners.

*Levelling.*—In bedding bricks, great care should be taken
to keep all courses perfectly level.   To do this, the footings and
the starting course should be carefully levelled through, using a
spirit level with a stock at least 10 feet in length, commencing
at one end and levelling towards the other, and taking care to
reverse the level each time at each forward step, and completing
the length to be levelled in an even number of steps.   A piece
of slate or iron is left projecting from the lowest course, and
from this all other courses at the corners can be levelled by
using the gauged rod, which is usually about 10 feet in length,
with the courses marked on it.   The work should then be again
tested by the level, and the operation repeated.   The levelling
of brickwork is most efficiently accomplished with a surveyor's
level, particularly where the walls are long.

*Boning Method of Levelling.*—Boning rods are used for
levelling trenches, ground work, paving, etc.   There are three
rods in a set, two of these are levelled at a distance of about
10 feet apart ; a third rod is then levelled at a similar distance,
taking care to reverse the long level.   The centre rod is then
removed, and the level transmitted to any point along the line
by sighting or boning over the first and third rods.

Figure 208 shows the method of using boning rods and setting a kerb-stone.

*Toothing.*—The usual method of leaving a brick wall which is to be continued at some future time is to tooth it, which consists in leaving each header projecting 2¼ inches beyond the stretching courses above and below to allow the new work to be bonded to the old as shown in figure 21.

The usual practice in joining new cross walls to old main walls is to cut out a number of rectangular recesses in the main walls equal in width to the width of the cross wall, three courses in height, and half a brick in depth, a space of three courses being left between the sinkings (as shown in figure 20); the new cross wall is then bonded into the recesses with cement mortar to avoid any settlement. It is necessary that the sinkings should not be less than 9 inches apart, as in the cutting, the portion between is likely to become shaken and cracked.

*Racking.*—Racking is the term applied to the method of arranging the edge of a brick wall, part of which is unavoidably delayed while the remainder is carried up. The unfinished edge must not be built vertically or simply toothed, but should be set back 2¼ inches at each course, to reduce the possibility and the unsightliness of defects caused by any settlement that may take place in the most recently built portion of the wall.

*Stability of Brickwork.*—The stability of brickwork is affected in three general ways :—1st, By loading a given area of ground beyond its ultimate resistance, by the irregular concentration of great pressures on a soft soil, or by the tendency of the substratum to slide; the result of any of the above causes is that the walls are thrown out of the upright, crack, or disintegrate (see Chapter on Foundations, *Advanced Course*). 2ndly, By bad bonding, resulting in disintegration. 3rdly, By side thrusts; these may be distributed or concentrated, and their tendency is to overturn the walls; they are provided for

Fig. 20.

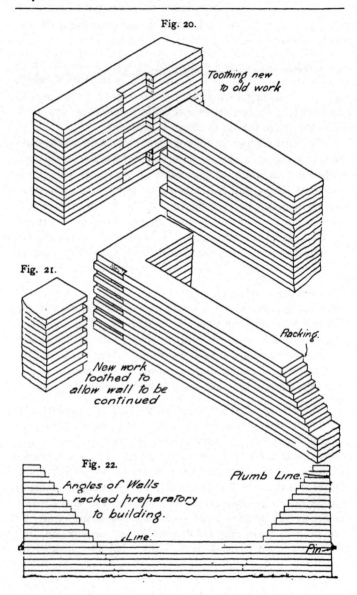

Toothing new
to old work

Fig. 21.

Racking.

New work
toothed to
allow wall to be
continued

Fig. 22.

Plumb Line.

Angles of Walls
racked preparatory
to building.

Line.

Pin

by designing the walls of a sufficient thickness, or by disposing buttresses at the salient points, so that the weight of wall or buttress when compounded with the overturning thrust shall give a resultant that shall fall a sufficient distance within the base of the wall, and also if any tensile stresses are anticipated by building the wall with a tenacious cement mortar. (See *Advanced Course.*)

*Wall Classification.*—Walls are constructed to enclose areas and to support the weight of floors, roofs, earth or water. They are classified as follows :—

    (*a*) Walls to resist vertical pressures (continuous or concentrated).

    (*b*) Walls to resist oblique thrusts (continuous or concentrated).

Under the first section of heading (*a*) would be included all house walls, solid or hollow, supporting single floors, and couple close raftered roofs. The second section would include all walls carrying the girders of framed floors and the trusses of framed roofs.

Under the first section of heading (*b*) would be included all walls supporting continuous barrel vaulting, retaining walls, dams, etc. The second section would include walls supporting groined vaulting and untied roof trusses.

Figures 23 to 28 illustrate solid and hollow walls; figures 32 to 34 illustrate walls designed to carry girders of framed floors, etc. They are arranged as a series of piers at intervals joined by thin curtain walls to complete the enclosure. Where such walls are on a yielding stratum the piers should be connected at their bases by inverted arches, to distribute the pressures uniformly over the foundations, as shown in figure 32. Figure 31 shows a wall supporting an untied roof truss; the latter transmits an inclined thrust on the wall at some point above the ground level. The wall depends for its stability upon its weight, which, when compounded with the inclined thrust of the vault, should be sufficient to give a resultant that shall fall a safe distance within the base (see

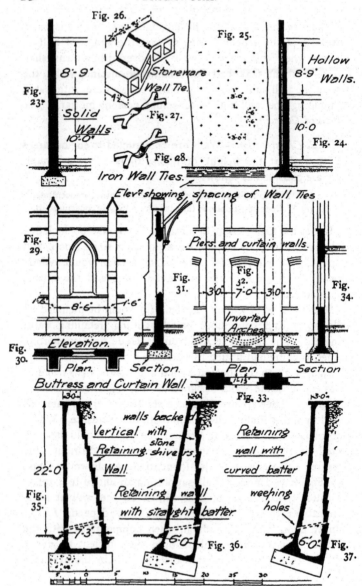

Fig. 26.

Stoneware Wall Tie.

Fig. 25.

Hollow 8'-9" Walls.

Fig. 23.

8'-9"

Solid Walls. 10'-0"

Fig. 27.

Fig. 28.

Iron Wall Ties.

10'-0"

Fig. 24.

Elevⁿ showing spacing of Wall Ties

Fig. 29.

1'-6" 8'-6" 1'-6"

Fig. 31.

Piers and curtain walls.

Fig. 32.

3'-0" 7'-0" 3'-0"

Fig. 34.

Inverted Arches.

Fig. 30.

Elevation.

Plan.

Section.

Plan

Section

Buttress and Curtain Wall.

1'-1½"

Fig. 33.

3'-0"

walls backed Vertical with stone Retaining shivers. Wall.

2'-0"

Retaining wall with straight batter

3'-0"

Retaining wall with curved batter

weeping holes

22'-0"

Fig. 35.

7'-3"

6'-0"

Fig. 36.

6'-0"

Fig. 37.

*Advanced Course*). The thickness of such walls is dependent upon the magnitude of the inclined thrust and its leverage from the base, and secondly upon the weight of the wall material. Figures 35 to 37 show three simple forms of retaining walls subjected to continuous pressure, and consequently of a uniform thickness. Figure 35 shows a wall with a vertical face. Figure 36 with a battered face; this is a more economical and stable section. Figure 37 illustrates a wall with a curved batter; this more nearly approaches the theoretical section for such walls, but is not often used, as the labour in forming curved surfaces is more costly. Provision should be made in such walls to drain the retained earth by inserting drains, known as weep holes, through the walls near their bases. Gutters at the feet of such walls should be constructed to carry off all surplus water, and thus prevent the softening of the earth about the foundations.

Figures 38 to 43 show projections of retaining walls with counterforts. Where it is necessary to have a wall with a plain face the counterforts are constructed at the back of the wall, as shown in figures 38 to 40 ; there is a slight saving in brickwork over the ordinary method of building retaining walls by adopting this arrangement. Figures 41 to 43 show the counterforts arranged as a series of buttresses on the faces of the walls, the counterforts being connected at their backs by a series of horizontal arches ; this is a far more economical method than those last described. The top ends of the buttresses are connected by arches, as shown in figures 41 and 42. Figures 44 and 45 show a method of supporting earth by a series of relieving arches, having their axes at right angles to the earth face. The earth is banked up behind the structure and allowed to assume its natural slope, as shown in figure 44. By this method no pressure is transmitted to the face of the wall.

Figures 29 to 31 show the method of constructing walls to resist concentrated thrusts. Buttresses are arranged in the

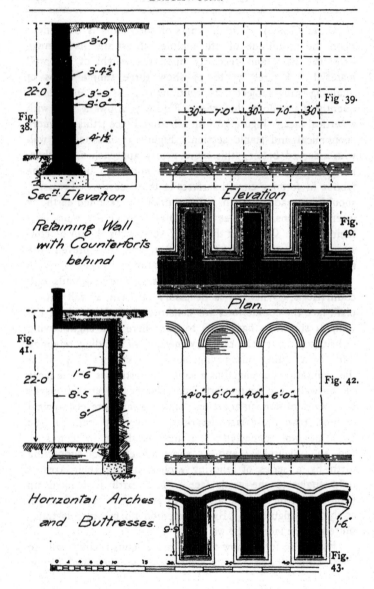

3'-0"

3'-4½"

22'-0"

3'-9"

8'-0"

4'-1½"

Fig. 38.

Sec.ᵗˡ Elevation

Retaining Wall
with Counterforts
behind

Fig 39.

3·0"   7·0"   3·0"   7·0"   3·0"

Elevation

Fig. 40.

Plan.

Fig. 41.

22'-0'

1'-6"

8·5

9'

Horizontal Arches
and Buttresses.

Fig. 42.

4·0"   6·0"   4·0"   6·0"

Fig. 43.

8·5                 1·6"

plane of the thrust to maintain equilibrium, and are connected by thin curtain walls to complete the enclosure.

*Bond* (that is, to bind), the name given to any arrangement of bricks in which no vertical joint of one course is exactly over the one in the next course above or below it, and having the greatest possible amount of lap, which is usually one-fourth the length of a brick.

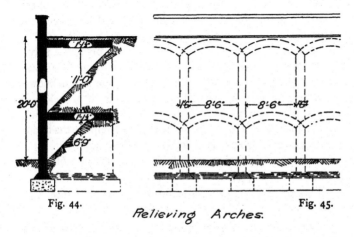

Fig. 44.

Fig. 45.

*Relieving Arches.*

*Bonds in Brickwork.*—To ensure good bond the following rules should be rigidly adhered to :—1st, the arrangement of the bricks must be uniform ; 2ndly, as few bats as possible be employed ; 3rdly, the vertical joints in every other course to be perpendicularly in line on the internal as well as the external face ; 4thly, stretchers are only to be used on the faces of the wall, the interior to consist of headers only, except in footings and corbels ; 5thly, the dimensions of bricks should be such that when bedded the length should equal twice the width, plus one mortar joint.

There are several kinds of bond used in brickwork. Those described in this chapter being as follows :—1st English,

2nd Double Flemish, 3rd Single Flemish, 4th English Cross, 5th Dutch, 6th Stretching, 7th Heading, 8th Garden Wall, 9th Facing, 10th Raking, 11th Hoop-Iron, 12th Miscellaneous, 13th Broken.

*English Bond.*—When the bond is arranged as shown in elevations and plans, figures 46 to 56, it is known as English bond, and sometimes old English bond. It consists of one course of headers and one course of stretchers alternately. In this bond bricks are laid as stretchers only on the boundaries of courses, thus showing on the face of the wall, and no attempt should be made to break the joints in a course running through from back to front of a wall. That course which consists of stretchers on the face is known as a stretching course, and all in course above or below it would be headers with the exception of the closer brick, which is always placed next to the quoin header to complete the bond, and these courses would be called heading courses.

It may be noticed that in walls the thickness of which is a multiple of a whole brick, the same course will show either—

(*a*) Stretchers in front elevation and stretchers in back elevation.

(*b*) Headers      „      „      „ headers      „      „

but in walls in which the thickness is an odd number of half bricks, the same course will show either—

(*a*) Stretcher in front elevation and header in back elevation.

(*b*) Header      „      „      „ stretcher „      „

In setting out the plan of a course to any width, draw the quoin, or corner brick ; then next to the face (which in front elevation shows headers) place closers to the required thickness of wall, after which set out all the front headers, and if the thickness is a multiple of a whole brick, set out headers in rear ; if the thickness is an odd number of half bricks, set out stretchers in rear : the intervening space, if any, is always filled in with headers.

*Double Flemish Bond* has headers and stretchers alternately

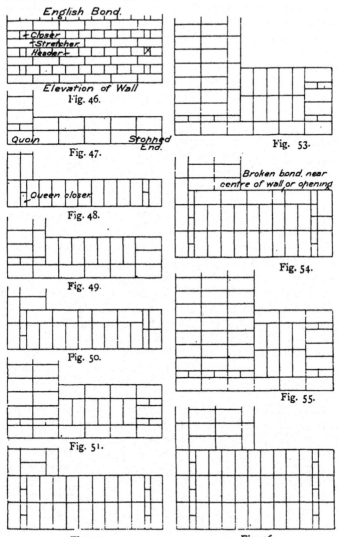

English Bond.

+Closer
+Stretcher
Header

Elevation of Wall
Fig. 46.

Quoin　　　　　　Stopped End
Fig. 47.

Queen closer
Fig. 48.

Fig. 49.

Fig. 50.

Fig. 51.

Fig. 52.

Fig. 53.

Broken bond. near centre of wall or opening

Fig. 54.

Fig. 55.

Fig. 56.

Figs. 46–56.　Examples of English Bond.

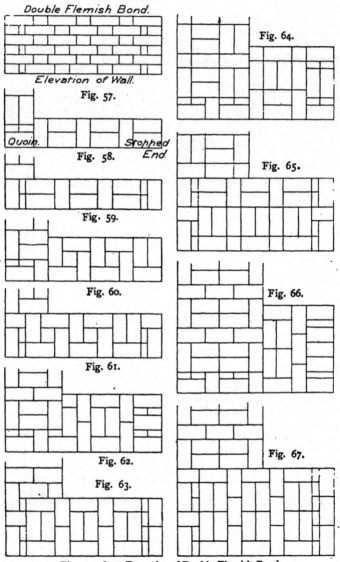

Double Flemish Bond.

Elevation of Wall.

Fig. 57.

Quoin.                    Stopped End.

Fig. 58.

Fig. 59.

Fig. 60.

Fig. 61.

Fig. 62.

Fig. 63.

Fig. 64.

Fig. 65.

Fig. 66.

Fig. 67.

Figs. 57-67.    Examples of Double Flemish Bond.

in the same course, .both in front and back elevations, as shown in figures 57 to 67. It is weaker than English bond, owing to the greater number of bats and stretchers, but is considered by some to look better on the face. It is also economical, as it admits of a greater number of bats being used, so that any bricks broken in transit may be utilised. By using double Flemish bond for walls one brick in thickness, it is easier to obtain a fair face on both sides than with the English bond.

*Single Flemish Bond* consists in arranging the bricks as Flemish bond on the face, and English bond as backing. This is often done on the presumption that the strength of the English bond as well as the external appearance of the double Flemish is attained, but this is questionable. It is generally used where more expensive bricks are specified for facing. The thinnest wall where this method can be introduced is $1\frac{1}{2}$ brick thick. Plans of alternate courses are given, figures 68 to 76. The front elevations are the same as in double Flemish bond.

*English Cross Bond.*—A class of English bond. Every other stretching course has a header placed next the quoin stretcher, and the heading course has closers placed in the usual manner.

*Dutch Bond.*—In every alternate stretching course a header is introduced as the second brick from the quoin, three-quarter bricks are used in the remaining stretching courses at the quoins, and the closers are dispensed with in the heading courses, as shown in figures 77 to 82; the longitudinal tie becomes much greater, and the appearance of the elevation is certainly superior to much of the inferior work one is accustomed to see as examples of the modern bricklayer's skill in bonding. Should there be a fracture, it is supposed to throw it more obliquely.

*Stretching Bond* is used for walls half brick thick, as for

D

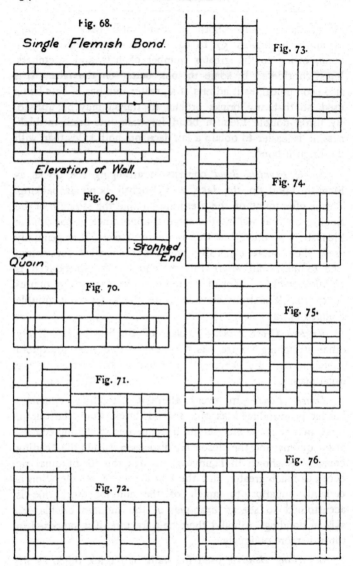

Fig. 68.

Single Flemish Bond.

Elevation of Wall.

Fig. 69.

Quoin

Stopped End

Fig. 70.

Fig. 71.

Fig. 72.

Fig. 73.

Fig. 74.

Fig. 75.

Fig. 76.

Figs. 68-76. Examples of Single Flemish Bond.

partition walls, bricknogging in partitions and in half-timbered work. All bricks are laid as stretchers upon the face.

*Heading Bond.*—All bricks show as headers on face. Used chiefly for rounding curves, for footings, corbels and cornices.

*Garden or Boundary Wall Bond* is the name given to walls built with three stretchers and one header in same course, constantly recurring, as shown in elevation, figure 83. This method is used for walls one brick thick that are seen on both sides, as it is easier to adjust the back face by decreasing the number of headers, the lengths of which usually vary.

*Facing Bond.*—Modifications of the rules given for bonding are necessary, first, where the thicknesses of the facing and backing bricks vary ; secondly, where the facing bricks are expensive and it is necessary to economize.

In the first case, one heading course only to several stretching courses, the distance between the heading courses being the least common multiple of the backing and facing bricks ; there should be at least one course of headers to every foot. In the second case the usual practice is to have three courses of stretchers to one of headers, as shown in figure 84.

*Raking Bonds.*—Walls as they increase in thickness increase in transverse strength ; but become proportionally weaker in a longitudinal direction, owing to the fact that stretchers are not placed in the interior of a wall. This defect is remedied by using raking courses at regular intervals, of from four to eight courses in the height of a wall. The joints of bricks laid in this position cannot coincide with the joints of the ordinary courses directly above or below, the inclination to the face usually being determined by making the longitudinal distance between the opposite corners equal to the length of a brick. It is not advisable

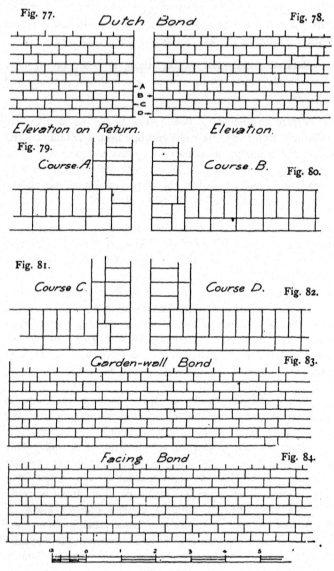

Fig. 77.

*Dutch Bond*

Fig. 78.

Elevation on Return.      Elevation.

Fig. 79.

Course. A.

Course. B.   Fig. 80.

Fig. 81.

Course C.      Course D.   Fig. 82.

*Garden-wall Bond*   Fig. 83.

*Facing Bond*   Fig. 84.

Figs. 77–84.

to use one raking course directly above another, as there is always a weakness with the face bricks at the junction of the raking.

Raking bonds are always placed in the stretching courses in walls of an even number of half bricks in thickness, in order that they may be effective over a greater area than would be the case if they were placed in the heading courses.

The alternate courses of raking bonds should be laid in different directions, in order to make the tie as perfect as possible.

There are two varieties of raking bonds, viz. diagonal and herring-bone.

*Diagonal Bond.*—This is used in the thinner walls, *i.e.* between two and four bricks in thickness. The operation is

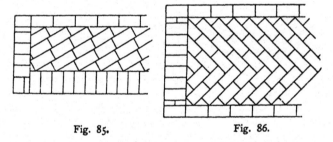

Fig. 85.    Fig. 86.

as follows:—The face bricks are laid, one or more bricks (in the latter case placed end to end) are bedded between the face bricks, so that the opposite corners touch the latter; this determines the angle that the bricks should be laid, the triangular spaces at the ends of the bricks being filled up with small pieces of brick cut to shape, as shown in figure 85.

*Herring-bone Bond.*—The bricks in this method are laid at an angle of 45°, commencing at the centre line and working towards the face bricks. Herring-bone bond is used for walls four bricks and upwards in thickness. Figure 86 shows this method.

Diagonal and herring-bone patterns are often used to form ornamental panels in the face of walls, and also in floors paved with bricks.

*Hoop-Iron Bond.*—An additional longitudinal tie, termed "*hoop-iron bond*," is often inserted in walls, being usually pieces of hoop iron 1″ by $\frac{1}{16}$″, one row for every half brick in the thickness of the wall, as shown in figure 87. It should be carefully tarred and sanded or galvanized before using to prevent oxidation. It is hooked at all angles and junctions, as shown in figures 88 and 89. If bedded in two courses in cement, additional strength is gained; pieces of hoop iron may be used with advantage where the bond at any part of the wall is defective.

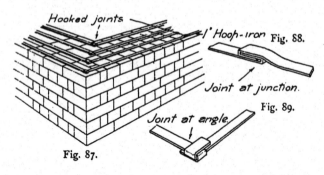

Hooked joints

1″ Hoop-iron  Fig. 88.

Joint at junction.

Fig. 89.

Joint at angle.

Fig. 87.

*Piers.*—Piers in brickwork, as shown in figures 90 to 98, are rectangular pillars constructed to support heavy weights transmitted to them by beams and girders, or to receive the thrusts of two or more arches, the resultant of the thrusts of which falls in a vertical line.

Figures 90 to 94 show plans of square piers built in English bond, one to three bricks square. It is only necessary to draw the plan of *one* course, as the adjoining courses have the same arrangement of bricks, but placed in such a manner that those in the front elevation of one course are in the side elevation of the next, above or below.

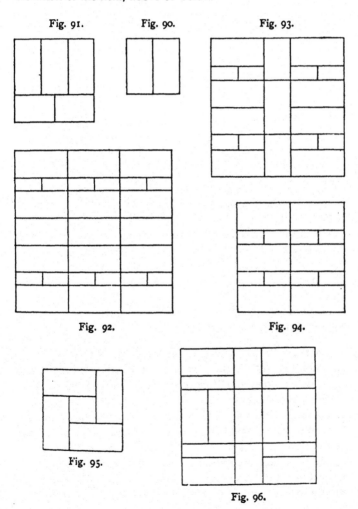

Fig. 91.

Fig. 90.

Fig. 93.

Fig. 92.

Fig. 94.

Fig. 95.

Fig. 96.

Figures 95 to 98 show plans of square piers built in double Flemish bond one-and-a-half to three bricks square.

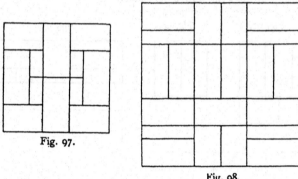

Fig. 97.

Fig. 98.

*Junctions of Cross Walls.*—The bond is obtained in cross or party walls abutting against main walls by placing a closer, $4\frac{1}{2}''$ from the face in every alternate course in the main wall, thus leaving a space $2\frac{1}{4}''$ deep and of a length equal to the thickness of the cross wall for the reception of the $2\frac{1}{4}''$ projection in every other course of the cross wall, as shown in figures 99 to 104.

Figures 105 and 106 illustrate the junction of one brick English bond with one-and-a-half brick Flemish bond.

*Jambs.*—The vertical sides of door and window openings are known as jambs; they are of three kinds:—1st plain, 2nd rebated, 3rd splayed. First, plain jambs are where the sides of the perforations in walls are taken through in one plane surface as in internal door openings, as shown in figure 89o. Secondly, the rebated jambs; here the sides are of more than one surface, as in figure 117. Jambs of this description are employed largely for external door and window openings; the projecting jamb that gives the sight line of the opening is known as a reveal. Thirdly, in window openings in thick

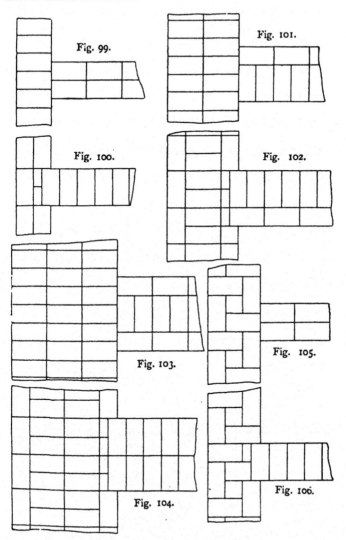

Figs. 99-106.    Junctions of Cross Walls.

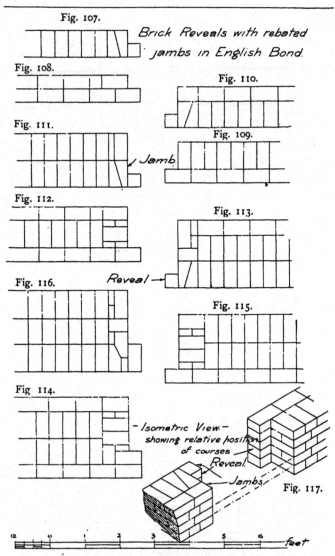

Fig. 107.

Brick Reveals with rebated jambs in English Bond.

Fig. 108.

Fig. 110.

Fig. 111.

Fig. 109.

Jamb

Fig. 112.

Fig. 113.

Fig. 116.

Reveal →

Fig. 115.

Fig 114.

— Isometric View —
showing relative position
of courses

Reveal

Jambs

Fig. 117.

feet

Figs. 107-117.   Brick Reveals with rebated jambs in English Bond.

walls, the internal jamb is frequently splayed at angles varying between 60° and 45° to the face of the wall in order to better diffuse the light in the interior. Figures 125 and 126 show the method of bonding a splayed jamb of a three-brick wall.

*Squint Quoins.*—External angles other than a right angle in plan are called squint quoins. They are of two kinds, acute and obtuse. Two general rules should be kept in view, viz., (1) no birdsmouth joint in plan should be employed; they would be useful in the interior in some cases, but sufficient care is not usually taken in cutting the re-entering angle where the brick is not exposed to view, the latter generally becoming cracked or broken, as bricks do not lend themselves to be easily cut in this manner. (2) The number of small pieces of bricks should be reduced to a minimum. Closers are usually employed in obtuse angles. Special bricks are cut showing three-quarters of a brick and one-quarter of a brick on the other; the three-quarter face is placed in the stretching course, the quarter brick face in conjunction with a closer is placed in the heading course. This arrangement is shown in figures 118 to 124. In acute angles the quoin brick shows three-quarter brick on one face and a half-brick on the other, the three quarter being placed in the stretching course and the half-brick in the heading course, no closer being used; this necessitates the quoin brick being reduced in width, the amount depending upon the angle of the wall. Squint piers are constructed on similar principles to squint quoins, a modification sometimes being necessary where it is desirable to construct rebated jambs.

*Projecting Courses.*—There are three cases in which it is necessary to enlarge the horizontal areas of walls: 1st, to increase the area of the base to distribute the pressure over a greater area of earth as in footings; 2ndly, to form a projection to afford a bearing area to support the ends of girders or joists; and 3rdly, for the purpose of obtaining architectural

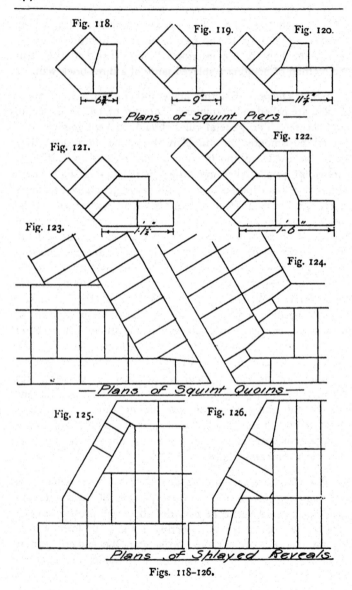

Fig. 118.

Fig. 119.

Fig. 120.

— *Plans of Squint Piers* —

Fig. 121.

Fig. 122.

Fig. 123.

Fig. 124.

— *Plans of Squint Quoins* —

Fig. 125.

Fig. 126.

*Plans of Shlayed Reveals*

Figs. 118–126.

effect, as in the construction of strings or cornices. The following two rules must be complied with in order to obtain the greatest efficiency. First, the ratio of the projection to the depth of the bed in any brick should not be more than one-fourth. This is to ensure the bricks from overturning provided they are properly weighted at their back ends. Secondly, all bricks as far as possible should be laid as headers; this renders the bricks more secure from being drawn from the wall.

*Footings.*—These are the wide courses placed at the base of a wall to distribute the pressure over a greater area of ground. Beneath the footings of all brick walls it is usual to place a bed of concrete of sufficient width and thickness to prevent the ground being loaded beyond its ultimate resistance, and of a thickness sufficient to bridge over any soft places in the soil, and to prevent the concrete beam from failing under the load. For methods of determining thicknesses of concrete in foundations, see *Advanced Course.*

The course coming immediately upon the concrete, to comply with the bye-laws, should be twice the required width of the wall; thus, in a two-brick wall this course would be four bricks wide. Offsets of $2\frac{1}{4}''$ are then made on each side of each successive course till the desired thickness is obtained. Walls of two or more bricks in thickness frequently have their bottom courses of footings doubled, as shown in figures 132 and 133.

To comply with the second rule, care should be taken that the bricks in footings should be laid as far as possible as headers, but if stretchers are required in any course they should be laid near the centre of the wall.

Figures 127 to 133 give sections of footings for walls from one to three bricks in thickness.

*Corbelling.*—It is sometimes necessary to support loads by the method of brick corbelling, which consists of one or

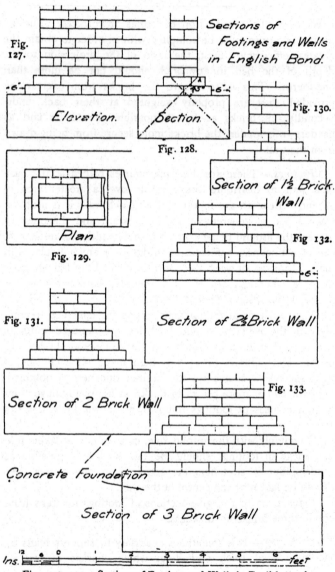

Fig. 127.

Elevation.          Section.

Sections of Footings and Walls in English Bond.

Fig. 128.

Fig. 130.

Section of 1½ Brick Wall

Plan

Fig. 129.

Fig 132.

Fig. 131.

Section of 2⅖ Brick Wall

Section of 2 Brick Wall

Fig. 133.

Concrete Foundation

Section of 3 Brick Wall

Ins.          feet

Figs. 127–133.  Sections of Footings and Walls in English Bond.

more courses projecting a distance sufficient to afford the
required bearing area for the load.

The two conditions given for projecting courses must be
rigidly adhered to. Corbelling renders the walls less stable
by bringing the centre of gravity of the mass nearer the
internal edge of the wall. This is compensated for, in the
case of girders and floor joists, by the lateral support given by
these members. Figures 134 and 135 give two examples.
Stone and iron corbellings are shown in the chapters on
Masonry and Floors.

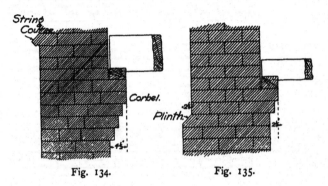

Fig. 134.       Fig. 135.

*Brick Cornices.*—Brick cornices are carried out on the
principles of corbelling, the length of bricks being 9 inches ;
no cornice made entirely of bricks should project more than
that amount. This being accepted, bricks are not suitable for
the large projecting cornices of buildings treated in the
Classic styles. Wherever bricks are employed in the latter
styles metal cramps of sufficient substance and suitable
section to obtain the necessary transverse resistance to
support the upper parts of cornice, and to tie in the mass,
should be employed ; if the cornice has modillions, the latter
are usually of stone of a colour resembling the bricks and
well tailed into the wall to form a support for the crowning
courses as shown in figure 142. Figure 141 shows the brick

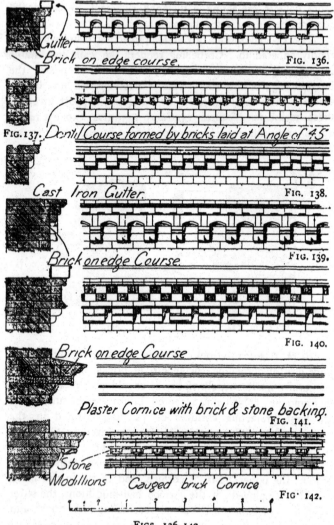

Gutter

Brick on edge course. FIG. 136.

FIG. 137. Dentil Course formed by bricks laid at Angle of 45°

Cast Iron Gutter. FIG. 138.

Brick on edge Course. FIG. 139.

FIG. 140.

Brick on edge Course

Plaster Cornice with brick & stone backing.
FIG. 141.

Stone
Modillions Gauged brick Cornice
FIG. 142.

FIGS. 136–142.

backing for a plastered cornice ; the large projection is also here obtained by the use of stone. Bricks are more suitable for cornices of buildings of the Gothic styles, which usually resolve themselves into a moulded band supported by a corbel table as shown in figures 136 to 140. In either variety there is no detriment in placing the bricks on edge wherever the dimensions of the members or disposition of the parts render that arrangement necessary.

*Plinth.*—A horizontal and usually projecting course built at the bases of walls, as in figure 135, to protect walls from injury and give additional strength.

*String Course.*—The name given to horizontal courses, sometimes projecting and moulded, built in the faces of walls, anywhere between plinth and cornice, as in figure 134, to act as a tie, and architecturally to emphasize the horizontal divisions of a building.

*Damp-Proof Courses.*—Walls are liable to become damp in three ways. First, by dampness rising up the wall from the ground and by being forced through the substructure into the basement story ; secondly, by dampness passing through from the faces of walls ; thirdly, by dampness passing down through from the tops of walls. Walls constructed on damp sites become wet for a certain height above the ground level, due to the action of capillarity. This is objectionable both from the constructional and hygienic standpoints. This is prevented by inserting an impervious material between the source of dampness and the portion of the wall necessary to be kept dry. Damp-proof courses are used horizontally and vertically.

There are four materials in common use for damp-proof courses. First, two courses of slate set in cement mortar ; secondly, a layer of asphalte ; thirdly, a course of vitrified stoneware ; fourthly, sheet lead.

Good slates are practically impervious to moisture; where used as damp-proof courses they are bedded in Portland cement mortar and laid in two courses, so that the joints in the two layers do not coincide, for the purpose of resisting the ready flow of the water through the joints. This system is used both vertically and horizontally. Special slates are prepared for the varying thicknesses of brick walls to avoid longitudinal joints. In vertical sheets the slates are usually bedded on the outside face of the wall.

Asphalte is laid both vertically and horizontally. For horizontal courses it is usually applied hot and screeded in layers of half inch in thickness over the whole width of the wall. Where applied vertically, it is usually placed $4\frac{1}{2}$ inches from the face of the wall. It is all essential that there should be no faults in the vertical sheets; to ensure this, when building the walls a board the thickness of the required asphalte sheet is bedded in the cavity between the outside and inside portion of the wall to ensure a clean cavity; this board is withdrawn at every third course and the asphalte is run in hot, being worked to prevent the formation of air bubbles; as asphalte melts at a low temperature, at each fresh application it fuses together and thus ensures a perfect joint.

Vitrified stoneware are blocks prepared to the standard widths of brick walls from a refractory earth, which, while being burnt in the kiln, is glazed by the application of chloride of sodium, as described in the *Advanced Course*. They are manufactured from 2 to 3 inches in thickness, and to prevent warping in burning have a series of perforations extending from front to back; these are valuable as air ducts. It is usual to apply vitrified stoneware in horizontal courses only.

Lead is used vertically and horizontally as a damp-proof course. A very effective horizontal damp-proof course is obtained by placing a layer of sheet lead through the whole thickness of the wall, either directly below the coping stone, or from 6 inches to 1 foot above the ground level. There is an objection that lead in the latter position, where it occurs under a heavy building, is liable to squeeze out to a certain extent; this is not sufficient to injure the lead, but the surface of the latter having become compressed and adapted to the rough surface of the stone in its movement about the outside edges, has the tendency to cause small pieces of the adjacent stone to spall. Lead is useful for vertical courses owing to the large continuous areas that can be obtained. Any joints required between the sheets are welted. The sheets of lead are placed $4\frac{1}{2}$ inches from the outer face of the wall. Lead is subject to being acted upon by the $CO_2$ from the air and the lime in the mortar, and so converted into a basic carbonate. This action would take an enormous time to accomplish, and may, therefore, not be regarded as a great objection to the use of lead.

*Generally.*—The whole of the site enclosed by the external walls of any house should have the vegetable mould removed, and a layer of cement concrete 6 inches thick should be spread over the whole area; this is effective for sterilizing the ground, not for the purpose of preventing the dampness rising from the ground, unless the concrete is fine and close as is specified for ferro-concrete in the *Advanced Course.*

*Floor level above ground.*—The damp-proof course is placed throughout the whole thickness of all walls at a height of 6 inches above the ground level. All woodwork should be placed above the damp-proof course. Efficient ventilation should be provided by means of air bricks or perforated stoneware for the enclosed space between the concrete and floor, as shown in figure 143.

Where the floor level is below the level of the ground it is necessary to have two horizontal damp-proof courses, the first 6 inches above ground, to comply with the bye-laws, the second beneath the wood sleeper plates of the floor. Under these conditions it is necessary to prevent the dampness passing through the vertical portion of the wall between the horizontal damp-proof courses ; this may be done as in figures 144 to 146 by placing a damp-proof material

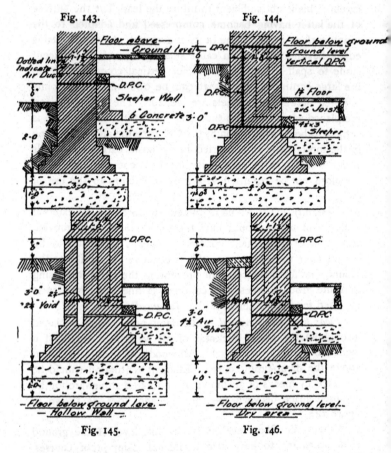

Fig. 143.

Fig. 144.

Fig. 145.

Fig. 146.

against the outside face of the wall or behind the front $4\frac{1}{2}$ inches of brickwork, or as shown in figure 145 by constructing a cavity $2\frac{1}{4}$ inches wide in the section of the wall, or as shown in figure 146 by forming a small external retaining wall,

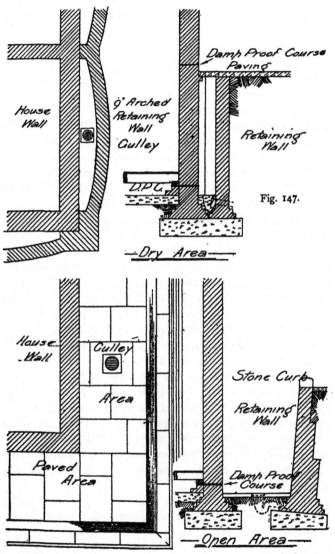

House Wall

9° Arched Retaining Wall

Damp Proof Course
Paving

Gulley

Retaining Wall

D.P.C.

Fig. 147.

—Dry Area—

House Wall

Gulley

Area

Stone Curb

Retaining Wall

Paved Area

Damp Proof Course

—Open Area—

Fig. 148

leaving a cavity; this leaves the total thickness of the wall unimpaired. The two latter methods are not to be recommended, as the spaces are too small to be conveniently cleaned out and cannot be air-flushed.

Where the floor level is much below the ground level, the pressure on the retaining wall becomes great; to avoid the construction of unnecessarily thick retaining walls, the latter are built segmental in plan and thus act as arches to resist the horizontal pressure. Figure 147 shows this method. The open area thus formed must be sufficiently large to enable it to be cleansed easily and also must be drained. Figure 148 shows an example with a wide open area, the earth being supported by an independent retaining wall.

The open area is often necessary for lighting the basement stories.

Basement stories constructed in water-logged soils may be rendered dry by treating them as tanks lined with asphalte, the latter being applied in existing buildings by constructing a $4\frac{1}{2}$-inch wall, $\frac{3}{4}$ inch distant from the existing wall, and the joints of which have been raked out, and filling the space between the two walls with asphalte in a heated and liquid condition. In constructing the $4\frac{1}{2}$-inch wall, three courses at a time should be built, and a board the thickness of the cavity being kept between the two walls to keep the space clear; the back joint of the $4\frac{1}{2}$-inch wall must also be kept back for about 1 inch. When the three courses are erected the board is withdrawn, and the asphalte is poured into the space left between the walls. This method of obtaining a key and binding the two walls together is sufficient to render the $4\frac{1}{2}$-inch wall stable enough in most cases to resist the pressure of water from the outside.

In the method known as Callender's Bitumen Damp Course, the material is placed in sheets against the old wall, in which a half-brick is taken out in every square yard of its face; into this a bitumen pocket is fitted, the edges of which lap over the face sheets. The 4½-inch wall is then built up against the latter, the pockets making it possible to form a tie by using 9-inch heads through the 4½-inch wall, as shown in figure 149.

The water may be prevented from passing through the floor by laying asphalte on the 6-inch concrete covering the site. This must be weighted down by bricks on edge or by 6 inches of concrete over the whole area. In spaces where the span between the walls is great, the upward pressure may be resisted by forming a concrete invert, as shown in figure 150, and bedding a one-ring inverted arch between the walls, and filling up to a level with concrete, or by omitting the arch and filling up to the level with concrete, as shown in figure 150, or by laying the asphalte level and bedding steel joists of sufficient section to resist the upward pressure and embedding in concrete.

*Prevention of Damp passing through Faces of Walls.*— Dampness is prevented from passing through in three ways, first, by rendering in Portland or Roman cement; secondly, by hanging slates or tiles on the front faces of the walls; thirdly, by constructing hollow walls.

*Rendering.*—The exposed surfaces of the walls are covered with two coats of Portland or Roman cement; of the two the Roman has the closer texture, and working fatter gives a smoother surface, but does not adhere with the same tenacity as Portland. The latter is mostly used.

*Slate and Tile Hung.*—For the purpose of resisting damp- ness, the exposed surfaces are covered with slates or tiles in a manner similar to roof surfaces. Battens are nailed to plugs in the wall, and each slate or tile secured with two strong nails. By this method a cavity is formed between the inner and outer surfaces, which causes the internal temperature to

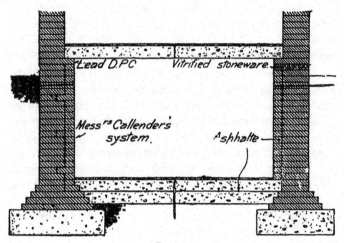

Lead D.P.C  Vitrified stoneware.

Mess.ʳˢ Callender's
system.

Ashphalte

Fig. 149.

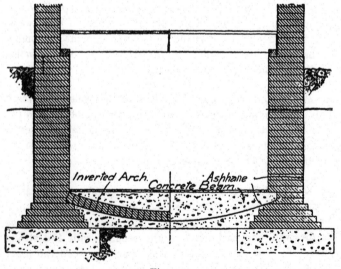

Inverted Arch.  Ashphalte

Concrete Beam.

Fig. 150.

be more equable for this reason ; thin brick or half-timbered walls are frequently slate or tile hung.

*Hollow Walls.*—Double walls are formed, one usually half brick in thickness with a cavity from 2¼ to 3 inches between. The thin wall in brickwork is placed on the outside. The two walls are connected by means of ties of stoneware or iron as shown in figures 24 and 25 ; these are bent or twisted to prevent water being conducted by them to the internal wall. The ties are placed 3 feet apart horizontally and every 18 inches in height, making two to every square yard. All wood-work, such as sash or door frames, should be protected by sheet lead placed over the top, as shown in figure 151. The cavity prevents the conduction of dampness, and makes the internal temperature more equable, but it is objected to on hygienic grounds, as all cavities that cannot be subjected to light and air flushing are not good.

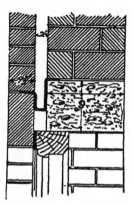

Fig. 151.

*Prevention of Dampness percolating from above.*—This object is attained by placing a damp-proof course either on top of the wall or immediately under the top courses. First a coping, as shown in figure 319, which is often a highly stratified stone, such as the York stone, bedded on the tops of the walls, projecting at least two inches on both sides of the walls, and having a throating worked on the under side to prevent the water trickling down the wall. The upper side of the stone should be weathered to throw off the rain rapidly. There are a number of specially moulded terra-cotta vitrified stoneware or blue-brick sections in the market, as shown in figures 152, 154, and 156, admirably adapted to be used as copings. A damp-proof course of two courses of slates or tiles set in cement is bedded beneath the top course of brickwork; these are termed creasing courses. The slates are given a projection of at least two inches, to throw the water clear of the wall. The projecting portion of the slates is weather flaunched in cement, as shown in figure 153. Special tiles are manufactured for creasing courses, as shown in figure 154. Asphalte and sheet lead are frequently used beneath the top courses of brick-

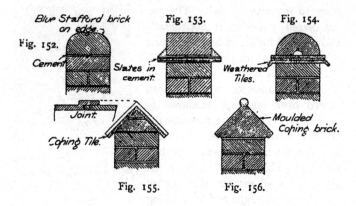

Fig. 152.

Blue Stafford brick on edge.

Cement

Slates in cement.

Fig. 153.

Fig. 154.

Weathered Tiles.

Joint.

Coping Tile.

Fig. 155.

Moulded Coping brick.

Fig. 156.

work or masonry to prevent the percolation of water from above.

*Arches.*—An arch is an arrangement of wedge-shaped blocks mutually supporting each other, built about the form of some curve, and being supported at both ends by abutments or piers. They are designed to support the weight of a wall over an opening. The important points to be observed in the construction of an arch are as follows :—

1st. Sufficient weight and strength in the abutments to safely resist overturning moment of the arch thrust.

2ndly. Depth enough in the arch to prevent the weight from above crippling it.

3rdly. Sufficient area in the pier and arch to prevent failure by crushing.

4thly. All bed joints should be perpendicular to the line of least resistance. In practice they are made normal to the curve of the arch, in which position they nearly approximate to normals of the line of least resistance. (See Article, *Advanced Course.*)

*Technical Terms.*—The following are the technical terms used in connection with arches :—

*Voussoirs.*—The courses of bricks or stones which compose the arch.

*Springers.*—The extreme or lowest voussoirs of arches.

*Skewbacks.*—The bricks or stones forming the sloping abutments on which the extreme or lowest voussoirs rest, of flat or segmental arches.

*Key.*—The uppermost or central brick or stone of the arch.

*Intrados* or *Soffit.*—The under or concave side of the arch.

*Extrados* or *Back.*—The upper or convex side.

*Abutments.*—The outside supports of an arcade, as shown in figure 157, are piers designed to resist the inclined thrust from one or more arches ; they are therefore made of greater weight than ordinary piers by increasing the height or by increasing the width in the direction of the thrust.

*Piers.*—The intermediate supports of an arcade. They are rectangular pillars constructed to support heavy weights transmitted to them by beams and girders, or to receive the thrusts of two or more arches, the resultant of the thrusts of which falls in a vertical line.

*Capital.*—The name given to the moulded or carved cornice of a column, as shown in figure 157.

*Abacus.*—The moulded slab crowning the capital of a column, by which the arch or lintel is supported, as shown in figure 157.

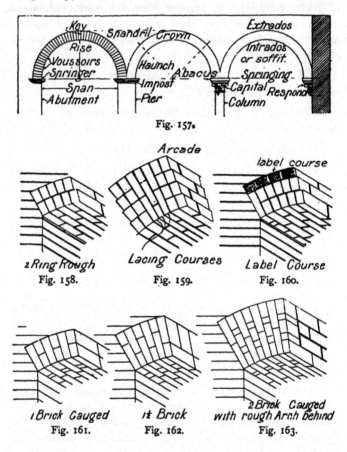

Fig. 157.

Arcade

2 Ring Rough
Fig. 158.

Lacing Courses
Fig. 159.

label course

Label Course
Fig. 160.

1 Brick Gauged
Fig. 161.

1½ Brick
Fig. 162.

2 Brick Gauged
with rough Arch behind
Fig. 163.

*Impost.*—The upper part of a pier or abutment on which an arch rests or from which it springs, generally finished with a moulded cap, as shown in figure 157.

*Arcade.*—A series of arches carrying a wall, above, and supported by columns or piers, as shown in figure 157.

*Respond.*—The semi-column or corbel forming the support of the end arch of an arcade, usually projecting from a flank wall, as shown in figure 157.

*Springing Points.*—The points from which the curves of the arch commence, as seen in elevation.

*Span.*—The horizontal distance between the springing points.

*Crown.*—The highest point of extrados of arch.

*Rise.*—The vertical distance between the highest point of the soffit and the level of the imposts or springing points.

*Haunch.*—The name given to the lower half of the arch, from the springer or skewback, midway to the crown.

*Spandril.*—The irregular triangular space between two arches, or the space enclosed by vertical lines drawn from the springing of the extrados and the horizontal line tangent to the crown.

*Ring Courses of Arches.*—The name given to those courses of brickwork that partake of the circular form, as seen on the face of the wall. Figures 158 to 163 give examples of $4\frac{1}{2}''$ rings, whilst figures 161 to 163 give examples of $9''$, $14''$, and $18''$ rings respectively.

*Lacing Courses.*—The continuity of the ring courses of arches is sometimes broken by a bonding or lacing course, as shown in figure 159. The object of the lacing courses is to distribute the pressure more evenly over the sectional area of the arch.

*Label Course.*—The name given to a course of bricks laid flatwise on an arch and deepening it, as shown in figure 160. These courses usually project and are moulded and weathered on their upper surface to throw all rain-water clear of the arch, and they are frequently made of a different colour to the arch to enhance the effect.

*Classification of Arches.*—There are two kinds of arches, rough and gauged: the latter are subdivided into axed and rubbed.

*Rough Arches.*—Arches, as shown in figures 164 and 165, constructed of the ordinary uncut bricks are known as rough arches. Owing to the bricks being rectangular in section, the the mortar joints are wedge-shaped. To prevent the thick end of the mortar joint becoming too large, no bricks are laid as stretchers on the face of the arch, but are built in half-brick rings. These are used for all purely constructional work where the appearance is of secondary importance. Rough arches are usually placed behind all gauged arches and over wood lintels; they are then generally known as relieving or discharging arches. They are built on centres or turning pieces where spanning a clear opening, or on a brick core when relieving a wood lintel. Coke breeze concrete lintels are usually shaped on their top surface to form the core. Where rough arches are used of several rings in thickness, they are sometimes tied together by three courses of bricks laid as stretchers, roughly axed to the wedge-shaped form and extending through the whole depth of the arch; these are known as lacing courses, as shown in figure 159.

The French, or Dutch arch, as shown in figures 166 and 167, is one form of a rough arch or lintel used for spanning openings with square heads in walls that are finished with a plastered surface. This is weak construction, and is only used for cheapness.

*Inverted Arches.*—These are rough arches inverted, as shown in figures 168 to 170, springing from piers or abutments upon which the bulk of the weight is concentrated and with which there would be a possibility of unequal settlement with the remainder of the wall due to the body of the wall or ground under the foundation being subjected to a non-uniformly

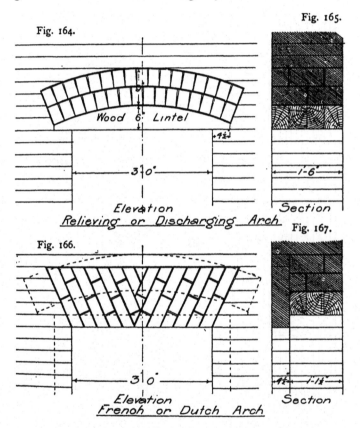

Fig. 164.

Fig. 165.

*Wood 6" Lintel*

*4½*

*3' 0"*

*Elevation*

*Section*

Relieving or Discharging Arch

Fig. 166.

Fig. 167.

*3' 0"*

*4½* *1'-1½*

*Elevation*

*Section*

French or Dutch Arch

distributed stress. Their effect is to distribute the pressure uniformly over the whole wall or along the whole length of the foundation. The segment of the arch should be such that the angle enclosed by producing the skewbacks subtends

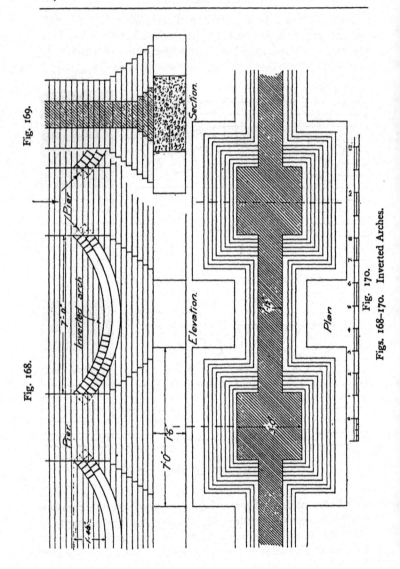

Fig. 169.

Fig. 168.

Section.

Elevation.

Plan

Pier

Pier

Inverted arch

7'·0"

7'·0"

Figs. 168-170. Inverted Arches.

Fig. 170.

120°, as this corresponds very closely to the curve of the line of least resistance produced by a uniformly distributed load.

*Trimmer Arches.*—These are a form of rough arch adopted for supporting hearths in front of chimney breasts, and having an abutment against the trimmer or trimming joists, as shown in the chapter on Floors.

*Gauged Arches.*—All arches in which the bricks are cut to definite sizes and shapes are known as gauged arches. There are two methods of cutting these, the first known as axed work, the second as cut and rubbed work.

*Axed Work.*—This labour is usually placed on the voussoirs of arches constructed of bricks too hard to be economically cut with the saw. It consists of reducing the brick to the required form by means of a bricklayers' axe or a scutch, the process being as follows:—A templet is prepared of wood about $\frac{3}{8}$ of an inch thick of the shape required. The templet is laid upon the brick which is marked and cut with a tin saw, the cut being sunk about $\frac{1}{8}$ of an inch in depth; the ends are squared over and cut, and the back of the brick treated in a similar manner. The required shape being now outlined, the superfluous portions of the brick are removed with a bolster and hammer, the brick is now placed on the chopping block, as shown on page 86, and the cut surface levelled with a scutch. The object of sinking with a tin saw is to preserve a sharp arris about the brick. Mouldings cannot be economically cut on hard bricks, there-fore these bricks are usually moulded before burning.

*Cut and Rubbed Work.*—In highly finished brickwork special bricks are employed known as cutters or rubbers (see *Advanced Course*) that can be easily cut and worked to the required forms. The bricks are cut with a bow saw, which latter is shown on page 86, and the surfaces of the brick are finished by rubbing them with a stone or with a file. The

F

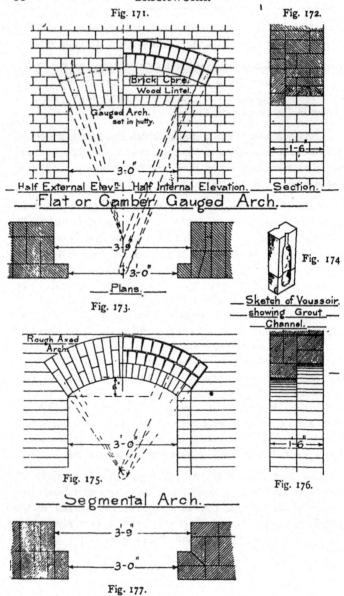

Fig. 171.

Fig. 172.

Brick Core.
Wood Lintel.

Gauged Arch.
set in putty.

3'·0"

Half External Elevn. | Half Internal Elevation. | Section.

Flat or Camber Gauged Arch.

3'·9"

3'·0"

Plans.

Fig. 173.

Fig. 174

Sketch of Voussoir.
showing Grout
Channel.

Rough Axed
Arch

6"

3'·0"

Fig. 175.

Fig. 176.

1'·6"

Segmental Arch.

3'·9"

3'·0"

Fig. 177.

processes vary slightly with the particular work in hand, but four general operations include most of the work in general practice, viz. :—1, squaring up as in ashlar work; 2, straight moulded work; 3, circular work, plain or moulded; and 4, work of double curvature. The object of rubbing gauged work is to produce regularity and extreme fineness in the joints, these being made from $\frac{1}{32}$ to $\frac{1}{16}$ of an inch in thickness. To produce such joints the substance of the bedding material must be reduced to a very fine state of division. Pure slaked lime known as putty is the material employed.

*Outline of Arches.*—The name by which an arch is known is generally that of the curve of which the arch is a segment.

The following figures (178 to 190) show the methods of

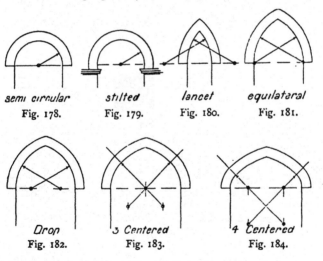

| semi circular | stilted | lancet | equilateral |
| Fig. 178. | Fig. 179. | Fig. 180. | Fig. 181. |

| Drop | 3 Centered | 4 Centered |
| Fig. 182. | Fig. 183. | Fig. 184. |

setting out the semi-circular, stilted, lancet, equilateral, drop, three-centred, four-centred, Venetian pointed, and Florentine forms of arches.

The setting out of the first seven is obvious.  The depth
of the arch at the springing in the Venetian pointed must be
given, the curve of the extrados and heading joints subtend at

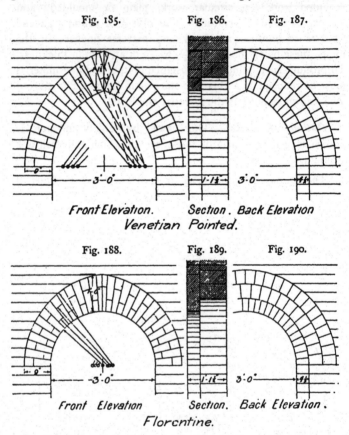

Fig. 185.　　　　　　Fig. 186.　　　　Fig. 187.

Front Elevation.　　Section. Back Elevation
　　　　　Venetian Pointed.

Fig. 188.　　　　　　Fig. 189.　　　　Fig. 190.

Front Elevation　　　Section. Back Elevation.
　　　　　Florentine.

their centres similar angles to that subtended by the arc of their
intrados, as shown in figures 185 to 187.  In the Florentine,
as shown in figures 188 to 190, the intrados is a semi-circle,
therefore the depth at the springing and crown must be given,

The centre will be upon the springing line at the intersection of the bisecting line of the line joining the crown and the springing of the extrados or heading joints.

*Elliptical Arches.*—The curve of the ellipse is usually struck for most practical purposes by the trammel. This is inconvenient for arches owing to the amount of labour in determining the bed joints, which should be all normals to the

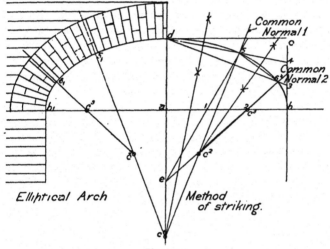

Fig. 191.

ellipse; and, further, if this were done each voussoir would be of a different shape, thus requiring a separate templet for each voussoir. This may be avoided by striking a curve from three centres, as shown in figure 191, which can be made a very close approximation to a true ellipse, and requiring only three templets to cut the voussoirs. To strike the semi-arch, let a rectangle be described having for its sides the semi-major and semi-minor axes, divide the lines *a b* and *b c* each into three equal parts at points 1, 2; 3, and 4. From the points 3 and 4, draw lines to the point $d$, and from $e$, which is distant from $a$ a length equal to $a\,d$, draw lines through points 1 and 2; the intersection of the lines drawn through 2 and 3 and 1 and 4 are points in the curve of a true ellipse. To find the first centre $c$, draw a line from $d$ to 5 and bisect it; at the intersection of the bisecting line and the minor axis produced will be the first centre. Join $c^1$ and 5, this will be the common normal between the curves $d.5$ and $5.6$, and must therefore contain centre No. 2, which may be found by bisecting a line joining 5 and 6 and producing till it cuts the common normal $c^1\,5$. From $c^2$ draw a line through point 6, this will be the second common normal and will contain $c^3$. The intersection of common normal 2 with springing line will give the third centre. All the bed joints of the voussoirs are struck from the centres of the segments in which they are situated.

*The Flat or Camber Arch.*—This arch, as shown in figures 171 to 174, requires to be accurately drawn on paper. The depth of the arch is invariably made some multiple of the thickness of a brick. The extrados is worked perfectly horizontal; the intrados is given a camber, that is, an upward curve to allow for any slight settlement and to correct the apparent sagging of a horizontal line, the usual allowance being $\frac{1}{8}$ of an inch rise in the centre for every foot of span. The amount of skewback should vary with the span of the arch; $1\frac{1}{2}$ inches for every foot of span in a depth of 12 inches gives good results. The centre of the arch will be at the intersection of the skewback lines produced; the voussoirs should be divided upon a segment drawn from the centre and tangent to the extrados as shown; an odd number of bricks is required to give a key brick. The lowest bricks in key and springers should be stretchers; to obtain this the number of voussoirs must be odd, and some multiple of four plus one, as, 5, 9, 13, etc. The heading joints should be all kept horizontal. A templet of $\frac{3}{8}$ of an inch pine is prepared, and on this the correct form of every voussoir is marked. The bricks are then prepared by rubbing one face and squaring one edge, they are arranged in pairs with the exception of the key. The heading joints are next formed, the bevils being taken from the templet, then marked on the bricks, cut with the saw and the joints finished accurately with the stone. The extrados and intrados of each voussoir is cut and finished in a similar manner; the templet is then applied to the face of the voussoir, the tapering is marked on and the voussoir placed in the cutting mould, adjusted and wedged in position, the superfluous portions of the brick are removed with a saw and the surface finished with a flat file. The voussoirs are jointed together on the surfacing slate, either into the complete arch or into convenient sections, the back of the joints being grouted with Portland cement. In setting the arch it can be placed *en bloc* on the turning piece and between the skewbacks prepared to receive it, or if large, the various sections are jointed together on the turning piece. The arch may be built upon the turning piece, in which case the exact position of each brick must be marked thereon.

Figures 192 to 194 show a gauged relieving arch treated by an ornamental arrangement of coloured bricks.

Figures 195 and 196 show a gauged relieving arch over a stone lintel. It is unnecessary for the relieving arch to spring from the extremities of the lintel, as in this case the lintel is of a material equally durable as the arch.

*Pointed Arch with Rere-arch.*—Figures 197 to 199 show an opening formed with splayed internal jambs, pointed arch, and stone cusps externally, and pointed barrel rere-arch internally, which is the method adopted in arcuated openings where splayed jambs are employed. The intersection of the rere-arch and splayed jambs shown in the figure 198 is obtained by drawing a number of vertical lines on the splayed jambs, shown in the internal elevation, figure 197, and drawing these

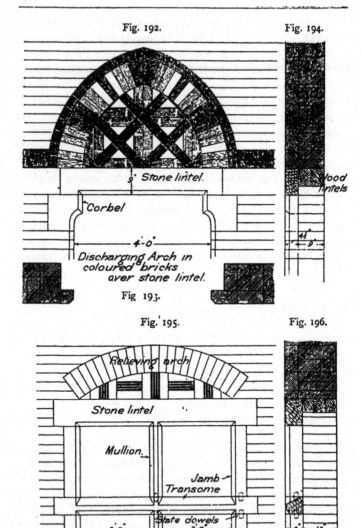

Fig. 192.        Fig. 194.

9ʺ Stone lintel.

Corbel

Wood lintels

4′·0″

Discharging Arch in
coloured bricks
over stone lintel.

4½ʺ

9ʺ

Fig 193.

Fig. 195.        Fig. 196.

Relieving arch

Stone lintel

Mullion.

Jamb
Transome

Slate dowels

2′·0″     4½ʺ     2′·0″

6ʺ   12ʺ

Relieving or Discharging arch over Stone lintel.

Figs. 192–196.   Arches with Coloured Treatment.

lines on the jamb in the sectional elevation and projecting
from the internal elevation ; the points of intersection of these
vertical lines with the barrel arch, joining these projected

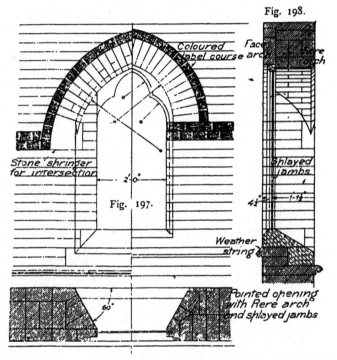

Fig. 198.

Fig. 197.

Fig. 199.

points in the sectional elevation will give the curve in this
projection.

*Arch Built with Orders.*—This is another treatment of the
arch when employed in the openings of very thick walls,
where the arch consists of a number of rings similar to a rough
arch. It is a common practice to diminish the width of each
ring from the outer (which is generally in width the full

thickness of the wall) to the inner ring, the width of which is very much smaller, thus forming a number of receding planes: each of these rectangular rims is termed an order. This

Fig. 200.          Fig. 201.

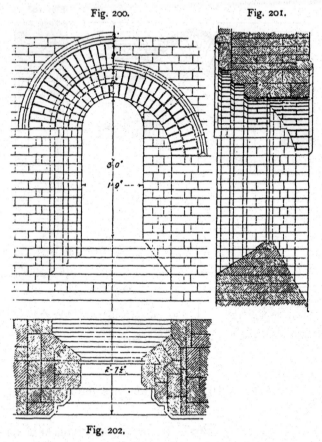

Fig. 202.

relieves the opening of the clumsy and heavy appearance that is attendant upon a square opening in such a position; it also admits of the distribution of light inside to a much greater extent than is the case with a square opening. The jambs inside such an opening are usually splayed; the inner edge of the splay may be carried round the outer arch in a concentric curve or curves, or the inner arch may be an ordinary barrel arch, with its crown at or near the same height as that of the

Fig. 203.

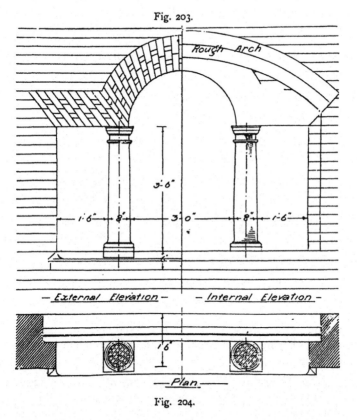

— *External Elevation* —    — *Internal Elevation* —

—*Plan*—

Fig. 204.

outer, and with the springing-line much lower, thus forming a curved line of intersection on the internal jambs and soffit. The edges of the external receding planes are often richly moulded. The orders may rest at the springing of the arch on the abaci of small collonettes, or they may be continued to the sill, as shown in figures 200 to 202.

*Three-Light Opening.*—This arch, as shown in figures 203 and 204, is sometimes known as a Venetian arch. It consists of an opening divided into three by light stone columns or small brick piers, the central opening usually being semi-circular. If the arch is formed of bricks, it is carried out in gauged work. The portions surmounting the two side openings have their bed joints at a constant angle, ranging between 50° and 60°. The bed joints for the central arch radiate from the point on the centre line at which lines drawn at the constant angle from the springing points intersect.

*Lintels.*—These are horizontal members placed over

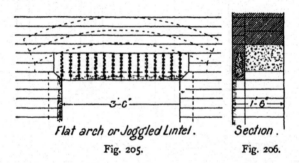

*Flat arch or Joggled Lintel.*

Fig. 205.

*Section.*

Fig. 206.

square openings to form the head and support the work above; they may be of brick, wood, iron, coke-breeze concrete, or stone.

*Flat Gauged Arch* or *Joggled Lintel*, as shown in figures 205 and 206. These are employed in openings in glazed brick-work and where it would be difficult to cut the bricks to the necessary wedged form. The bed surfaces of the bricks are rubbed to obtain a thin joint, they are then built upon a turning piece and the frogs (of which there are two on each brick) filled with Portland cement grout, which forms a joggle.

It is unwise to use these in openings greater than 4 feet. Above that span the bricks should be perforated and an iron bolt passed through.

*Wood Lintels.*—These consist of pieces of timber from $4\frac{1}{2}$ by 3 inches upwards, according to the span of the opening; they are generally assisted by a rough brick arch as a precautionary measure in the event of the timber being destroyed. The space between the lintel and the arch is filled with a brick core, see figure 171. The lintel sometimes has its upper surface curved, thus forming a turning piece for the arch, see figure 164.

*Iron and Steel Girders.*—These are generally used where the openings are very wide, and where there is no space for an arch above, as shown in the figure illustrating the shop front in the *Advanced Course.*

*Coke-breeze Concrete.*—This is formed of 4 parts by volume of coke-breeze to one of Portland cement, and is often employed, being to a certain extent fire-resisting and admitting of joinery being fixed to it, as it will hold nails and screws with great tenacity. These lintels are often curved on their upper edge, in which case they may be relieved by an arch above, or they are made straight on their upper surface and their depth is increased. Where used for wide openings, tensional strength is obtained by embedding a mild steel bar for each half brick in the thickness of wall, supported by lintel $\frac{1}{2}$ inch thick and having a depth of $\frac{1}{2}$ inch for each foot of the span.

*Stone Lintels.*—Stone is unsuitable to withstand transverse stresses, and should never be used for openings above 3 feet in span unless relieved of the weight above by a discharging arch or save stones, as shown in figures 346 and 349; otherwise, they would require to be abnormally deep.

*Fixing Joinery to Brickwork.*—The following materials

are generally built in or inserted into brickwork to assist in the fixing of joinery thereto :—

*Breeze Bricks.*—These are frequently built into the jambs of openings and other parts of walls to which joinery is to be fixed for securing the backings, as the bricks admit of nails and screws being driven into them and hold to same with tenacity.

They have the advantage of not shrinking or perishing to the same extent as wood, but are inferior to wood in that nailed backings cannot be adjusted after having once been fixed. They are made of a similar thickness and width to ordinary bricks, and can be built in the walls in a similar manner.

*Wood Bricks.*—These are made of similar dimensions to ordinary bricks. They have been used to form a fixing surface in brickwork, but they are objectionable because they shrink, become loose and are liable to rot.

*Wood Plugs* are wedges cut on the twist. They are usually driven between the joints of brickwork to obtain a fixing for joiners' work, as shown in Part II., but are not to be recommended in jambs or edges of walls, as they are very likely to damage such parts.

*Wood Joints.*—These are lengths of wood, $4\frac{1}{2}$ inches by $\frac{3}{4}$ inch, built into walls at intervals, or wherever fixing is required. The shrinkage in thickness in such thin pieces is inappreciable ; it is the best mode of fixing in most situations, but has to be inserted while the brickwork is being carried up.

*Joints on Face.*—The joints on the face of work are finished in a variety of ways, as shown in figures 207 *a* to *l*, to increase the effect, and to resist the weather ; they may be finished as the work proceeds, or as the scaffold is taken down on the completion of the building ; the former is the stronger and more durable, the latter is cleaner and has a better appearance, and is rendered necessary when the work has been built during frosty weather; where the latter method is employed, the joints should be raked out for at least $\frac{1}{2}$ inch in depth as the work proceeds. The joints in new work should be clean, sharp and regular. Figures 207 $a$ to $l$ show the forms of joints applied to brickwork.

*Flat or Flush Joint.*—This is formed (as shown in figure 207$a$) as the work proceeds by pressing with the trowel the wet mortar that protrudes beyond the face, flat and flush with the wall. This method is used for internal work, where the brick surface is the finished face. The edges are neatly trimmed with a trowel and straight-edge. It has the advantage that it affords no lodgment for dust.

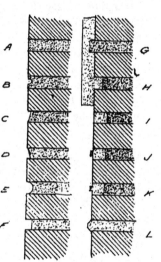

Fig. 207 $a$ to $l$.

*Flat Joint Jointed.*—This is formed similarly to the above, as shown in figure 207 $b$, but has in addition to the previous joint a semi-circular groove run along the centre of each joint with a jointing tool and straight-edge. This has the effect of making the mortar more dense.

*Struck Joint.*—This is formed by pressing the mortar with the trowel along the upper edge of the joint slightly below the surface of the brickwork, as shown in figure 207 $c$. This is a good joint, as the upper edge of the mortar is protected, and any water is thrown off with facility; its appearance is good, as it presents a sharp shadow at every horizontal joint, and forms the method of finishing new work; it is sometimes called a weather struck joint.  The mortar is often ignorantly struck back on the lower edge, as shown in figure 207 *d*, under the impression that the appearance is enhanced thereby, the idea being that a sharp line is presented on the upper edge of the bricks, but as no shadow is formed, the effect is lost at a few feet above the eye, a ledge is formed on which the water lodges, which freezes in the winter, and rapidly destroys the upper edges of the bricks and the joint.

*Keyed Joint,* as shown in figure 207 *e*, is formed by drawing a jointing tool with a curved edge, the same width as the joint, along the latter ; it has the effect of making the mortar dense at this part, and improves the appearance by making the joints distinct.  It is not much used.

Keyed joints of the form shown in figures 207 *g* and *h* are employed where the wall is to be rendered.  In the first case, the mortar in the joints is left protruding ;  in the second, they are raked out.

*Recessed Joint.*—This is used to obtain a pleasing and deep shadow, but care must be taken that the bricks are hard and unlikely to be damaged by the weather.  It is the joint employed in the new Church House, Westminster.  Figure 207 *f* gives this joint.

*Pointing Old Work.*—This operation consists in raking out the decayed mortar from the joints to a depth of at least ¾ inch, and in filling the same with cement, or some hard-setting mortar, as shown in figure 207 *i*.  The joints may be finished in any of the methods stated, or by one of the two methods known as tuck and bastard tuck pointing, which are fancy forms adopted by bricklayers to increase the effect by forming sharply defined joints.

*Tuck Pointing*, as shown in figure 207 *j*, consists in filling up the raked-out joints flush with a stopping of cement or some hard mortar.  The joints in this condition generally appear very wide, owing to the edges of the bricks being ragged, this being due to the frost or to the clumsy method in which the joints have been raked. The whole front, joints included, is then coloured with a compound of copperas, and a pigment of the colour required, or the front is rubbed with a piece of soft brick till the bricks and joints are of one colour. White lime putty, made as described in the article on Plastering, is pressed on to the joints in straight lines, with a jointer worked on a bevelled edge straight-edge; and before the latter is removed, the edges are trimmed with a tool called a frenchman, which usually consists of an ordinary table knife with the end of the blade turned up at right angles to the remainder. The edge of the knife cuts the putty, and the turned-up end drags off the superfluous stuff, leaving a white joint $\frac{1}{4}$ inch in width, and $\frac{1}{16}$ inch in thickness on the face of the work. This is not the best method of pointing if the bricks are sound and their edges sharp and regular; but if the edges are broken, the joint, when stopped, appears very wide and irregular, and is thought by some not to look well if the above process were not adopted.

*Bastard Tuck Pointing* is the name given when a ridge $\frac{1}{4}$ inch to $\frac{3}{8}$ inch is formed on and of the stopping itself, as shown in figure 207 $k$.

*Masons' V Joint.*—Figure 207 $l$ shows the usual joint used for masons' work.

*Paving.*—External areas in towns, or about houses where not required for garden purposes, are usually covered with some form of stone or brick to give a hard surface to walk upon and to facilitate drainage. The following materials and methods are employed, brick, stone or concrete slabs, concrete slabs formed *in situ*, tar paving, asphalting, tiling and mosaic. Bricks for paving should be hard, well burnt and wear with a rough surface. Specially hard bricks, chequered on surface

G

to afford a firm foothold, are manufactured for use in stables and similar positions. Bricks may be laid flat or on edge. Where laid flat, the area is usually covered with 6 inches of concrete, the bricks being then bedded on this in lime or cement mortar; where laid on edge, the ground is either concreted as before, or the surface of earth is levelled and the bricks are bedded in sand, the side joints of the bricks being then filled up with lime or cement grouting.

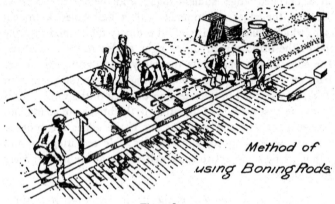

*Method of using Boning Rods*

Fig. 208.

*Stone or Concrete Slabs.*—Slabs of highly stratified stone that can be easily rent into approximately uniform thicknesses, tooled on surface, squared on edges, are largely used for paving. The method of laying is as follows. The ground is first levelled, and the surface covered with sand or sifted earth; the stones are then laid, being bedded in lime mortar about their edges. They are punned to their requisite levels with a tool known as a beedle, which also consolidates them upon their bed. Figure 208 illustrates the laying of stone paving. Concrete slabs have of late years been largely used for paving; they have the advantage of being cast in uniform superficial dimensions and thicknesses, which presents a uniformity in the appearance of the joints upon completion which greatly enhances the appearance. To obtain a similar effect in stone would be expensive, due to the waste in cutting.

*Concrete Paving in Situ.*—This method is largely adopted, in preference to the slab method, on account of the uniformity of the pavement and the fewness of the joints, which renders them clean and sanitary. They are usually formed in sections about 6 feet square. About 3 inches of coarse concrete is first deposited over section, and finished with about $\frac{1}{4}$ inch of fine stuff floated to the correct levels. While the surface is still soft, a metal roller is passed over it, producing a chequered face. The disadvantage of forming large concrete pavements is, that if any part settles and cracks the fracture is likely to star and spread in all directions. This effect is minimised by the formation of sheets not larger than 6 feet square, any damage being localised within this area.

*Tar Paving.*—The material is composed of limestone and coal tar. The paving is formed with a bottom layer equal to one-half of the whole thickness of paving, broken to pass through a sieve with a mesh of $1\frac{1}{2}$ inch, a middle layer equal to one-fourth of the whole thickness of paving, broken to pass through a $\frac{3}{4}$-inch mesh sieve, and the remaining one-fourth broken to pass through a sieve with a $\frac{1}{2}$-inch mesh. The compound is spread over the site by means of an iron rake, and is finished by being well rolled to an even surface. The compound is in the proportion of 12 gallons of coal tar to a cubic yard of broken limestone. Pitch is frequently added, to thicken the coal tar. The limestone and tar are heated before being mixed. Before being rolled, finely divided limestone, sand or grit is thrown over the surface and rolled into it whilst still soft. The total thickness of tar paving by this method varies from $2\frac{1}{2}$ to 4 inches. Another method is to lay a coat of tar paving $1\frac{1}{2}$ inch thick, composed of fine particles, upon a prepared surface of concrete 6 inches thick.

*Asphalte Paving.*—The various methods of asphalte paving are fully dealt with in my *Advanced Course.*

*Tile Paving.*—These are thin slabs of finely prepared and well-burnt clay. They are manufactured in various colours and shapes up to about 12 inches square. The area to be tiled is covered with a bed of 6 inches of concrete; it is then floated with a coat about ¾ inch thick, composed of one of cement and one of sand screeded to an even surface. The tiles are then placed in position and pressed into the cement bed, the latter oozing up between the side joints.

*Mosaic Paving.*—There are two varieties of mosaic paving, Roman and Venetian. In both it is necessary to have a prepared ground of 6 inches of good concrete, on which a coat of ¾ inch of neat cement is floated.

The Roman variety consists of pressing into this floating coat small approximate cubes of marble of various colours about ¼ inch length of side. These may be arranged in various patterns. Pictorial designs are frequently worked. The design is first worked and coloured on paper, the stones required are selected, fitted and stuck on to the drawing, the whole pattern is then placed in position, stones lowermost on the prepared bed of cement. When the cement floating coat has set, the paper is damped and removed; the small cubes forming the ground colours are then bedded singly up to the design, when the whole surface is covered; the face is levelled by rubbing with a stone and finished by polishing.

The Venetian variety is carried out by sprinkling small irregularly shaped chips of marble on to the floating coat of cement previously described; these are pressed into the surface with a hand float, and the whole is consolidated by rolling with a heavy iron roller. Patterns in various designs may be formed in this method by first cutting out the design in wood, and bedding the pieces of the latter in position. The ground work is then bedded about the patterns, and when set, the wood patterns are removed and the voids filled with cement coloured with pigments and with chips of different coloured marbles pressed in as described. The whole is then levelled and polished as in the Roman variety.

In floors composed of steel and concrete the rigid surface formed by the mosaic is subject to cracking, owing to the expansion and contraction of the joists. The ill effect of this is minimised by arranging the floor in sections, so that border lines, and consequently natural joints, are formed along the lines in which the variable expansion takes place.

It is not advisable to make any dimension of these sections greater than ten feet.

*Excavators' and Labourers' Tools.*—The following tools as shown in figures 209 to 238 are used by the excavator and labourer: 1, the pick for breaking up hard ground; 2, the grafting-tool for digging out earth, such as stiff clay; 3, the shovel for lifting earth; 4, the line and pegs, for outlining trenches; 5, the boning rods for levelling; 6, the level, which consists of a spirit-level mounted in a straight-edge at least 10 feet in length; 7, the hod, this consists of three sides of a rectangular box mounted on a staff, and used for carrying bricks and mortar, and capable of carrying 12 bricks, or about $\frac{2}{3}$ of a bushel of mortar; 8, the larry, which is a steel blade bent at right angles, and fixed to a long handle and used for the mixing of mortar; 9, the beedle is a large wood mallet with a circular pine head, with rounded ends about 18 inches and 15 inches in diameter, with a handle about 3 feet long. It is used by the pavior for punning paving stones into their position when bedding, as shown in figure 236; 10, the rammers are of two kinds—1st, that used for ramming granite sets in roadways, and consists of a cylindrical piece of wood about 3 feet 6 inches long, with a vertical handle at top and a horizontal handle about half-way up, as shown in figure 237; 2nd, that used for the bottoms of trenches and for consoli-

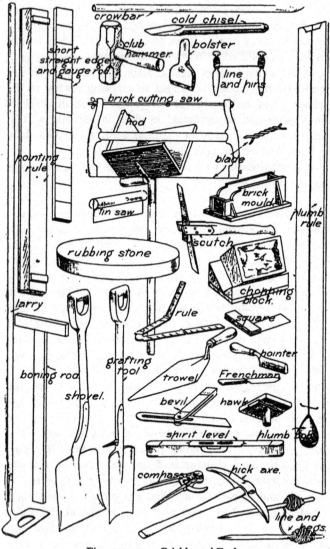

Figs. 209-235. Bricklayers' Tools.

dating ground. They are of the shape shown in figure 238, the head being of iron and about 10 lbs. in weight; the handle is of ash about 10 feet long.

*Bricklayers' Tools.*—These may be described under the following heads—1, setting out tools; 2, cutting tools; 3, laying tools; and 4, pointing tools. The setting out tools include

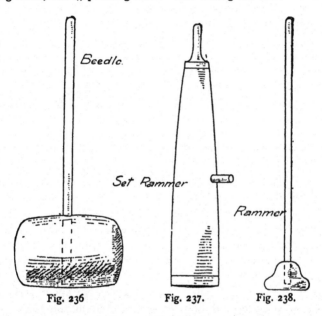

Fig. 236.    Fig. 237.    Fig. 238.

the 2-foot rule for taking small measurements, and the 10-foot rod for the setting out of buildings, the 66-foot metallic woven tape also used for the same purpose; more accurate work, however, can be done with the rigid wooden rod than with the flexible tape. The large wood square for setting out right-angles; templets for setting out work curved or shaped in plan; the plumb-bob and rule for testing the verticality of work; the straight-edge, usually three or four feet in length, for trying the accuracy of plane surfaces; the pocket spirit-level usually employed with the short straight-edge for testing the horizontality of points not far apart; the long spirit-level with a stock about 10 feet in length for determining the horizontality of points great distances apart; the telescopic or surveyor's level for a similar purpose to the preceding tool; the level of points a great distance apart may be more accurately and expeditiously compared by means of this instrument than with the long level. Lines and pins for the building of straight lengths of walls between given corners; the square with a steel blade and wood stock and the bevil for marking bricks to be cut; the compasses for the division of lengths; the radius-rod for describing segments of large circles, and a large drawing-board for the setting out of arches or shaped work.

*Brick-Cutting Tools.*—Rough cutting : 1, the large trowel ; 2, the club hammer and bolster, for cutting with greater exactitude than with the trowel; 3, the cold chisel for the cutting of chases and for general work.

Fair cutting, *Hard Bricks:* 1, the tin saw for making an incision $\frac{1}{8}$ of an inch deep, preparatory to cutting with bolster; 2, the chopping-block, which is an arrangement of two blocks of wood so fixed as to support a brick in an angular position convenient for cutting ; 3, the scutch consists of a stock and a blade, the latter generally formed of a flat file about 10 by 1 inches, sharpened at both ends and fixed in the stock by means of a wedge. This displaces, and is an improvement on, the old brick-axe, as the blade can be removed and sharpened readily; it is used to hack away the rough portions on the side of a brick after the edges have been cut by the tin saw and bolster.

Fair cutting, *Soft Bricks:* 1, the saw consists of a frame holding the blade, which is of twisted soft steel or malleable iron wire, No. 18 B.W.G., and is used for cutting soft rubbing bricks; 2, the rubbing-stone is a circular slab of gritty York stone 20 inches diameter, for rubbing the faces of bricks to a true surface; a few hand stones for finishing small surfaces and a few long files, parallel, flat and round, and a few rifflers for working returned and stopped mouldings; a surface plate of slate for testing the accuracy of surfaces and the correct adjustment of the joints of contiguous blocks; 3, the mould is a wood box enclosing bricks that are to be cut to a shape, the sides of the box being formed to that shape, and the edge over which the saw-blade works is protected by a strip of zinc. A few pieces of thin sheet steel for working moulded circular work.

*Laying and Pointing Tools.*—1, the large trowel for the spreading of mortar when laying bricks; 2, small trowels for filling up joints of new brickwork; 3, the pointing rule, which is a feather-edged straight-edge, with two small pieces $\frac{3}{8}$ inch thick nailed at each end to keep the rule away from the wall and allow the trimmings to fall through; 4, the frenchman for trimming joints consists usually of an old table-knife, with the end ground and turned up, as shown in plate containing figures 209 to 235; 5, the jointer, used for tuck pointing in old work.

# CHAPTER III.

# MASONRY.

---

*Definition.*—Masonry is the art of building in stone. Brickwork is built with uniform-sized blocks, admitting of a number of definite systems of laying the bricks; whereas in stone, owing to the expense in working the material, the face stones usually are squared, and the interior or hearting is filled up with smaller stones roughly fitted with a hammer. The stones are in the great majority of cases of varying dimensions, thereby making it a matter of great skill to obtain a proper bond in the work; and owing to the irregular shape of the material the walls have to be made considerably thicker than walls of the same height in brick, with the exception where the walls are built of coursed stones properly squared, in which case the thickness may be even less than that of brick walls.

The great dimensions in which stone may be obtained, lends itself to a much greater degree than bricks for buildings of architectural pretensions, rendering it possible to have cornices and corbelled work of great projection which is impossible in brickwork.

*Technical Terms.*—The following is a list, and also an explanation, of the labours in stonework :—

*Apex or Saddle Stone.*—Highest stone of a gable end cut to form the termination of two adjacent inclined surfaces, as shown in figure 247.

*Architrave.*—In classic architecture, the beam or lintel, as.

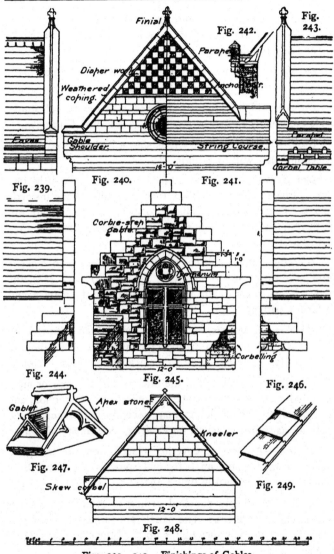

Finial

Fig. 242.

Fig. 243.

Parapet

Diaper work

Weathered coping.

Anchor bar.

Eaves

Gable Shoulder.

String Course.

Parapet

Corbel Table

Fig. 239.

Fig. 240.

Fig. 241.

Corbie-step gable

Tympanum

Corbelling

Fig. 244.

Fig. 245.

Fig. 246.

Gablet

Apex stone

Kneeler

Fig. 247.

Skew corbel

Fig. 249.

Fig. 248.

Figs. 239—249.  Finishings of Gables.

shown in figure 355, spanning the opening between two columns. Generally it is the moulding in wood or stone immediately surrounding the opening.

*Archivolt.*—The moulding worked about the rim of an arch, as shown in figures 375 and 376.

*Baluster.*—The uprights in a fence. In a stone balustrade these are usually turned or moulded, as shown in figure 375.

*Base.*—The lowest member, as shown in figure 375, usually projecting on a wall or column.

*Bed Surface.*—The bed surface must be worked in one plane surface. Masons to form thin joints often make the beds hollow. This is bad, as it is liable to spall; all the pressure will be thrown on the outer part, which is liable to spall the edge of the stone.

*Blocking Course.*—A course of stones erected to make a termination to the cornice (see figures 327 and 328), the object being to gain extra weight to tail down the cornice, and to form a parapet.

*Bonders.*—Long stones placed through from front to back of a wall, as shown in figures 309 and 310, to tie the wall transversely. These may be either headers or through stones.

*Boss.*—A stone placed at the intersection of three or more moulded members to enhance the effect of the junction, and to avoid where the mouldings vary in section and irregular inter-sections, as shown in article on Vaulting; see *Advanced Course.*

*Broach.*—A form of covering placed over the triangular surface formed where an octagonal spire is supported by a square tower, as shown in figure 292.

*Coffer.*—A deeply panelled ceiling of rooms or porticoes having classic cornices. The panels are formed by traversing the ceiling with the crowning members of the cornice, as shown in figure 379.

*Column.*—A pillar of circular section, as shown in figure 375.

*Console.*—A decorated bracket or corbel used in classic architecture, as shown in figure 360.

*Copings.*—The highest and covering course of masonry, forming a waterproof top, to preserve the interior of wall from wet, which in frosty weather might burst the _wall. Figures 248 and 249 show copings flat on the top surface, which should be used only for inclined surfaces, as on a gable or in sheltered positions. Figures 311 and 312 give section and elevation of feather-edged copings, the upper surface of which is inclined to throw water off on one side of the walls, and are necessary for parapet walls of crowded thoroughfares. Saddle-back is the name applied when the upper surface is inclined or weathered both ways, as in figures 315 and 324; and segmental, when the section of coping shows the upper surface to be a part of a circle, as in figures 316 and 317.

*Corbel.*—A piece of stone projecting from a wall to support a projecting feature. Figures 244 to 246 show methods of supporting a projecting angle of a building by means of corbels.

*Corbie Step Gables.*—A common method of finishing gables by constructing a number of steps formed of some hard stone squared, the top surfaces being slightly weathered, as shown in figures 244 and 246, and known as corbie or crow-step gabling.

*Skew Corbel.*—As shown in figure 248, is a projecting stone at the lowest part of the triangular portion of the gable end of a wall supporting the starting piece of coping, and resisting the sliding tendency of the latter. The skew corbels are often tied in to the wall by long iron cramps.

*Corbel-table.*—A system of corbelling supporting a parapet, and often forming an architectural feature, as shown in figure 243.

*Cornices.*—The moulded course of masonry crowning buildings, generally having a large projection, as shown in figures 327 and 328, to throw off the rain.

*Cramp.*—A slate or metal connection used in stone-work. See Joints.

*Crocket.*—A fourteenth-century hip or angle ornament of roofs, spires, etc., as shown in the article on Fleches; see *Advanced Course.*

*Cupola.*—Used now as a synonym for a dome, but strictly, the under surface; see figure in the article on Vaulting in the *Advanced Course.*

*Diaper Work.*—Is the name given to bands, surfaces and panels in the stone work formed by square stones and similar squares, filled in with brick or flint work, giving a chequered appearance, as shown in figures 240 and 241. The term is also applied to any ornament arranged in squares upon the surface of ashlar masonry.

*Dome.*—A roof usually circular or polygonal in plan, and hemi-spherical or any other curve or compound curve in elevation; see *Advanced Course.*

*Dormer.*—A small window built in a pitched roof, as shown in figure 292.

*Dressings.*—Stones are said to be dressed when their faces are brought to a fair surface; but any stones which are cut or prepared and are used as finishings to quoins, window and door openings, and may be applied as an ornamental feature, such as quoin stones, window and door jambs, as shown in figure 366, are described as dressings.

*Drip-stone.*—A projecting stone moulding, as shown in

the hood moulding, figures 389 to 392, having a throated under-surface, to throw water off clear of walls, and usually placed over arches of doors, windows, etc.

*Entablature.*—The combination of the architrave, frieze and cornice in a classic order ; see Portico, figure 375.

*Entasis.*—The convex swelling in a classic column.

*Finial.*—The aspiring ornament of an apex stone, often richly foliated, as shown in figures 239 to 241.

*Footings.*—The object of footings is the same as in brick walls. Stone footings should be large rectangular through stone blocks, as shown in figure 328. Square stones in plan are not so good as oblong. All stones in the same course must be of the same height, but all courses need not necessarily be of the same depth. The breadth of set-offs need not exceed 3 or 4 inches.

If the expense of stone is an objection, footings may be made of bricks or deep beds of concrete, as already described in the chapter on Brickwork.

*Frieze.*—The central division or a classic entablature, often used as a field for sculpture.

*Gable Details.*—The tops of stone walls are protected by coping, and these, where placed on steep gables, need support at their lower ends and at intervals ; this may be done by constructing a shoulder at the foot as shown in figures 240 and 241, or by the use of skew corbels as shown in figure 248 ; the intermediate supports are obtained by kneelers, which consist of stones having a part worked as a coping, the remainder tailing well into the wall, as shown in figure 248.

*Gablets.*—Many skew corbels are constructed with a small gablet as shown in figure 247, which gives extra weight to the skew corbel, thus rendering it more efficient for resisting the outward thrust of the coping stones. The apex stones are often treated in a similar manner.

*Galleting.*—The term given when sharp bits of flint, or pebbles, are pressed into the face joints of rubble walls to preserve the mortar and to give a pleasing effect, as shown in figure 313.

*Gargoyle.*—A stone water-spout, as shown in figure 250, employed in buildings of Gothic character to carry off the rain from the gutters, and projecting sufficiently far to throw the water clear of the building. At present down pipes are employed, but the gargoyle is often retained as an overflow in lieu of a warning pipe.

*Grout.*—This is a thin mortar, which is poured over the stones when brought up to a level surface, to fill up any interstices between the stones in the hearting of walls or other positions as necessity requires.

*Headers.*—-These are bonding stones, which extend across not more than three fourths of the thickness of the walls, as shown in figure 310.

*Hood-mouldings.*—These are similar to drip-stones, but are of necessity throated, their chief function being to accentuate the outline of the arch, as shown in figures 389 to 392.

*Jambs, Window and Door.*—For purposes of strength these should be of cut stone, attention being given that each course is securely bonded. For that reason it would not be advisable to build them of rubble. Figures 346 to 348 show alternate courses, one showing the stones forming the window jambs acting as stretchers—these are termed outbands, as shown in figure 361 ; and the other a through stone or header forming the face of reveal and the face of rebated jamb, and is termed an inband, as shown in figure 361.

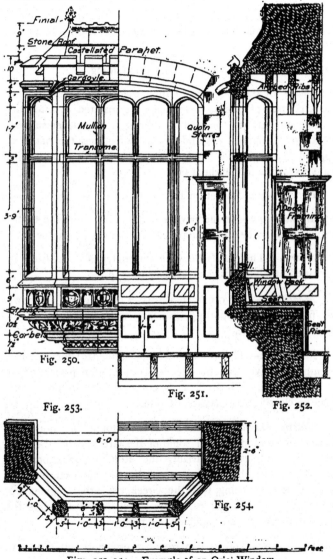

Fig. 250.

Fig. 251.

Fig. 253.

Fig. 252.

Fig. 254.

Figs. 250–254. Example of an Oriel Window.

H

*Kneeler.*—This is a long stone, tailing well into the gable wall, as shown in figure 248, and resists the sliding tendency of the coping.

*Label.*—Another name for a hood moulding.

*Lacing Course.*—Owing to the absence of bond in flint walls, courses of bricks, three deep, are inserted at intervals, to give strength to the wall and bring it to a level surface, as shown in figures 314 and 315. Sometimes the name is applied to a horizontal band of stone placed in rubble or flint walls to form a longitudinal tie.

*Lintels.*—See chapter on Brickwork.

*Machiolation.*—In mediæval military architecture the spaces between the corbels that supported the projecting parapet.

*Metope.*—The panels between the triglyphs in the frieze of a Doric entablature, as shown in figure 375.

*Modillion.*—The ornamental trusses or brackets, as shown in figure 355, supporting the crowning mouldings in a Corinthian cornice.

*Mouldings.*—The ornamental forms, applied to constructional members to give æsthetic effect. They may be projecting, as in the Classic, or incised, as is often in the Gothic. They may be classified as Classic or Gothic.

*Classic Mouldings.*—Most of the groups of mouldings consist of combinations of some of the Greek or of the eight Roman examples. Invariably the Roman mouldings are found to have their prototype in the Grecian examples, the chief difference being, that the Greek are either segments of some of the conic curves or are struck freehand, while the Roman curves are all segments of circles.

(1.) *Fillet.*—This is a narrow flat projection often used to divide individual or groups of mouldings in any composition; it is similar in both Greek and Roman work, as shown in figure 259.

(2.) *Astragal.*—A small semicircular moulding, as shown in figure 260, often used in combinations of mouldings, but chiefly to mark the division between the shafts and caps of columns. It is also largely used in the joints of straight members in joinery that are not glued, such as the joints of match-boarding or the meeting styles of doors, sashes, etc., so that if the wood shrinks the shadow cast by the moulding will detract from the unsightliness of the joint. This member is similar in Greek and Roman.

(3.) *Cavetto.* — The cavetto is a hollow moulding, as shown in figure 256, consisting in the Greek of a quarter of an ellipse and in the Roman of a quadrant.

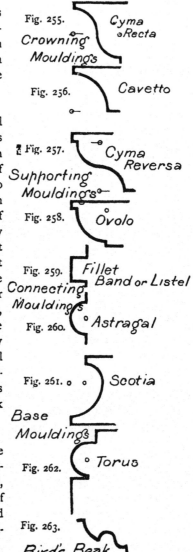

Fig. 255.

Crowning Mouldings

*Cyma Recta*

Fig. 256.

*Cavetto*

Fig. 257.

Supporting Mouldings

*Cyma Reversa*

Fig. 258.

*Ovolo*

Fig. 259.

Connecting Mouldings

*Fillet*

*Band or Listel*

Fig. 260.

*Astragal*

Fig. 261.

*Scotia*

Base Mouldings

Fig. 262.

*Torus*

Fig. 263.

*Bird's Beak.*

(4.) *Ovolo.*—This moulding in the Greek consists of a segment of an inclined ellipse, having a fillet at the top and bottom, and forming at the top a quirk. In Roman work, as shown in figure 258, it is a quarter circle, rounded at top and bottom by a fillet.

(5.) *Cyma Recta.*—This is a double curve, formed in the Greek of two quarter ellipses whose minor axes are in the same straight line and bounded top and bottom by a fillet. The Roman example, as shown in figure 255, is similar, but consisting of two quarter circles. This moulding has a concave portion of its surface above the convex, and is generally used as a crowning member.

(6.) *Cyma Reversa.*—As its name implies, is the reverse of the preceding moulding, slightly modified in the Greek by having a quirk above, between the same and the fillet, and the hollow portion slightly more concave. The Roman is an exact reverse, as shown in figure 257.

(7.) *Scotia.*—The scotia in the case of the Greek is formed of an inclined ellipse, having a fillet above and below. The Roman is struck from two centres on a common radial line, as shown in figure 261.

(8.) *Torus.*—The torus is a base moulding, the Greek form being the reverse of the scotia. Many Greek examples are, however, similar to the Roman, consisting simply of a large semicircle with a quirk below and a fillet above, as shown in figure 262.

(9.) *Bird's Beak.*—This moulding only occurs in the Greek mouldings; it consists of a quarter ellipse (with the major axis horizontal), in the lower side of which a small hollow has been worked, as shown in figure 263, and is used as a supporting moulding.

In the designing of groups of mouldings for cornices, strings, etc., reference should be made as to the suitability of the mouldings for their intended position, and for this purpose they may be divided into base mouldings, connecting mouldings, supporting mouldings, and crowning mouldings. The base mouldings would include such mouldings as the torus, the scotia or the inverted cyma recta, and any combination of such mouldings that would tend to broaden the base and distribute the weight of the mass supported.

*Connecting.*—These include the fillet and the astragal.

*Supporting.*—The supporting mouldings include such members as the ovolo, bird's beak and the cyma reversa, mouldings that do not have their hollow members near their upper edge, and such as have their mass in a position to strengthen the same, and are fitted to act as corbels. These mouldings are used to form the bed mouldings or lower parts of combinations, such as cornices which are divided into two parts, the bed mouldings and the crowning mouldings.

*Crowning Mouldings* are those mouldings which are not expected to carry anything above, such as the cyma recta and the cavetto, the top members of which are small and delicate.

The above ideas are not always rigidly adhered to, and successful departures from the same are often made with good effect ; but it is prudent to bear these principles in mind when designing any groups, as if too widely departed from confusion ensues.

*Mouldings.*—Figures 264 to 291 give a selection of the mouldings commonly used in the Gothic periods, combinations in archivolts, also for strings, wall bases, bases and capitals of columns.

*Order.*—In classic architecture the combination of the column and entablature without other members is known as an order.

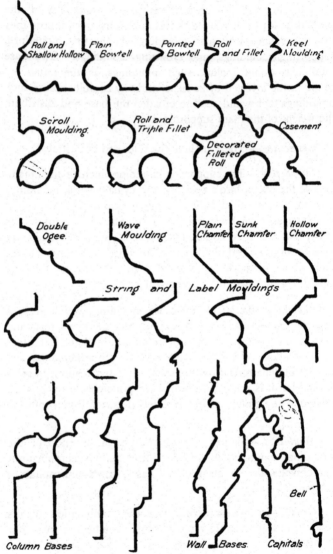

Roll and Shallow Hollow · Plain Bowtell · Pointed Bowtell · Roll and Fillet · Keel Moulding

Scroll Moulding. · Roll and Triple Fillet · Decorated Filleted Roll · Casement

Double Ogee. · Wave Moulding · Plain Chamfer · Sunk Chamfer · Hollow Chamfer

String and Label Mouldings

Column Bases · Wall Bases · Capitals · Bell

Figs. 264–291.   Types of Gothic Mouldings.

*Oriel.*—A projecting window supported by corbelling, as shown in figures 250 to 254.

*Parapet.*—The fence wall in front of the gutter at the eaves of a roof, as shown in figure 242. The castellated parapet is formed by a number of embrasures similar to the parapets used in ancient military buildings, as shown in figure 250, and much used in the later Gothic work as an ornamental feature.

*Patera.*—A small concave ornament, usually circular and decorated inside, used in classic architecture at the junction or to break the continuity of mouldings.

*Pedestal.*—An elevated base, used for elevating the orders in Roman architecture, as shown in figure 375.

*Pediment.*—The gable over a window, door, portico or end of a classic building usually surmounted by the crowning members of the cornice, as shown in figure 372.

*Pendant.*—A hanging ornament, accentuating the junction of constructional members.

*Pilaster.*—A wall pillar, rectangular in section, as shown in figures 375 to 378.

*Pillar.*—A vertical support, square or rectangular in section, and always independent of walls.

*Pinnacle.*—The pyramidal termination or apex to a buttress or spire, as shown in figure 294.

*Plinth.*—A horizontal projecting course or courses built at the base of a wall, as in figures 363 and 372.

*Porch.*—An external construction forming a covered approach to a building, as shown in figure 378.

*Quoins.*—The angle stones of buildings. In rubble and

inferior stone walls, quoins are built of good blocks of ashlar stone to give strength to the wall. These are sometimes worked to give a pleasing effect, and where hammer-dressed and chamfered, as in figure 327, are said to be rusticated. They are, at times, merely built with a rough or quarry face, as in elevation figure 324, only having the four face edges of each stone lying in one plane.

*Respond.*—A semi-column or corbel forming the support of the end arch of an arcade, as shown in figure 157.

*Scontions.*—The stones forming the inside angle of the jamb of a door or window opening, as shown in figure 365. These are often cast in concrete to effect a saving in labour.

*Sills.*—These are the lower horizontal members of openings, and those in stone are usually of one length, being pinned in cement to both sides of the opening. They should be fixed after the carcase of a building has been finished, and any settlement that was likely to occur through a number of wet mortar joints has taken place. They may be plain and square, as for door-sills, or sunk, weathered, moulded with drip and with properly formed stools and grooved for metal water-bar, or sunk, weathered, throated and grooved for metal water-bar, all as illustrated in chapter on Joinery; or moulded, grooved and weathered, as shown in figures 363 to 365.

*Spalls or Shivers.*—These are broken chips of stone, worked off in the dressing.

*Spire.*—A steep pyramidal roof placed over the towers, as shown in figures 292 to 295, in 13th and 14th century Gothic work.

*Springer.*—The lowest stone in an arch.

*Squinch-arches.*—The arches placed across the angle of a square tower, to form a springing for the sides that traverse the angles, as shown in figure 295.

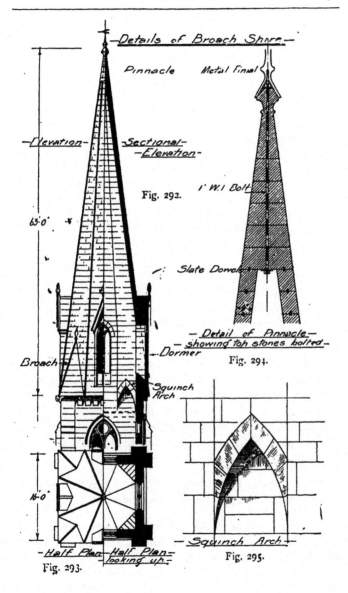

Details of Broach Spire.

Pinnacle     Metal Finial

Elevation     Sectional
              Elevation

Fig. 292.

1" W.I Bolt

63·0"

Slate Dowel

Broach

Dormer

Squinch
Arch

— Detail of Pinnacle —
— showing top stones bolted —

Fig. 294.

16·0"

Half Plan    Half Plan
             looking up.

Fig. 293.

— Squinch Arch —

Fig. 295.

*String Courses.*—Horizontal projecting bands of stone, as in figures 327 and 328, often carried below windows, of architectural importance, imparting a feature to the building.

*Stylobate.*—The elevated base or platform used in Greek work, consisting of a series of stepped layers upon which the building is supported.

*Tailing Irons.*—These are formed of H, L or T irons for holding down the ends of corbels in Oriel windows, as shown in figure 252.

*Templates.*—Pieces of stone placed under the end of a beam or girder to distribute the weight over a greater area, as in figures 445 and 446.

*Tenia.*—The fillet or band in a Doric order, dividing the architrave and frieze.

*Throatings.*—Grooves on the under surfaces of copings, sills, string courses, etc., as in section, figure 319, forming a drip to prevent the water that would otherwise trickle down and disfigure the walls.

*Through Stones.*—Stones which extend through the entire thickness of wall, to tie or bond it, as shown in figure 312. These are considered objectionable, for the reason that if any projection has to be taken off at the back, to present a fair face, it would disturb the setting of the adjoining masonry, also damp is more likely to show on the interior of walls where the continuity of the material is uninterrupted.

*Tracery.*—The intersecting mullions or bars in the tympanum of a Gothic window opening, as shown in figure 393.

*Tympanum.*—The masonry filling in between the relieving arch and the head of a door or window, also the wall surface in a pediment of a Classic gable. This is often taken advantage of to form a ground for carved ornament, as shown in figure 245.

*Vaulting.*—The arched stone ceiling of any apartment; see *Advanced Course.*

*Weathering.*—The top face of a stone worked to a plane surface inclined to the horizontal for the purpose of throwing off the water is said to be weathered, as in sills, cornices, etc.

*Zoccolo.*—The platform forming a step placed upon the pedestal of a Roman order upon which the columns are placed.

*Labours.*—The following are the chief labours recognised in the preparation of stonework :—Scabbling, hammer-dressing, self-faced, half-sawing, chisel-draughting, plain work, which includes a combed or dragged face over a Bath or soft limestone, a tooled stroke over the plane faces of a hard limestone and sandstone, such as Portland or York, or a smooth-axed face on granite ; sunk work, moulded work, rubbed and polished work. These labours may be put upon the straight, circular, or circular circular work.

*Scabbling or Scappling.*—That is, taking off the irregular angles of stone ; is usually done at the quarry, and is then said to be quarry pitched, hammer faced, or hammer blocked ; when used with such faces the stone is called rock or rustic work.

*Hammer-Dressing.*—Roughest description of work after scabbling, as shown in figure 296.

*Self-Faced.*—The term applied to the quarry face, or the surface formed when the stone is detached from the mass in the quarry ; also the surfaces formed when a stone is split in two.

*Half-sawing.*—The surface left by the saw. Half the cost

of the sawing being charged to each part of the separated stone.

*Chisel-Draughted Margins.*—To reduce an irregular surface to a plane surface a rebate about an inch in width is worked at two opposite edges of the surface; the parallelism of these rebates is ensured by testing with the winding strips and straight-edge. A similar rebate is worked on the two remaining edges, connecting those first made. A continuous margin or rebate is thus formed about the four edges of the stone, every portion of which lies in the same plane surface. If the stone be small, the irregular excrescence is then removed with the chisel to the level of the rebate. For large surfaces subsidiary draughts are formed traversing the stone between the rebates. In walls the stones in which are furnished with hammer-dressed or rusticated surfaces, chisel-draughted margins are sunk about the four edges to ensure the accuracy of the work. This work is shown in figure 297.

*Plain Work.*—This is divided, for purposes of valuation, into half plain and plain work. The former term is used when the surface of the stone has been brought to an approximately true surface, either by the saw or with the chisel. Plain work is the term adopted for surfaces that have been taken accurately out of winding with the chisel, and are finished with a labour usually to form an exposed surface. These labours are usually placed upon the bed and side joints of stones in walling.

For sandstone or hard limestone, plain work includes a tooled stroke and chisel-draughted margins, for soft limestones a combed or dragged surface, and for granite a smooth-tooled face.

*Combed or Dragged Work.*—This is a labour employed to work off all irregularities on the surfaces of soft stones. The drag or comb is the implement used, and consists of a piece of steel with a number of teeth like those of a saw. This is drawn over the surface of the stone in all directions, after it has been roughly reduced to a plane with the chisel, making it approximately smooth, as shown in figure 300.

*Boasted or Droved Work.*—This consists in making a number of parallel chisel marks across the surface of the stone by means of a chisel, termed a boaster, which has an edge about 2½" in width. In this labour, the chisel marks are not kept in continuous rows across the whole width of the stone, as shown in figure 298.

*Tooled Work.*—This labour is a superior form of the above, care being taken to keep the chisel marks in continuous lines across the width of the stone, as shown in figure 299. The object of this and the preceding is to increase the effect of large plane surfaces by adding a number of shadows and high lights.

*Axed Work.*—Axed worked and tooled work are similar labours. The axe is employed for hard stones, such as granite, but the mallet and chisel for soft stones, as being more expeditious.

The method of preparing the hard stones after being detached from their beds in the quarry is as follows :—The stones are roughly squared with the spall hammer; the beds are then prepared by sinking a chisel draught about the four edges of the bed under operation, the opposite draughts being out of winding, and the four draughts in the same plane surface; the portions projecting beyond the draught are then taken off with the pick. After the pick the surface is wrought with the axe, the latter being worked vertically downward upon the surface, and taken from one side of the stone to the other, and making a number of parallel incisions or bats, the axe is worked in successive rows across the stone, the incisions made being kept continuous across the surface. In axed work there are about

four incisions to the inch. This labour is used for the beds
of stones for thresholds and kerbstones, and in this state
the pick marks are easily discernible. Fine axed work is
a finer description of axed work, and is accomplished with
a much lighter axe having a finer edge. In fine axed work
there would be eight incisions to the inch.

*Pointed Work.*—The bed and side joints of stones are
often worked up to an approximately true surface by means of
a pointed tool or punch. This labour, as shown in figure 303,
is often employed to give a bold appearance to quoin and
plinth stones, and where so used it usually has a chisel-
draughted margin about the perimeter.

*Sunk Work.*—This term is applied to the labour of making
any surface below that originally formed, such as chamfers,
wide grooves, the sloping surfaces of sills, etc., as shown in
figure 306. If the surface is rough, it is known as half-sunk;
if smooth, sunk; and any other labour applied must be added
to the same, such as sunk, rubbed, etc.

*Moulded Work.*—Mouldings of various profiles are worked
upon stones for ornamental effect, as shown in figure 304.
Mouldings are worked by hand as well as by machine.
In the former case, the profile of the moulding is marked
on the two ends of the stone to be treated by means of
a point drawn about the edge of a zinc mould, cut to the
shape of the profile. A draught is then sunk in the two
ends to the shape of the required profile. The superfluous stuff is then cut away with the chisel, the surface
between the two draughts being tested for accuracy by
means of straight-edges. The machines for moulded work
somewhat resemble the planing machines for metal work.
The stone is fixed to a moving table. The latter has imparted
to it a reciprocating rectilinear motion, pressing against a fixed
cutter of the shape of required profile, or some member of it.

The cutter is moved nearer to the stone after each journey, thus gradually removing the superfluous stuff till the profile is completed. Moulded work is, strictly speaking, the name given to profiles formed with a change of curvature, and, therefore, should not be applied to cylindrical sections, such as columns.

The weathering properties of stones moulded by hand labour are considered by some far superior to those worked by machinery, as in the latter method the moulding irons, being driven continuously, become heated and partially calcine the surfaces of the stones, thus rendering the same peculiarly susceptible to atmospheric deterioration.

*Rubbed Work.*—This labour consists in rubbing the surfaces of stones until perfectly regular, and as smooth as possible. The work is accomplished by rubbing a piece of stone with a second piece. During the first stages of the process, water and sand are added, gradually reducing the quantity of sand up to the finish. Large quantities of stone are rubbed by means of large revolving iron discs. The stones are placed on, and kept from revolving with the disc by means of stationary timbers fixed across the table a few inches above the stone. No pressure is applied other than that obtained by the weight of the stone. Water and sand are added to accelerate the process. Only plane surfaces can be rubbed in this way.

*Polishing.*—Dense stones, such as the marbles and granites, after being worked to a smooth surface, are often polished. These processes are fully described in the *Advanced Course.*

*Circular Work.*—Labour put upon the surface of any convex prismatic body, such as the parallel shaft of a column or large moulding, is termed circular work, as shown in figure 307.

*Circular Sunk Work.*—Labour put upon the surface of any

concave prismatic body, such as a large hollow moulding or the soffit of an arch, is termed circular sunk work, as shown in figure 306.

*Circular Circular Work.*—The labour placed upon columns with entases, spherical or domical work, as shown in figure 308.

*Circular Circular Sunk.*—The labour worked upon the interior concave surfaces of domes, etc., as shown in figure 308.

*Moulded Work, Circular.*—This term is given to mouldings, as shown in figure 305, stuck upon circular or curved surfaces in plan or elevation.

*Internal Mitres.*—The name given to the intersection of two mouldings making an angle less than 180°.

*External Mitres.*—The name given to the intersection of two mouldings making an angle greater than 180°.

*Returned Mitred and Stopped.* — The name given to a moulding returned in itself, and stopping the same against an intersecting surface.

*Long and Short Work.*—This work is usually used for quoins and dressings in rubble walls, and is especially notice-able in old Saxon work. It consists in placing alternately a flat slab, which serves as a bonder, and a long stone approxi-mately small and square in section.

This arrangement in modern work is sometimes known as block and start work.

*Vermiculated Work.*—This labour is placed chiefly on quoin stones to give effect. The process is as follows :—A margin of about ¾″ is marked about the edge of the stone, and in the surface enclosed by the margin a number of irregularly shaped sinkings are made. The latter have a margin of a

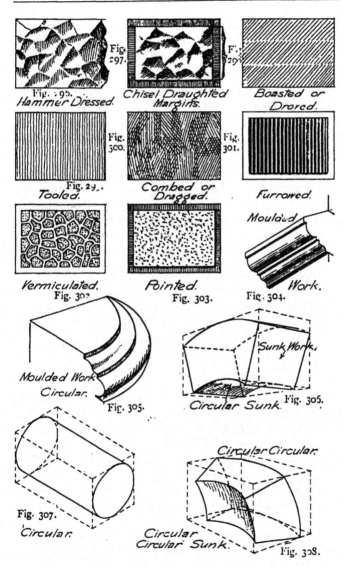

Fig. 297.
Fig. 298.
F. 299.
Fig. 295.
*Hammer Dressed.*
*Chisel Draughted Margins.*
*Boasted or Drored.*

Fig. 300.
Fig. 301.
Fig. 29..
*Tooled.*
*Combed or Dragged.*
*Furrowed.*

Fig. 302
*Vermiculated.*
Fig. 303.
*Pointed.*
Fig. 304.
*Moulded Work.*

*Moulded Work Circular.*
Fig. 305.

*Sunk Work.*
*Circular Sunk.* Fig. 306.

Fig. 307.
*Circular.*

*Circular Circular.*

*Circular Circular Sunk.*
Fig. 308.

I

constant width of about $\frac{3}{8}''$ between them. The sinkings are made about $\frac{1}{4}''$ in depth. The sunk surface is punched with a pointed tool to give it a rough pock-marked appearance, as shown in figure 302.

*Furrowed Work.*—This labour, used to emphasise quoins, consists in sinking a draft about the four sides of the face of a stone, leaving the central portion projecting about $\frac{3}{8}''$, in which a number of vertical grooves about $\frac{5}{8}''$ wide are sunk, as shown in figure 301.

*Foundations and Footings.*—Stone walls may be provided with concrete foundations. Where built on rocky sites, the concrete is dispensed with or only used to level up irregularities. Stone footings usually consist of two courses of long stones, rectangular in section, and the least dimension being about 9 inches ; on these the walls are erected, as shown in figure 328.

*Classification of Stone Walling.*—This is classified as follows :—

(1) *Rubble.*—Flint ; random rubble set dry ; random rubble set in mortar ; Kentish rag ; random rubble built in courses ; uncoursed, squared or snecked rubble ; squared rubble built up to courses ; regular coursed rubble.

(2) Block in Course.

(3) *Ashlar.*—Ashlar facing with brick backing ; ashlar facing with rubble backing.

Rubble walls are those built of thinly-bedded stone, generally under 9 inches in depth, of irregular shapes as in random rubble, or squared as in coursed rubble.

Block in course is composed of squared stones usually larger than coursed rubble, and under 12 inches in depth.

Ashlar is the name given to stones, from 12 to 18 inches deep, dressed with a scabbling hammer, or sawn to blocks of given dimensions and carefully worked to obtain fine joints.

The length of a soft stone for resisting pressure should not exceed three times its depth ; the breadth from one-

and-a-half to twice its depth; the length in harder stones 4 to 5 times its depth, and breadth 3 times its depth.

## RUBBLE WALLS. .

*Flint Walls.*—These are largely employed in the chalk districts, where large quantities of flints are found beneath the beds of chalk. The flints are small and of irregular form. The larger stones are selected for the facing, the exposed surface being chipped to present the silica face. For the best work they are squared up to form small briquettes ; they are laid in regular courses, as at St. Saviour's Cathedral, Southwark. The quoins, windows and door dressings are always built with squared stone or brick to give strength and obtain regularity in these parts. The facing flints are then laid, being separately and carefully bedded in mortar ; the hearting is then inserted from every 6 to 9 inches in height, and bedded either by grouting or by a process analogous to larrying in brickwork. Owing to the smallness of the material, flint walls are liable to be deficient in longitudinal and transverse strength. To remedy this, lacing courses, consisting of large, long flat stones, or of three courses of plain tiles or bricks, are built in at intervals of about every 6 feet in height, as shown in figures 314 and 315. In mediæval work, a very effective wall decoration was obtained by building an unmoulded stone tracery and filling this in with flint work ; this is known as flint tracery, as shown in figure 322.

*Random Rubble.*—The name given to walling built of stones that are not squared.

*Random Rubble set Dry.*—In the stone districts boundary walls are built of rubble set without mortar in courses about 12 inches high, provided with a waterproof top, as shown in figures 309 and 310, to keep water from getting into the body of the work, and bursting it in frosty weather.

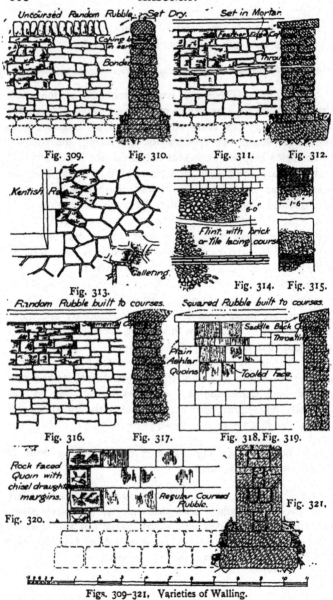

Uncoursed Random Rubble. Set Dry.    Set in Mortar.

Coping
in sand
Bonder

Feather Edge Co
Throat

Fig. 309.     Fig. 310.     Fig. 311.     Fig. 312.

Kentish Ra

6-0"

1-6

Flint with brick
or tile lacing cours

Galleting.

Fig. 313.     Fig. 314.    Fig. 315.

Random Rubble built to courses.    Squared Rubble built to courses.

Segmental

Saddle Back C
Throating

Plain
Ashlar
Quoins    Tooled Face.

Fig. 316.     Fig. 317.     Fig. 318. Fig. 319.

Rock faced
Quoin with
chisel draught
margins.

Regular Coursed
Rubble.

Fig. 320.         Fig. 321.

Figs. 309-321, Varieties of Walling.

*Uncoursed Random Rubble set in Mortar.*—The stones are used as they come from the quarry, care being taken to obtain them as uniform as possible, the bond being obtained by fitting in the inequalities of the stone, and by using one bond stone every super yard on face; any openings between stones to be pinned in with spalls, as shown in figures 311 and 312. If good mortar is used, walls built of random rubble should be made one-third thicker than the thickness necessary for brick walls.

*Kentish Rag.*—Walls of this type, as shown in figure 313, are built of an unstratified sandstone largely found in Kent. The blocks are roughly dressed to a polygonal form and being fitted with a hammer as they are bedded in the wall. In this work all quoins and dressings are built with properly squared stones. The face stones are first carefully bedded, the hearting being afterwards filled in partly by bedding and partly by grouting.

*Random Rubble built in Courses.*—This consists of stones forming horizontal beds at intervals of 12 to 18 inches, every stone being bedded in mortar. The object being to ensure that there shall be no continuous vertical joints, as illustrated in figures 316 and 317. To save expense in bedding each stone in mortar, masons bed only the stones on faces of wall, and at these levels pour a pail of thin mortar, called grout, to fill up any cross joints between stones, taking care not to let any run over the face of wall.

*Uncoursed, Squared or Snecked Rubble.*—This type of work, as shown in figures 323 and 324, is built in districts where a highly stratified stone is abundant; such stones lend themselves to be easily squared. With these stones, horizontal bed joints are readily obtained, but as the stones are of irregular depths there is a great tendency to obtain long vertical joints; to prevent this, small stones, termed snecks, are inserted where required.

Fig. 322. Elevation.

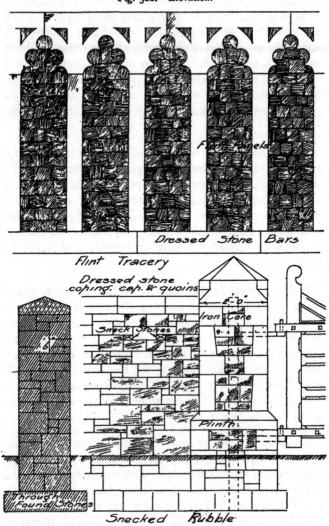

Flint Panels

Dressed Stone | Bars

Flint Tracery

Dressed stone
coping. cap. & quoins

Sneck Stones

Iron Core

Plinth

Through Found Stones

Snecked Rubble

Fig. 323        Fig 324

For good work not more than four side-faces should form any continuous vertical face joint. To prevent long vertical joints, no stone for house walls should be more than eight inches in depth. One bonder should be inserted in every superficial yard of wall face.

*Squared Rubble built up to Courses.*—This work is built up to courses to prevent long vertical joints. Figures 318 and 319 show squared rubble brought up to level beds with hammer-dressed quoins and chisel-draughted margins.

*Regular Coursed Rubble.*—In this kind of work all stones in one course are squared to the same height, usually varying from 4″ to 9″, as illustrated in figures 320 and 321, and are generally obtained from thin but regular beds of stone.

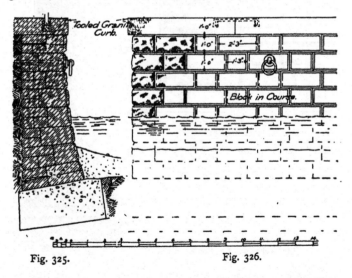

Fig. 325.        Fig. 326.

*Block in Course* is the name applied to stone walling, chiefly used by engineers in embankment walls, harbour walls, etc., where good work is required. The stones are all squared and brought to good fair joints, the faces usually being hammer-dressed, as shown in figures 325 and 326. Block in course closely resembles coursed rubble, or ashlar, according to the quality of the work put upon it.

Fig. 327.          Fig. 328.

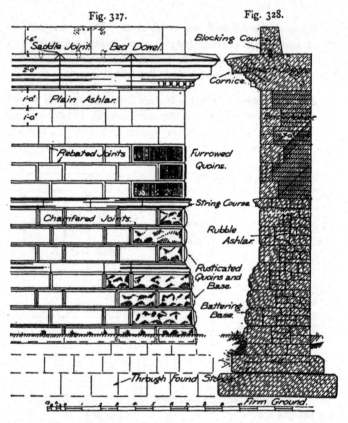

*Ashlar.*—Ashlar is the name applied to stones that are carefully worked, and are usually over 12″ in depth, and have joints not more than ⅛ inch thick.

As the expense would be too costly to have walls built

entirely of ashlar, they are constructed to have ashlar facing and rubble backing, or ashlar facing and brick backing, both shown in figures 327 and 328; but, as the backing would have a greater number of joints than the ashlar, the backing should be built in cement mortar, and brought to a level at every bed joint of the ashlar, to ensure equality of settlement.

The ashlar facing may be plain, rebated, or chamfered, as illustrated in figure 327, and looks best when laid similar to Flemish bond in brickwork.

*Joints.*—In arranging the joints of masonry the following general principles should be observed :—

(1) All the bed joints must be arranged at right angles to the pressure coming upon them.

(2) Joints should be arranged to prevent any members, such as sills, being under a cross-stress.

(3) The joint should be arranged so as to leave no acute angles on either of the pieces joined.

The first condition applies to all kinds of masonry. It is necessary to prevent any sliding tendency taking place between the stones.

The second condition applies chiefly in sills and lintels. In the case of sills, if bedded along their entire length, there is a liability to fracture due to the tendency to irregular settlement under the walls and the openings. To prevent cross-stress, the extremities of the sill are only bedded, the remaining portion of the under surface having no bearing. With lintels over small openings save stones, as shown in figure 349, are used, and with larger lintels other devices are employed to relieve the lintels of the load, as described in the article on Lintels. In brickwork the sills are frequently fixed after the work has been completed and had time to settle. In large stone openings a joint is arranged in the sill in the line of the reveal, as shown in figure 393.

The third condition applies chiefly to the joints in tracery
work, and any exposed joints in any other work. Stone being
a granular material, anything approaching an acute angle is
liable to weather badly; therefore in any tracery work having
several bars intersecting, a stone must·be arranged to contain
the intersections and a short length of each bar, as shown in
figure 393, and the joints should be (a) at right angles to the
directions of the abutting bars if straight, or (b) in the direction
of a normal to any adjacent curved bar. This not only prevents any acute angles occurring, as would be the case if the
joints were made along the line of intersection of the moulding,
but also ensures a better finish, as the intersection line can be
carved more neatly with the chisel, and is more lasting than
would be the case if a mortar joint occurred along the above
line. In no case, either in tracery, string courses, or other
mouldings, should a joint occur at any mitre line.

Joints and connections in stonework may be classified as
follows :—

        1.  Joggle joints.
        2.  Cramps.
        3.  Weathered joints.

Under the first head are included joggle joints, tabling,
cement joggles, dowels, and pebble joints.

*Joggles.*—A joggle is a form of joint in which a portion of
the side joint of one stone is cut to form a projection, and a
corresponding sinking is made in the side of the adjacent stone
for the reception of the projection, as shown in figures 329 and
330. It is chiefly used in landings to prevent any movement
between the stones joined and so retain a level surface, and
also to.assist in distributing any weight over every stone in
the landing.

*Tabling Joints.*—This is a form of joint that has been used
to prevent lateral displacement in the stones of a wall subjected
to lateral pressure, such as in a sea-wall. It consists of a joggle

in the bed joints, the projection in this case being about
1½" in depth and a third of the breadth of the stone in width,
as shown in figure 331. This joint is expensive to form, and
is often substituted by long slate joggles placed in a space to
receive same in the bed joint at the junction of side joints of
two stones and the top bed joint of another, as shown in
figure 332. These were used in the piers of the new Tower
Bridge.

*Cement Joggles.*—These are generally used in the side joints
of the top courses of masonry to prevent lateral movement in
the same, and consist of a V-shaped sinking in the side joint
of each adjacent stone in the same course, as shown in
figures 333 and 334.

*Dowels.*—Dowelling is another method of resisting lateral
movement. The dowels consist usually of pieces of hard
stone or slate about one inch square in section and varying
from about two inches to five inches in length, shown in
figures 335 and 336, being sunk and set in cement in corresponding mortices in the adjacent stones. They are used
in both the side and bed joints. They are generally employed
in the top courses of masonry where the weight on or of the
individual stones is not great, and also in the dressings, about
openings and in the bed joints of the drums of columns,
balusters, and in any position where lateral movement is
likely to occur. The united mass thus formed from the
connected stones renders any movement impossible under
normal conditions.

*Pebbles.*—Small pebbles used to be commonly employed to
connect stones before the introduction of slate joggles owing
to the ease with which the latter may be fitted. The pebbles
are still largely used.

Under the second head are included metal cramps, lead
plugs, slate cramps, anchor bolts, and rag bolts.

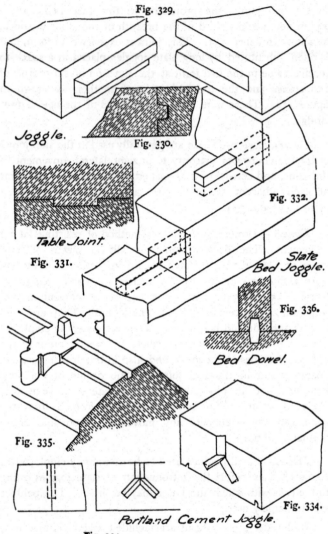

Fig. 329.

Joggle.

Fig. 330.

Table Joint.

Fig. 331.

Fig. 332.

Slate
Bed Joggle.

Fig. 336.

Bed Dowel.

Fig. 335.

Fig. 334.

Portland Cement Joggle.

Fig. 333

Figs. 329-336.

*Cramps.*—Metal cramps are used similarly to dowels to bind work together, but are more particularly adapted for positions in which there is a tendency for the stones to come apart, such as in copings covering a gable, or in face stones of no great depth, or cornices and projecting string courses to tie the stones to the body of the wall or to the steel skeleton supports. The cramps are made from thin pieces of metal of varying lengths and sectional area according to the work, bent at right angles about $1\frac{1}{2}''$ at each end. A chase with a dovetailed mortice at each end is made in the stones to receive the cramp, the ends of which are made rough and inserted, as shown in figures 337 and 338. The cramps are usually prepared from either wrought iron, copper, or bronze. If wrought iron is used, it is usually subjected to some preservative process, such as tarring and sanding or galvanising, to prevent oxidation. Iron is useful on account of its great tensile strength. Copper is valued for its non-corrosive properties under ordinary conditions, and its tensile strength, which is not much less than wrought iron; it is, however, comparatively soft. Bronze partakes of properties similar to those of copper, with the exception that it is much harder, which, under some conditions, is a desideratum.

The best bedding materials are Portland cement, lead, and asphalte. Care should be taken to completely envelop the cramp in the bedding material. Lead is at times objected to for external work owing to its liability to form a galvanic couple with the cramp in the presence of moisture, in addition to the oxidation.

*Lead Plug.*—Stones may be connected together by means of lead in the following manner, shown in figures 340 and 341 :—Dovetailed shaped mortices of the form shown are made to correspond in the side joints of two adjacent stones, into which, when placed in position, molten lead is poured, and when cool is caulked, thus completely filling the mortices and connecting the pieces.

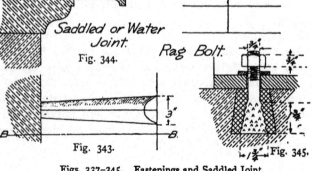

Figs. 337.

10½″

1¼

Section.

Cramp run in Lead.

Fig. 340.

½″

Lead Plug.

Plan.

1¼″ · 1½″

½″

Fig. 338.

Fig. 341.

2¼ · 1½ · 7″

Slate Cramp.

1″

Throating.

Fig. 339.

12″

Section B.B.

Saddled or Water Joint.

Fig. 344.

3″

3″

Elevation of Joint.

Fig. 342.

Rag Bolt.

¾″

¾″

¾″

Fig. 343.

B.                                    B.

¾″      Fig. 345.

Figs. 337–345.   Fastenings and Saddled Joint.

*Slate Cramps.*—These consist of pieces of slate about 7 in. × 2 in. × 1 in. cut to a double dovetail form; they are bedded in Portland cement in sinkings formed to receive them, and are generally used in flat coping stones, as shown in figure 339.

*Anchor Bolts.*—Long iron bolts, as shown in figure 780, are frequently employed at the backs of cornices that have great projection; in such cases the centre of gravity of the mass is dangerously near the edge of the wall. The bolts are passed through a hole drilled through the back of the cornice, or are inserted into a chase worked along the back face of the stone, and extend a sufficient distance down the back of the wall, being provided at their lower ends with large iron plates or washers; the effect is to give homogeneity to the whole mass, and thus bring the centre of gravity of the combined mass a distance back from the front of the wall sufficient for safety.

In pinnacles at the tops of spires and buttresses, where formed of small stones, it is usual to connect a sufficient number of them together with an iron bolt, which latter usually contains their common axis, thus increasing the stability by rendering the mass homogeneous.

*Rag Bolt.*—The ends of the bolts are often fixed by having the end that is let into the stone jagged, and run with lead, or Portland cement, the mortice being dovetail shaped, as in figure 345, to secure it from any upward pressure.

Where there is any probability of a great upward stress, the rag bolt is replaced by an anchor bolt and plate. The bolt is passed through a hole drilled through the stone.

*Weathered Joints.*—Under the third head is included all joints or precautions taken to prevent the deterioration of the joints of cornices or other exposed parts of masonry due to the percolation of water into the joints.

*Saddled or Water Joint.*—To protect the joints of cornices and other exposed horizontal surfaces of masonry the sinking is sometimes stopped before the joint and weathered off, as shown in figures 342 to 344. Any water passing down the weathered surface is guided away from the joint.

*Rebated Joints.*—These joints are used for stone roofs and copings to obtain weather-tight joints. There are two kinds: (1) when both stones are rebated; (2) when the upper stone only is rebated. In the first case the stones are of the same thickness throughout, their upper surface being level when the joint is made. In the second case the stones are thicker at the bottom edges than at the top, the bottom edge having a rebate taken out equal to the thickness of the upper edge of the stone below it, over which it fits. The part that laps over should not be less than $\frac{3}{4}''$ thick. The under surfaces or beds of the stones should be level, as shown in figures 396 and 398.

The upper exposed surfaces of all masonry built of soft or porous stones should be protected by a lead covering.

*Stone Lintels.*—Square openings in buildings are frequently bridged with stone lintels. Stone, owing to its low tensile resistance, is not well adapted to act as a beam, and in wide openings every care must be taken to relieve these members as far as possible of the superincumbent weight.

Lintels are of two kinds: 1st, where the openings are small, each opening can be bridged in one piece of stone; 2nd, where the openings are large, each lintel is built up with several stones.

The first class is divided into (*a*) those bridging narrow openings, or where the lintel is made of great depth, rendering any precautions for relieving the pressure unnecessary; (*b*) lintels with a relieving arch over, usually employed in rubble walls, as shown in figures 346 to 348; (*c*) lintels in rubble or ashlar work, where it is inconvenient to use a relieving arch. A flat

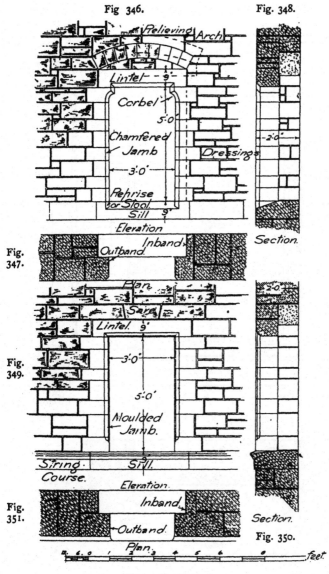

Fig 346.

Fig. 348.

Relieving Arch

Lintel 9

Corbel

5:0

Chamfered Jamb

Dressings

3:0

Reprise or Stool.

Sill 9

Elevation

Section.

Fig. 347.

Inband

Outband

Plan.

Fig. 349.

Saip

Lintel. 9

3:0

5:0

Moulded Jamb.

Fig. 351.

String Course.

Sill.

Elevation.

Inband

Outband.

Section.

Plan.

Fig. 350.

2:0

2:0

feet.

Figs. 346–351.

K

Fig. 352.

Fig. 353.

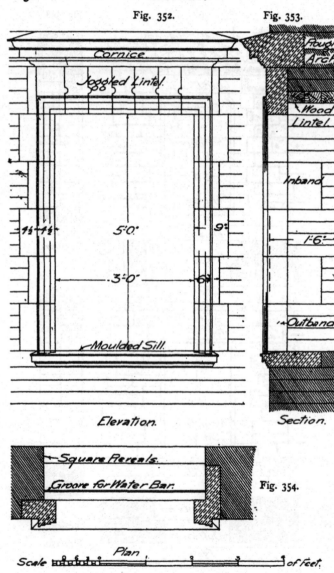

Cornice.

Joggled Lintel.

Rough Arch

Wood Lintel

Inband

4½·4½

5'·0"

9"

1'·6"

·3'·0"

6"

Outband.

Moulded Sill.

Elevation.

Section.

Square Rereals.

Groove for Water Bar.

Fig. 354.

Plan

Scale

of feet.

Figs. 352-354.

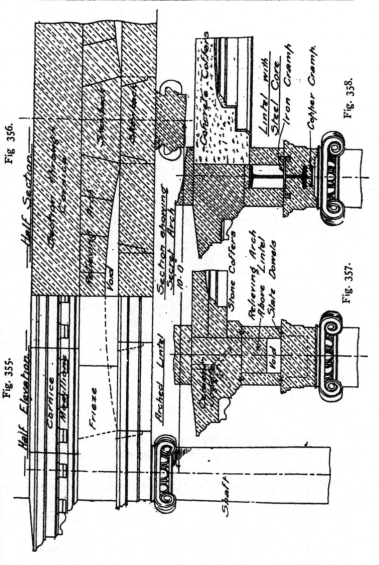

Fig. 356.

Half Section

Section showing Cornice

Relieving Arch Void

Section showing Secret Arch 10'·0

Coffered Coffers

Lintel with Steel Core

Iron Cramp

Copper Cramp

Fig. 358.

Fig. 355.

Half Elevation

Cornice

Modillions

Frieze

Arched Lintel

Shaft

Stone Coffers

Relieving Arch Above Lintel

Slate Dowels

Void

Fig. 357.

arch of three stones is constructed above the lintel, as shown
in figures 349 to 351 ; the centre stone or key is termed the
save.   In bedding the save stones no mortar is placed on the
lintel, but the stones are supported in their position by means
of small wood wedges.   After a sufficient mass of the wall has
been built to tail down the side saves the wedges are removed.

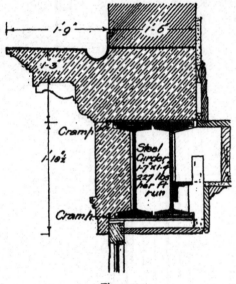

Fig. 359.

In finishing the wall, the joint between the saves and the lintel
is pointed only, thus no weight from the wall above is brought
to bear on the lintel.

Where it is inexpedient to employ relieving arches or save
stones, and where the openings are too great to admit of the
lintel being formed of· one length, the lintels are often formed
either with a joggle or with an iron core.

Figures 352 to 354 show a joggled lintel in this example

the joints are vertical, the intermediate stones being kept in their place by a semicircular joggle from front to back of every pair of bed-joints in the centre of each block. This form is also largely used for terra-cotta work.

(2) The method now frequently adopted is to build the lintel up of a number of pieces with vertical joints and in two thicknesses, the front and back portions being made, as shown in figure 358, to envelop the flanges of a steel girder, which bridges the whole span and takes its bearing on the columns. The back and front pieces are connected on the soffit by small copper cramps, the latter being bedded in cement mixed with dust from the stones to be united. The upper surface is connected by cramps of iron or copper, extending from front to back either over the top of the steel girder or through its web. The whole soffit is finally rubbed over with a piece of stone similar to the lintels, to render the joint as nearly as possible invisible. Care must be taken to protect the steel girder from the danger of oxidation by applying one of the preservative processes employed for iron and steel.

The stone entablatures built over shop fronts are formed in this way, but have the stone on one side only of the girder, being connected to the latter with cramps, as shown in figure 359.

*Arched Lintels.*—Lintels are often constructed as flat arches, and are divided as follows: (*a*) With radiating joint and joggle; (*b*) with radiating joint stepped; (*c*) with vertical joint, secret and relieving arch.

(*a*) The joints are radiated to a common centre, and have a joggle or secret arch, or a Portland cement joggle, as shown in figures 360–362.

(*b*) In this example the joints radiate to a common centre and are stepped, a method usually considered necessary where a moulded band traverses the lintel; the step renders any

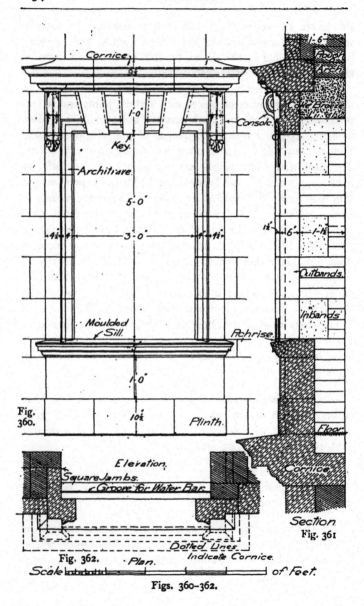

Cornice.

1'-0"

Console.

Key.

Architrave.

5'-0"

4½"+4"   3'-0"   4"+4½"

Moulded Sill.

Rchrise.

1'-0"

Fig. 360.

10½"   Plinth.

1'-6"

Rough Arch.

Cold Bros [inch]

1½"   6"   1'-1½"

Outbands.

Inbands.

Floor.

Cornice.

Section.
Fig. 361

Elevation.

Square Jambs.
Groove for Water Bar.

Fig. 362.   Plan.

Dotted Lines
Indicate Cornice.

Scale |⊢⊢⊢⊢⊢⊢|                    | of Feet.

**Figs. 360–362.**

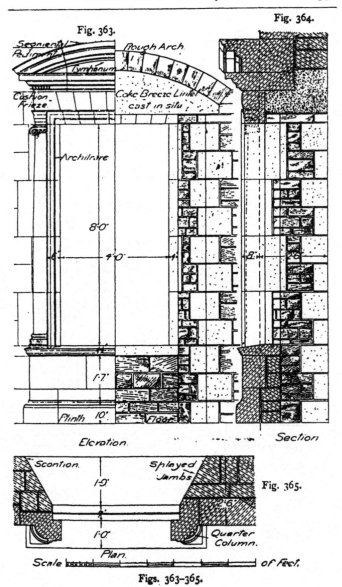

Fig. 363.

Fig. 364.

Segmental
Pediment

Tympanum

Rough Arch.

Cushion
Frieze

Coke Breeze Lintel
cast in situ

Architrave

8'0"

6"     4'0"     4"     2'4"

1'1"

1'7"

Plinth  10"     Floor

Elevation.                                          Section.

Scontion.                          Splayed
                                   Jambs                    Fig. 365.

1'0"

2'6"

1'0"                               Quarter
                                   Column.

Plan.

Scale ⊢⊣⊢⊣⊢⊣                                of feet.

Figs. 363–365.

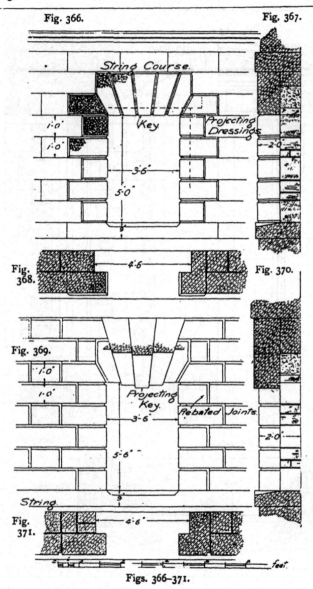

Figs. 366–371.

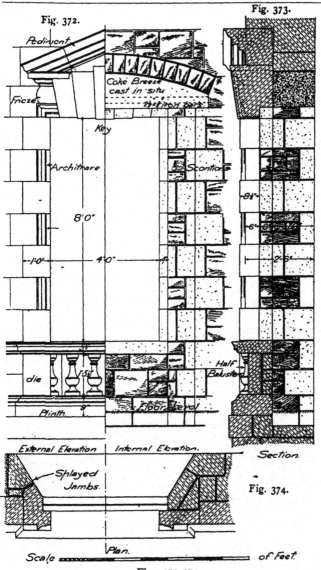

Fig. 372.

Pediment.

Frieze.

Coke Breeze
cast in situ

flat iron bars

Key

Architrave

Scontions

8'0"

1'0"    4'0"

8½"

6"

2'5"

Half
Baluster

die

Plinth.    Floor level.

External Elevation    Internal Elevation.    Section.

Splayed
Jambs.    Fig. 374.

Scale ▰▰▰▰▰▰    Plan.    of Feet.

Fig. 373.

Figs. 372–374.

Fig. 383.

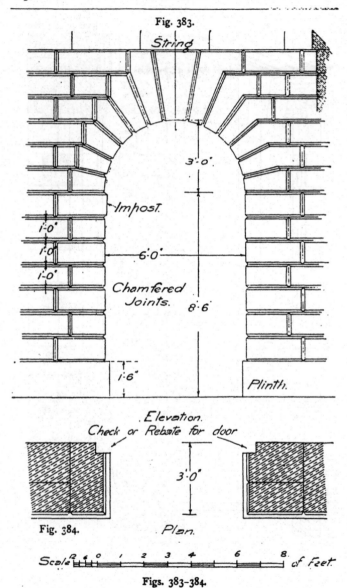

String

3'·0"

Impost.

1'·0"

1'·0"

6·0"

1'·0"

Chamfered
Joints.

8·6

1'·6"

Plinth.

Elevation.

Check or Rebate for door

3'·0"

Fig. 384.

Plan.

Scale

of Feet.

Figs. 383-384.

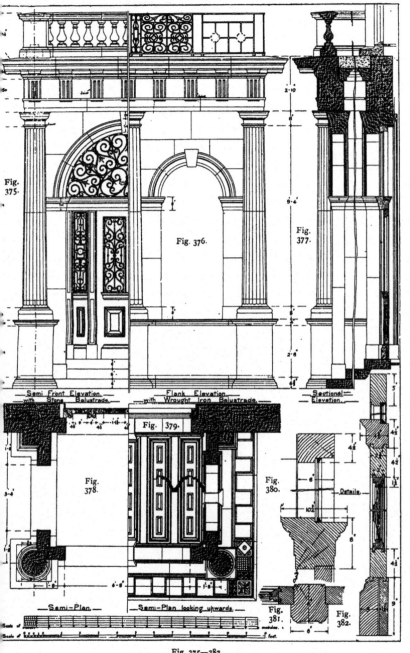

Fig.
375.

Fig. 376.

Fig.
377.

Semi Front Elevation,
with Stone Balustrade.

Flank Elevation,
with Wrought Iron Balustrade.

Sectional
Elevation.

Fig. 379.

Fig.
378.

Fig.
380.

Details.

Semi-Plan.

Semi-Plan looking upwards.

Fig.
381.

Fig.
382.

Scale of
Scale of

Fig. 375—382.

[*Between pages* 138 and 139.

dropping of the central voussoirs impossible. Figures 363–365 illustrate this method.

(c) This method is usually employed in the formation of large classic entablatures, as shown in figures 355 to 357. The lintel has vertical face points, a secret arch being formed in the central portion of the joint; this being a weak form of arch, the bulk of the weight is relieved by the employment of relieving arches or steel joists above.

*Stone Arches.*—These may be built with horizontal intrados where square openings are required, or they may be built to any of the curved forms shown in figures 178 to 191. These include arches with plain radiating bed joints as shown in figure 366. In the Palladian style the central voussoirs and the keystone often have a considerable projection beyond the face of the work, as shown in figures 369 and 372. The large voussoirs in stone arches are usually provided with cement joggles to prevent any sliding on their beds. Stone arches built to a curved form may have curved or stepped ex-

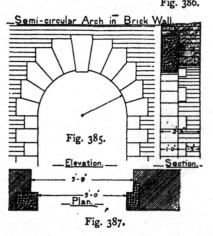

Fig. 386.

Semi-circular Arch in Brick Wall.

Fig. 385.

Elevation.

Section.

5'-9"

5'-0"

Plan.

Fig. 387.

trados. Figure 375 shows an arch with a curved extrados. In ashlar masonry built in regular courses or in brick walls the extrados are often stepped, each voussoir having a vertical point to correspond with the depth of the courses. Arches of this description are extremely useful, where the opening is near to the end of the wall; the tendency for the arch to spread is greatly reduced as the superimposed walling has no tendency to slide on the back of the arch.   Figures 383 and 387 show examples of this type of arch.   Figure 383 also shows a further development of this idea in which the voussoirs are cut to bond with the courses; this further strengthens the arch. A peculiarity of this arch is that it has a greater depth at the crown than at the springing.

Openings in buildings of a Gothic character are spanned

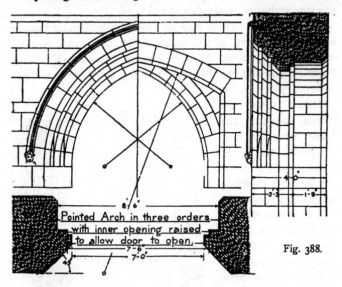

Pointed Arch in three orders with inner opening raised to allow door to open.

Fig. 388.

with pointed arches of the curved-back description, although the stepped extrados would be a most useful constructional expedient to reduce outward thrust.  The principle of constructing arches with suborders is essentially Gothic, and is to be found in most openings in this style of early date, and in thick walls of both early and late work.  Figure 388 shows a door opening with suborders and splayed jambs intersecting with the arch orders.  Where the doors were hung near the middle of the wall in pointed openings it necessitated an

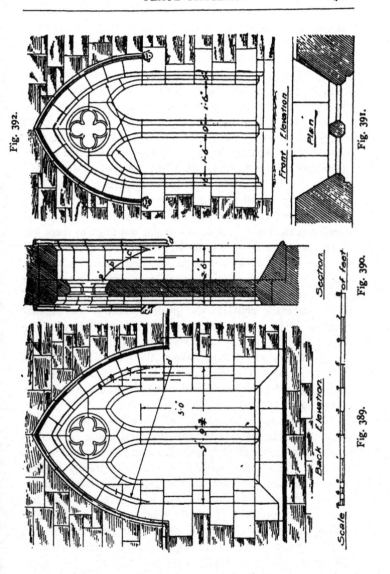

Fig. 392.

Front Elevation.

Plan.

Fig. 391.

Section.

Fig. 390.

Scale of feet.

Back Elevation.

Fig. 389.

opening of a different form at the rear to allow the doors to open. Window openings in this style in early work were employed singly, but in later work these were grouped, the walling between gradually developing into plate tracery, as shown in figures 389 to 392. This consists of a thin plate of the wall relieved of all weight from above, by a large enclosing relieving arch. The required openings were built up or pierced in this plate, leaving a series of massive mullions. The glass plane in these windows was usually near the external surface of the wall; this in thick walls left very deep jambs, which were usually splayed to admit of a better dispersion of the light internally. In small openings this splay was frequently carried round the intrados of the internal opening, forming a conical opening at the head; but in the larger openings, as there was no advantage to be gained by having an arch with a splayed soffit, and only difficulties of construction introduced by the employment of such openings, a rere-arch was used consisting of an ordinary barrel arch with a horizontal axis, the span of which was made equal to the greatest width of the internal opening. The splayed jambs were carried up till they cut the intrados of the arch, thus giving an elliptical intersection. The jamb and arch stone about this line should be in one. The method of finding the line of intersection with an example of plate tracery and rere-arch is shown in figures 389 to 392.

As the Gothic styles progressed there was a tendency for the walls to become thinner and the glass plane in window openings to be placed at or near the centre of the thickness of the wall. The thick mullions of the plate tracery became more attenuated, and the massive plate above the mullions, with its one or more foliated piercings, developed into bar tracery, the members of which were arranged in the fully developed style into forms of great complexity and beauty. Great care and judgment are required in the arrangement of the joints in bar tracery and in the selection of the stones to

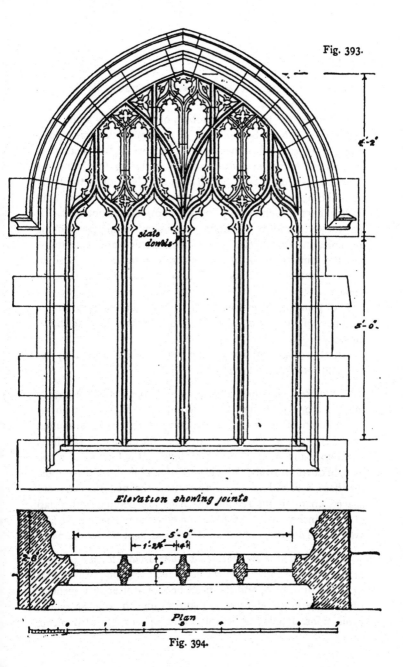

Fig. 393.

4′-2″

5′-0″

slate
dowts

Elevation showing joints

5′-0″

1′-2½″   4″

2′-8″

0″

Plan

Fig. 394.

prevent weathering. A window with bar tracery is shown in figures 393 and 394.

Wherever the moulded members of the tracery admit of it the practice should be followed of designing the tracery and fitting it in rebated stone reveals, similar to the method of fixing wood frames in reveals, as it is easier to fix the tracery after the opening is built, and it also affords an opportunity for the building to settle, and thus prevents fracture of the light tracery bars. Figures 375 to 382 show semi-front and side elevations, sectional elevation, semi-plan, semi-plan looking upwards, and details of a stone Doric portico and balustrade, and the joiner's details necessary for an entrance door, also an alternate method for balustrade showing ironwork suitable for same.

The principal dimensions of the masonry are given in modules of 30 parts, also in the ordinary English lineal measures.

In all wrought stonework the voussoirs are accurately set out with templets and cut.

*Stone Roof.*—Figures 250 to 252 show the method of arranging the stone roof over an oriel window. Figure 395 shows the method of forming a stone-covered roof over a vaulted chamber, such as was frequently used during mediæval times in military and monumental buildings. It is formed of stone flags bedded on rubble filling over the vault. In these roofs the flags are laid in two systems: in the first the flags are spaced apart, in the second the flags are bedded with a lap of 2 or 3 inches over the top sloping edges of the flags in the first system. The whole upper surface has a slight fall for drainage.

*Niches.*—These are recesses in walls usually for the purpose of receiving statues. They are generally polygonal, semicircular or elliptical in plan, and the head or roof is usually a segment of a sphere or ellipsoid. Figures 399 to 403 show the projections, details and templets of a spherical-headed niche.

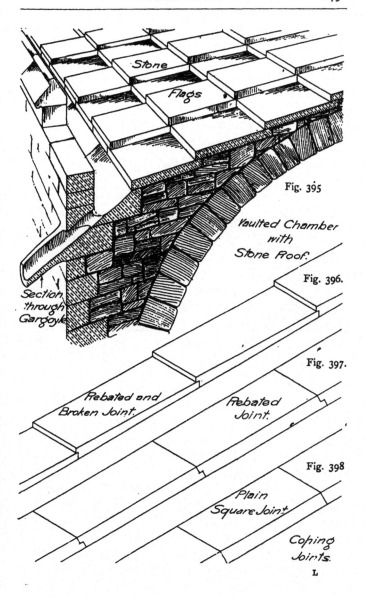

Stone Flags

Fig. 395

Vaulted Chamber
with
Stone Roof.

Fig. 396.

Section
through
Gargoyle

Fig. 397.

Rebated and
Broken Joint.

Rebated
Joint.

Fig. 398

Plain
Square Joint.

Coping
Joints.

L

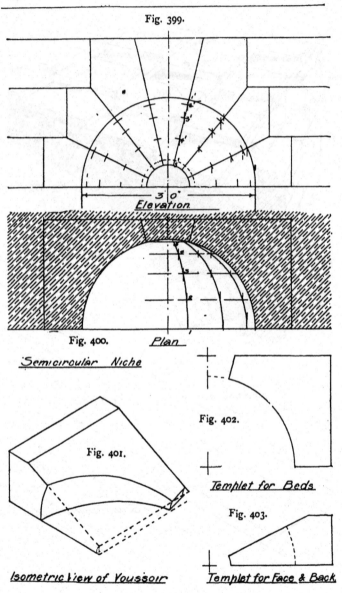

Fig. 399.

3' 0"
Elevation.

Fig. 400.          Plan

Semicircular Niche

Fig. 402.

Templet for Beds.

Fig. 401.

Isometric View of Voussoir

Fig. 403.

Templet for Face & Back

The stones comprising the head consist of a conical eye or key and a series of stone voussoirs, cut to form sectors of a sphere. In regular coursed masonry the voussoirs are arranged to form regular steps on the extrados. Figure 401 shows a detail of one of the voussoirs or sectors preparatory to bedding. Figures 402 and 403 show the templets for marking the correct shape of the voussoirs.

*Tools.*—Figures 404 to 444 show the tools commonly used by masons.

*Classification.*—These tools may be classified under the following heads :—1. Picking and surfacing tools ; 2. Setting out and setting tools ; 3. Cutting tools ; 4. Saws worked by hand ; 5. Hoisting apparatus.

*Picking and Surfacing Tools.*—1. The mallet is formed of a truncated cone of hard wood with handle, and is used for striking the mallet-headed chisels. 2. The dummy is similar in form, but with a smaller head, and generally used in carved work. 3. Iron hammer : this has a head similar to a carpenter's mallet, but made of iron ; this is used for carved work. 4. The mash hammer : this has a heavy head and short handle, and is used upon the hammer-headed chisels. 5. The waller's hammer, used for roughly squaring stones in rubble work. 6. The pick : this has a long head, pointed at both ends, and weighs about 16 lbs. ; used for roughly dressing granite and also for splitting other stones in the quarry. 7. The axe has a head shaped like a double wedge, and of about 12 lbs. in weight, used for forming a labour, similar to tooled work upon granite. 8. Patent axe : this is also used for fine axed granite work, and consists of a number of plates bolted together in a specially formed head. 9. Spalling hammer, used for roughly dressing stones in the quarry. 10. Scabbling hammer, has one end pick-pointed, and is used for roughly dressing hard stones in the quarry.

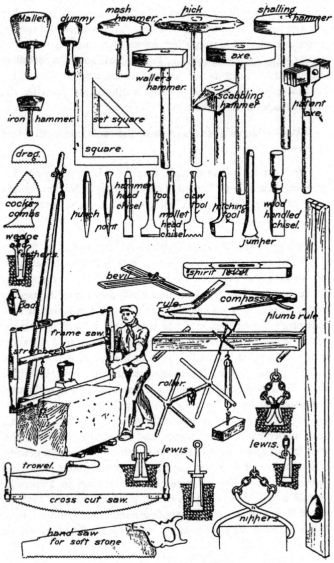

Figs. 404-444.   Mason's Tools.

*Setting Out and Setting Tools.*—1. The square, formed of iron, each arm being about 18 inches long. 2. The set square, formed of iron, set to 45° or other angles as required. 3. The bevil, formed of two blades of iron slotted and fastened with a thumbscrew. 4. The spirit level, which should be at least 18 inches long. 5. Compass for dividing. 6. The rule. 7. The plumb rule and bob for testing verticality. 8. The trowel for spreading mortar.

*Cutting Tools.*—Chisels are divided into two classes—those for use with the hammer, and those for use with the mallet. The hammer-headed chisels have the section of their striking end made smaller, to lessen the amount of burr. The mallet-headed chisels have their striking ends made broader to avoid injuring the mallet. 1. The punch has a cutting edge about $\frac{1}{4}''$ long; it is used with the hammer for removing superfluous stone in roughly dressing. 2. The point has an edge similar to that of the punch, but is used with the mallet for hard stones. Hammer and mallet-headed chisels are made with lengths of edges from $\frac{1}{4}''$ to $1\frac{1}{4}''$ wide; mallet-headed chisels from $1\frac{1}{2}''$ edge upwards are termed boasters and are used for tooling surfaces. The claw tool has an edge with a number of teeth from $\frac{1}{8}''$ to $\frac{3}{8}''$ in width, and is used for dressing surfaces of hard stones after the point or punch has been used. The pitching tool has a long edge with a thick point, and is used with the hammer for reducing stones. Jumpers are used for boring holes in quarrying granite, or for blasting operations; they are of varying lengths, and have the length of their edges greater than the diameter of the tool, so that they may be turned round in the hole being bored. The wood-handled chisel, similar to a carpenter's firmer chisel, is used upon Bath and other soft stones. The drag is a piece of steel plate with an edge cut similar to a saw tooth and is used for bringing or dragging the surfaces of soft stones to a uniform level. Cockscombs are specially shaped drags for moulded work in soft stones. The wedge and feathers, used generally for splitting granite, consists of a conical wedge and two thin curved plates; these are placed in a series of holes made by a jumper along a required line of section for severing the stone. Gads are small iron wedges used for splitting softer stones; these are inserted in chases made by the pick.

*Saws Worked by Hand.*—1. The hand saw, 28″ in length, is used for cutting soft stones, it is similar to a carpenter's half rip saw. 2. The double-handed saw, from 5 to 6 feet in length, is worked by two men. 3. The frame saw, used for cutting large blocks of stone, consists of a frame as shown in figure, the weight of which is supported by being slung on to a system of pulleys arranged to give the saw the required amount of vertical play. The blade of this saw consists of a piece of steel about $\frac{1}{10}$″ in thickness and 4″ in width, it is fixed to the frame by means of two steel pins. The cutting action is assisted by coarse sand and water, which latter dribbles into the cut from a tinware vessel placed conveniently by the side of the work. The saw should be about 2 feet 6 inches longer than the stone to be cut, for this purpose blades and stretchers of varying lengths are kept; and the coupling irons at the top may be increased in length by the addition of loose links.

*Hoisting Apparatus.* — 1. The nippers consist of two curved and pointed arms rotating on a pivot, and fitting into notches cut on the sides of thè stone; the tendency is to bite more tightly as they are lifted. These are convenient for manipulating stones too large to be easily handled. It is necessary where stones are too large to be easily lifted that they should be placed exactly in their position by the lifting apparatus, to avoid disturbing the green work underneath. For this purpose lewises, which are arrangements of dovetailed shaped blocks, are let into a special hole cut into the top bed. Four varieties are shown in the plate. 2. The scaffold jib, as shown in figure, is employed where a derrick or other crane is not used, and where the stones are first raised on the scaffold, and from there manipulated to their bed; it consists of two timbers, about 9" or 11" by 3", according to the weight to be lifted, bolted together at their ends, and leaving a space of about 4" between the timbers; it is supported by the ledgers of the inside and outside scaffolds at a convenient height above the work; it is supplied with an iron roller of about 1½" diameter with four arms at each end, the chain supporting the pulleys is passed over the roller, and the stone when raised can be moved in a transverse direction by turning the levers of the roller.

# CHAPTER IV.

# GIRDERS.

———

*Definition.*—Members of buildings which are in one piece, or built up to form one member, and for the purpose of spanning openings and supporting loads, and which are subjected to a transverse stress, are called girders, and may be either straight or curved and of various sections.

*Classification.*—They may be of (*a*) wood, (*b*) wood, halved, reversed and bolted, (*c*) flitched beams, (*d*) built-up beams, (*e*) trussed beams, and when entirely of metal may be of cast iron or steel. Wrought iron has been largely used, but it is now practically superseded by steel, owing to the latter being 30 per cent. to 40 per cent. stronger and only from 5 to 10 per cent. dearer than wrought iron. The steel sections may be rolled joists or rolled sections built up as plate, box, lattice, Warren girders, etc.

Girders fixed at one end and free at the other are called semi-beams or cantilevers; when supported or fixed at both ends, they are known as beams.

A load carried by a beam supported only at its ends would cause the upper fibres of the beam to be in compression, and the lower to be in tension. The neutral plane is an imaginary plane containing the fibres of that part of the beam which is neither in tension nor compression.

*Bressummer*, or *breastsummer.*—A girder with a wide supporting surface, acting as a lintel, as shown in figures 445 and 446; as in the case of a beam supporting a brick or stone wall over a shop front, bay window, etc.

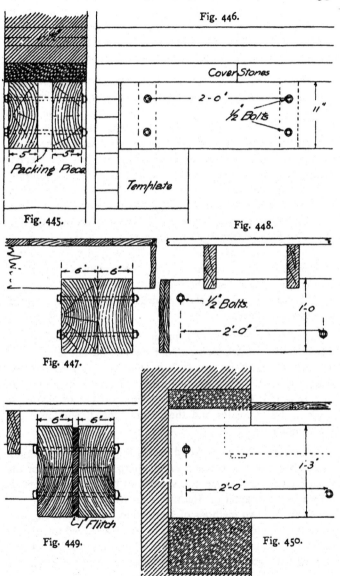

Fig. 446.

Cover Stones

2-0'

½ Bolts

11"

Packing Piece

Template

Fig. 445.

Fig. 448.

6' 6'

½ Bolts.

1-0'

2'-0'

Fig. 447.

6' 6'

½' Flitch

1-3'

2'-0'

Fig. 449.

Fig. 450.

Figs. 445-450. Examples of Girders.

*Wood Girders.*—These are usually employed in uniform rectangular sections, and to ensure against lateral bending, the minimum breadth should be ·6 of the depth.

*Flitch Girders.*—These are formed by sawing, reversing, and bolting timbers, as in figures 445 to 448. Where the dimensions are such that the whole section of the log is necessary to obtain the required scantling, it is usual to cut the timber depthways through the heart, to season the wood, for the purposes of inspection. The butt end of one piece is placed against the top end of the other, to ensure a uniformity of strength throughout, and to allow the timber to shrink without developing large shakes. The two flitches are placed together with their heart sides outwards : this operation is known as halved and reversed, after which it is bolted.

Figures 445 and 446 show a flitch girder, the flitches being blocked apart to carry a wall, the thickness of which is greater than the sum of the widths of the flitches.

Figures 447 and 448 show an ordinary flitch girder, halved, reversed and bolted.

*Steel Flitch Girder.*—To further strengthen a wood girder, a steel plate or flitch is inserted between the cut timber and bolted, as shown in figures 449 and 450. To prevent the girder riding on the steel flitch when the timber shrinks, the steel is kept $\frac{1}{4}''$ from the top and bottom surfaces, or chases are made in the templates and the cover stones if employed as a bressummer.

*Laminated Beams.*—Wood beams required of a length not obtainable in the market, or with a sharp curvature as for the ribs of a centre or dome, are sometimes built up in layers, as shown in the chapter on Joints in Carpentry.

There are two methods of lamination, vertical and horizontal. Where timbers are required to be of a curved form, such as curved beams, curved roof timbers or ribs of centres, the vertical laminated method is much used, as works of curvature

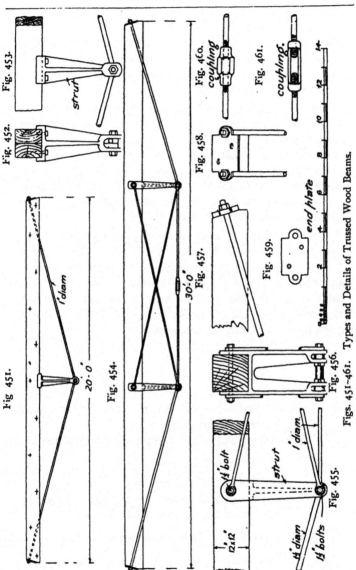

Figs. 451-461.   Types and Details of Trussed Wood Beams.

can be readily made by this method. The horizontal method
of lamination is applicable for curved beams of great span on
account of the comparative fewness of butt joints, but for
general purposes is rejected on account of the relatively
great expense in production.

*Trussed Wood Beam.*—To avoid clumsy and heavy sections
in timber, where great strength is required, steel, which offers
greater tensile resistance than timber, is often substituted for
the lower and tensional portion of the beam.

Figures 451 to 461 show the types and details which are greatly
employed for the beams of travellers, and for purlins of roofs.

Figures 460 and 461 show types of couplings used for
adjusting the lengths of the bars.

*Wood Pillars.*—Timber of rectangular section is largely
used for story posts and struts. The method of determining
the resistance of pillars, stanchions, struts, etc., is very complex,
and no formula has yet been determined to give the exact
value of the resistance. The most satisfactory formula, as
explained in the *Advanced Course*, is that based upon that
given by Rankine and Gordon, as follows:—

$$p = \frac{f}{1 + a\dfrac{l^2}{p^2}}$$

$p$ = pressure per unit of sectional area.

$f$ = intensity of pressure to crush a short column of one
  square inch sectional area.

$a$ = constant and is equal to the reciprocal of the limit
  of elasticity of the material $= 1 \div \dfrac{f}{4}$

$l$ = length of bar in same unit as $h$.

$h$ = least transverse dimension of a triangle or rectangle
  circumscribing the section of the bar.

$r$ = radius of gyration in a rectangular section $= \dfrac{h}{\sqrt{12}}$

If the ends are rounded, the value of $a$ is multiplied by 4.

The value of $f$ for dry northern pine is taken as $2\frac{1}{4}$ tons or 5040 lbs.

The following table has been calculated upon the above data and will be found useful in practice:—

$$\text{Radius of gyration in a square section} = \frac{h}{\sqrt{12}}$$

$$\text{Factor of safety} = 4 + \cdot07 \ \frac{\text{length in inches}}{\text{least diameter in inches}}$$

| $\frac{l}{h}$ | Value of $p$ or ultimate resistance to compression in lbs. per square inch. | factor of safety. | value of $p^1$ or safe working load under compression in lbs. per square inch. | $\frac{l}{h}$ | Value of $p$ or ultimate resistance to compression in lbs. per square inch. | factor of safety. | value of $p^1$ or safe working load under compression in lbs. per square inch. |
|---|---|---|---|---|---|---|---|
| 5 | 4070·8 | 4·35 | 935·81 | 28 | 595 28 | 5·96 | 99·87 |
| 6 | 3753·6 | 4·42 | 849·14 | 29 | 559·41 | 6·03 | 92·77 |
| 7 | 3414·2 | 4·49 | 760·40 | 30 | 526·57 | 6·10 | 86·32 |
| 8 | 3131·4 | 4·56 | 686·70 | 31 | 496·90 | 6·17 | 80·53 |
| 9 | 2845·2 | 4·63 | 614·51 | 32 | 468·73 | 6·24 | 75·11 |
| 10 | 2581·6 | 4·7 | 549·25 | 33 | 443·22 | 6·31 | 70·24 |
| 11 | 2341·6 | 4·77 | 490·90 | 34 | 419·67 | 6·38 | 65·77 |
| 12 | 2125·3 | 4·81 | 441·85 | 35 | 397·89 | 6·45 | 61·68 |
| 13 | 1931·4 | 4·91 | 393·35 | 36 | 377·73 | 6·52 | 57·93 |
| 14 | 1758·2 | 4·98 | 353·04 | 37 | 358·78 | 6·59 | 54·44 |
| 15 | 1603·7 | 5·05 | 324·20 | 38 | 341·64 | 6·66 | 51·29 |
| 16 | 1466·0 | 5·12 | 286·88 | 39 | 325·46 | 6·73 | 48·35 |
| 17 | 1339·7 | 5·19 | 258·14 | 40 | 310·38 | 6·80 | 45·64 |
| 18 | 1233·6 | 5·26 | 234·52 | 41 | 296·31 | 6·87 | 43·13 |
| 19 | 1135·6 | 5·33 | 213·06 | 42 | 283·15 | 6 94 | 40·79 |
| 20 | 1049 | 5·4 | 194 | 43 | 270·83 | 7·01 | 38·63 |
| 21 | 969·23 | 5·47 | 177·19 | 44 | 259·28 | 7·08 | 36·62 |
| 22 | 898·47 | 5·54 | 162·18 | 45 | 248·45 | 7·15 | 35·47 |
| 23 | 834·70 | 5·61 | 148·79 | 46 | 238·27 | 7·22 | 33·00 |
| 24 | 777·09 | 5·68 | 136·81 | 47 | 228·70 | 7·29 | 30·65 |
| 25 | 724·93 | 5·75 | 126·08 | 48 | 219·67 | 7·36 | 29·84 |
| 26 | 677·59 | 5·82 | 116·43 | 49 | 211·17 | 7·43 | 28·42 |
| 27 | 634·53 | 5·89 | 107·73 | 50 | 203·15 | 7·50 | 27·08 |

*Example.*—Determine the safe resistance of a dry northern pine story post 9 inches by 6 inches by 10 feet long, ends fixed. The ratio of length to least dimension is 20 to 1. From the table the value of $p^1$ is found to be equal to 194 lbs. Therefore, 194 multiplied by 54 (the number of square inches in area) = 10,476 lbs. = 4·68 tons nearly.

*Technical Terms.*—The following are some of the terms used in steel girder construction :—

*Clear Span.*—The horizontal distance between the abutments.

*Effective Span.*—The distance between the centres of the bearing surfaces of the girder on the supports, and is taken for purposes of calculation in the event of the camber of the girder preventing the bearing surface of the girder resting on the outer edge of the stone template.

*Effective Load.*—The effective span in feet multiplied by the weight of the distributed load per foot run.

*Bearing Surface.*—That part of the lower face of the girder which when loaded rests upon the support. The minimum length of the bearing may be obtained by dividing the span by thirty to forty.

*Feathers.*—Triangular or circular fillets inserted at the angles of cast-iron girders, etc., where triangular, usually making angles of 135° with the horizontal and vertical faces; right angles in cast-iron are a source of weakness.

*Camber.*—This is a vertical curve in an upward direction from the bearing points. Beams are cambered to allow for the deflection of a beam when loaded, and to avoid the apparent sagging that exists in all long horizontal lines. Cast-iron girders should have a camber of $\frac{3}{4}$ inch to every 10 feet of span; steel girders $\frac{1}{2}$ inch in 10 feet.

*Sections of Steel Girders.*—The most economical disposition of the material for stress-resisting purposes is theoretically to place the mass of the material as distant as possible from the neutral plane of the section; practically this condition is nearly satisfied in the ⊥ section. In plate girders the flanges are considered to resist the bending stresses and the web the shearing stresses. In cast-iron girders and rolled joists, the webs for practical reasons are made much larger than the necessary area to resist shearing; and in computing their strength, the value of the inertia of these webs should be taken into account, to obtain the maximum resistance of such girders.

The form and dimensions of girders are modified according to whether they are fixed or supported only.

Beams are said to be fixed when they are bolted at the ends, or if continued over a number of supports, the parts over the inner spans may be considered as fixed.

*Cast-iron Girders.*—The best shape for cast-iron beams is that of $\top$ section, the tensional flange being four to six times the area of compressional flange.

The depth is usually about $\frac{1}{13}$ the length, and the width of compressional flange $\frac{1}{30}$ to $\frac{1}{20}$ the length.

All castings should vary in thickness gradually, and stiffeners should be cast about every 3 feet apart.

*Specification for Constructional Cast-Iron.*—The metal should preferably be good soft grey cold-blast iron, sound and clean, cast from the second melting, free from injurious cold shuts or blow-holes, true to pattern, and of a workmanlike finish. Sample pieces, 1 inch square, cast from the same heat of metal in sand moulds, shall be capable of sustaining on a clear span of 4 feet 6 inches a central load of 500 lbs. when tested in the rough bar.

Sharp corners are a source of weakness, and should be well rounded; or triangular fillets, called feathers, cast at all internal angles.

Figures 462 and 463 show section and part elevation of a cast-iron girder used as a beam, supported at the ends, resting on a stone template, the top flange being in compression and the bottom flange in tension.

Figures 464, 465, and 467 are sections of cast-iron girders fixed at both ends, where each flange would at different parts be under tensional and compressional stresses. Of these there are two forms: (1) the $\top$ section, as shown in figure 465, and

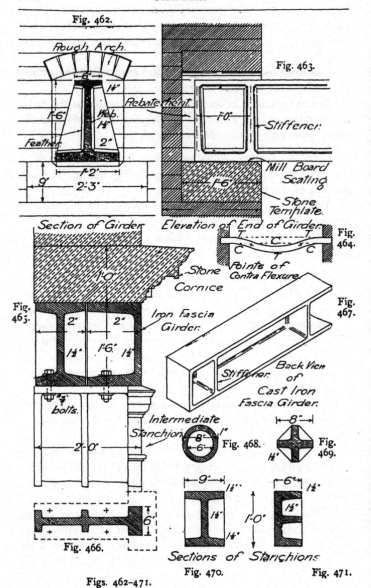

Fig. 462.

Rough Arch.

6"

14"

1·6"

Web.

14"

Rebate 4 ent

2"

Feather

9"

1·2"

2·3'

Fig. 463.

1·0"

Stiffener.

Mill Board Seating

1·6"

Stone Template.

Section of Girder.

Elevation of End of Girder.

Fig. 464.

Points of Contra Flexure.

Fig. 465.

1·0"

Stone Cornice

2"    2"

14"   1·6"   14"

3" bolts.

2·0'

Iron Fascia Girder.

Fig. 467.

Stiffener.

Back View of Cast Iron Fascia Girder.

Intermediate Stanchion.

8"

6"

Fig. 468.

8"

4"

Fig. 469.

Fig. 466.

6"

9"

14"

14"

1·0"

14"

6"

14"

14"

Sections of Stanchions

Fig. 470.

Fig. 471.

Figs. 462–471.

(2) the channel section; the latter is used extensively in Scotland as bressummers over shop fronts, as shown in figures 465 and 467.

Figure 467 is an isometric view of the cast-iron channel section girder, showing the position of the stiffeners.

Figure 464 is a diagram showing the points of contrary flexure and the undulating curved outline of the flanges of a girder, fixed at both ends when stressed by a central load.

*Cast-iron Stanchions.*—Figures 468 to 471 show sections of four forms of cast-iron stanchions. Figure 468 shows the

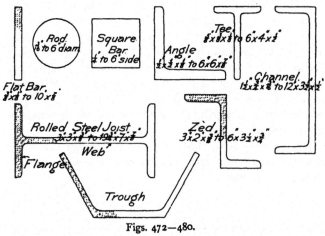

Figs. 472—480.

circular hollow section used where the column is exposed. This is the most economical form for any given area in section. Figure 469 is the cross section, used largely where stanchions are to be covered with faïence-ware or terra-cotta. Figure 470 shows the ⊥ sections, where stanchions are to be surrounded by brick; and figure 471 shows the ribbed channel form, used against party walls for supporting bressummers over shop fronts.

Cast-iron girders and stanchions are not considered as

M

reliable as those of rolled steel, as the former may contain serious flaws hard to detect, or internal stresses due to irregular cooling may exist, rendering the casting liable to fracture when subject to shocks or sudden changes of temperature. Cast-iron girders have snapped suddenly when cold water has been throw upon them while they have been in a heated state, as often happens in a burning building.

*Steel Girders.*—Owing to the improvements in the manufacture of steel, girders are now almost exclusively made of this material. Steel is rolled to numerous sections, as shown in figures 472 to 480, which can be readily adapted to any of the requirements of construction. Cast-iron girders were first largely displaced by wrought-iron girders, and now the latter have been almost entirely superseded by steel. Heat causes wrought iron and steel to expand considerably, for which reason girders of these materials should never be fixed at both ends ; heat also causes alterations in the strength and ductility of steel, this material rapidly losing strength when the temperature is raised above 500° F.

The following table shows the relative strengths in tons of cast iron, wrought iron, and steel, as most commonly used in ordinary construction :—

|  | Tension. | Compression. | Shearing. |
|---|---|---|---|
| Cast Iron ... ... | 7 to 8 | 40 to 45 | 12 |
| Wrought Iron ... | 22 | 16 to 17 | 22 |
| Steel ... ... ... | 30 | 30 | 24 |

Figures 472 to 480 show the usual rolled steel sections. Built-up girders of various forms are constructed of combinations of these sections riveted together. The I and [ sections are commonly used as joists for supporting light loads.

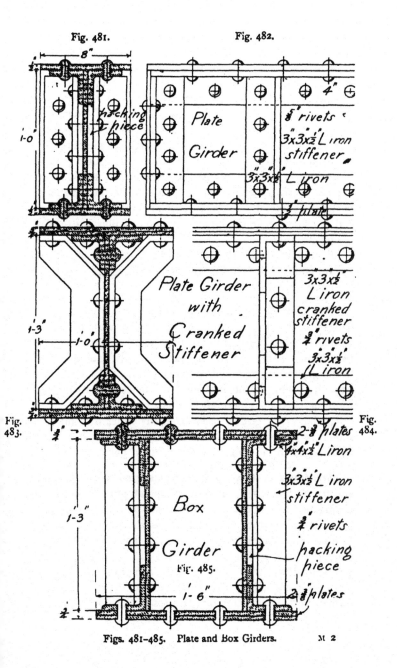

Fig. 481.

Fig. 482.

8"

1'-0"

packing
piece

Plate

Girder

⅜" rivets

3"x3"x½" L iron
stiffener

3"x3"x½" L iron

½" plate

Plate Girder
with
Cranked
Stiffener

3"x3"x½"
L iron
cranked
stiffener

¾" rivets

3"x3"x½"
L iron

1'-3"

1'-0"

Fig.
483.

½"

2-⅜" plates

4"x4"x½" L iron

Fig.
484.

3"x3"x½" L iron
stiffener

¾" rivets

packing
piece

Box

Girder

Fig. 485.

2-⅜" plates

1'-3"

1'-6"

½"

Figs. 481–485.   Plate and Box Girders.        M 2

*Plate Girders.*—It is not economical to use girders of uniform section to support heavy loads, but the area of the section is varied as the moment of the load; to accomplish this, girders are built up of plates, angles and T irons riveted together, and to strengthen the webs and flanges with stiffeners of T or angle irons. Stiffeners may be simply angle or tee irons, cut to fit and riveted between the flanges and against the longitudinal angle irons, a level surface being made with the latter; a vertical packing piece having a width equal to the stiffener and of the thickness of the longitudinal angle irons. Figures 481 and 482 show section and part elevation, and illustrate the fixing of end plates.

Figures 483 and 484 show section and part elevation of a plate girder with cranked stiffener, for use where wide flanges are necessary.

*Box Girders.*—Where girders are required with extra wide flanges, such as a girder carrying a thick wall, or in long girders to give lateral stiffness, they may be constructed with two webs. These are not so liable to twist laterally, owing to the greater support offered them by the webs, or to bend laterally, owing to the increased resistance due to the great width of flanges. Figure 485 shows a box girder.

In large box girders used externally either the end plates should be bolted to the body of the girder, or a manhole should be arranged in the web, for the purpose of painting the inside.

*Semibeams* or *Cantilevers.*—Beams fixed at one end only, and free at the other, are used to support projecting galleries. Owing to the tendency of these semibeams to act as levers and tilt or overturn the wall above, they are termed cantilevers. The overturning action may be resisted: 1st, by placing a great weight over the fixed end, or, better, by tying the end down with anchor-plates and ties as shown in figures 486 and 487. This latter method is preferable to the former, as it prevents internal stresses in the upper portion of

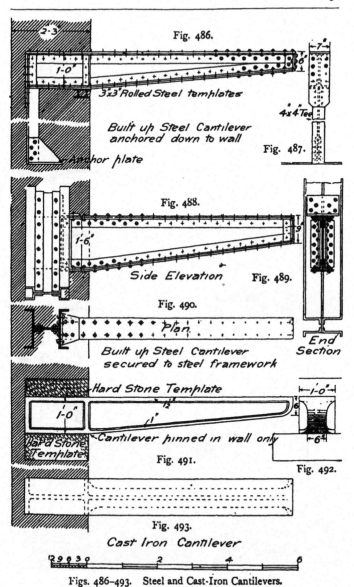

Fig. 486.

1'-0"

3x3" Rolled Steel templates

Built up Steel Cantilever
anchored down to wall

Fig. 487.

Anchor plate

7"

4"x4" Tee

Fig. 488.

1'-6"

Side Elevation    Fig. 489.

Fig. 490.

Plan

End
Section

Built up Steel Cantilever
secured to steel framework

Hard Stone Template

1'-0"

1'-0"

Hard Stone
Template

Cantilever pinned in wall only

Fig. 491.

6"

6"

Fig. 492.

Fig. 493.

Cast Iron Cantilever

Figs. 486–493.   Steel and Cast-Iron Cantilevers.

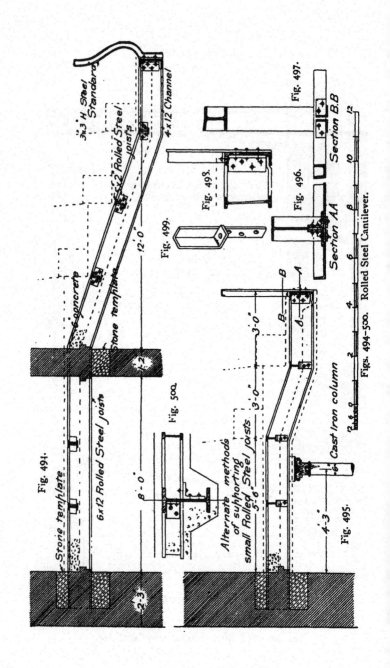

Fig. 494.

Stone template

6 x 12 Rolled Steel joists

3"x 3" H. Steel Standard

5 x 12 Rolled Steel joist's

4 x 12 Channel

Stone template

Concrete

Fig. 499.

12'-0"

2'-3"

Fig. 500.

8'-0"

Fig. 497.

Section B.B

Fig. 498.

Fig. 496.

Section A A

Alternate methods of supporting small Rolled Steel joists

3'-0"

3'-0"

B

B

A

A

Cast Iron column

5'-6"

4'-3"

Fig. 495.

Figs. 494-500.   Rolled Steel Cantilever.

the wall. The internal bearing edge is under a great com-
pressive stress and is liable to fracture, to avoid which the
ordinary stone template is now replaced by rolled steel joists of
small section, as shown in figure 486. Figures 488 to 490
show a built-up steel cantilever as used in modern steel
skeleton construction. The web of the cantilever is usually
carried right through, and in that part forms the web of
the stanchion. The cantilever is securely riveted to the
stanchion, which renders it independent for its stability upon
the stone or brick casing. Figures 491 to 493 show an
ordinary cast-iron cantilever depending on the superincumbent load for its stability. The bottom flange being under a
compressional stress is smaller in section than the top flange,
but must be widened out at the bearing edge of the wall
to avoid the crushing of the template beneath. It is necessary
to have a template above as well as beneath in cantilevers fixed
in this manner.

Figure 494 shows the arrangement of the rolled steel cantilever suitable for a public hall where there is an adjacent corridor;
the internal wall acting as the fulcrum and the outer and tailing
wall acting with the leverage of the width of the corridor.
The girder is cranked to enable the seats to be arranged in
tiers one above the other. The outer ends of all cantilevers
are connected by channel
iron; the standards for the
balustrading are formed of
H iron riveted to the
channel iron at the required
intervals, and the upper ends
of the standards are con-
nected by channel iron to
which the wood handrail is

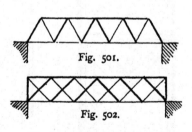

Fig. 501.

Fig. 502.

fixed. The floor is formed of 2" × 5" steel joists, riveted at
intervals to the main cantilevers, the spaces between being
filled with concrete.

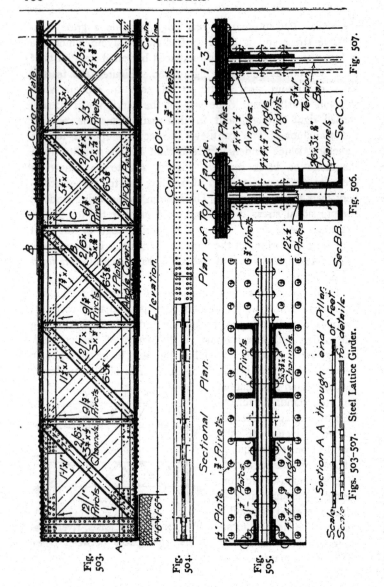

Fig. 503.

Fig. 504.

Fig. 505.

Elevation.

Sectional Plan.

Plan of Top Flange.

Section A A through end Pillar.

Figs. 503-507.   Steel Lattice Girder.

Fig. 505.

Fig. 506.

Fig. 507.

Sec. BB.

Sec. CC.

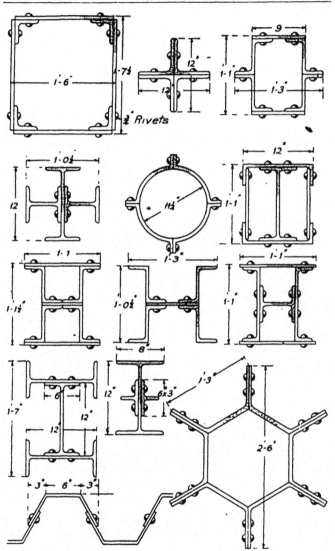

Figs. 508–520.  Sections of Stanchions.

Figures 495 to 500 show the arrangement where there is no adjacent corridor, and where it is desirable there should be no columns beneath the outer edge of the gallery to obstruct the view. A series of columns is placed under the cantilevers at a distance from the wall sufficient to form a convenient gangway.

*Warren and Lattice Girders.*—Where great depth in girders is no objection, as in bridges, they are often made on the principle of the truss, where by giving them increased depth

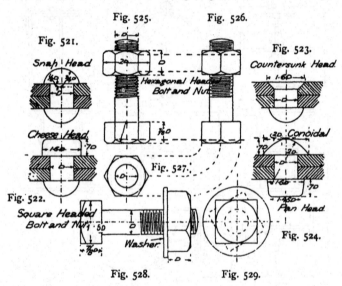

Fig. 525.  Fig. 526.

Fig. 521.

Snap Head

Cheese Head

Fig. 522.

Square Headed Bolt and Nut

Hexagonal Head Bolt and Nut

Fig. 527.

Washer

Fig. 528.

Fig. 523.

Countersunk Head.

Conoidal

Pan Head

Fig. 524.

Fig. 529.

the material may be economised. Figures 501 and 502 show line diagrams of types of the Warren and lattice girders, in which the web plates are replaced by bars and angle irons riveted together.

In these figures, when the girders are uniformly loaded on the upper flange, the thick lines indicate the members in compression and the thin those in tension. The compressional members are termed "struts," and are usually made of angle, tee or channel irons ; the tensional members are called "ties," and are of bar-iron. The vertical members joining the horizontal members, sometimes used in lattice girders between the apices of braces, serve to transmit the load in such a manner that one half of the load immediately above is carried by the top flange and the other half by the lower. These vertical members are formed by angle, tee, or channel irons, according to the size of the girder. The horizontal members of braced girders are termed booms, and are constructed of plates and angle irons in a manner similar to that of ordinary plate girders.

Steel sections framed as isosceles triangles, arranged in one system and bolted or riveted together to the form shown in figure 501, are known as Warren girders.

Figure 502 is an outline of a girder composed of two systems of triangles traversing each other, and is known as the lattice girder.

More than two systems of triangles may be arranged.

Figures 503 to 507 show a steel lattice girder of 60 feet span designed to support a distributed load of 3·6 tons per foot run, loaded on the bottom flange, and having upright bars to distribute the load equally upon both flanges, with details showing sections and connections of the various parts.

*Stanchions.*—Figures 508 to 520 show steel sections built up into stanchions. All systems of rectangular sections lend themselves to be encased with brickwork. The circular and polygonal sections are best adapted for being encased with terra-cotta or faïence ware to form circular columns. Open sections are best for exposed stanchions, as all of the parts are accessible for painting.

*Rivets.*—Figures 521 to 524 show forms of the following rivet heads :—snap, cheese, countersunk, conoidal and pan ; all parts of each rivet are given in terms of the diameter of the shank. For constructional purposes the snap or cup head is ordinarily used, the countersunk where a flush surface is required.

The rivet holes in steel plates up to $\frac{1}{2}$ inch in thickness are usually punched; plates $\frac{1}{2}$ to $\frac{3}{4}$ inch are punched and rymered or annealed afterwards, above $\frac{3}{4}$ inch they are drilled. Punched holes being slightly tapering afford a better grip for the rivets, but tend to weaken the plates, see Part II.

In constructing riveted plate girders or stanchions, it is more economical to drill than to punch the rivet holes. The parts of the girder are cramped together and the holes are drilled through the connected sections in one boring; this saves the setting out of each part and ensures the holes being concentric.

*Bolts.*—Figures 525 to 529 show the hexagonal and square headed bolts, all the parts being given in terms of the diameter of the shank.

# CHAPTER V.

# JOINTS IN CARPENTRY.

———

*Stress and Strain.*—Stress is the name given to an external force acting upon material, and is measured per unit area in gravitation units as lbs., cwts., etc.; strain is the amount of deformation of such material, and is measured either by the perpendicular distance of the axis from its original position, or by the proportionate reduction or elongation in length.

Timbers are said to be in tension when subjected to a length-stretching or pulling force, as in the case of a tie-beam, and all members wholly in tension may be considered as ties.

Timbers subjected to a crushing or pushing force are said to be in compression, as in a short pillar, and all members in compression may be considered as struts.

If a load or pressure be applied at right angles to the length of a beam it will cause a bending or cross strain, as when a floor joist is carrying weight. It may be noticed that the tendency of the upper part is to become hollow or concave; such fibres are in compression, and the lower in tension. Somewhere between the two will exist fibres neither in tension nor compression, and the plane containing these fibres is known as the neutral layer, or the layer under no stress, which in a theoretically perfectly elastic material would be in the centre of the section.

If a force acts with a twisting or wrenching tendency, the piece stressed is said to be in torsion; such stresses do not frequently occur in structural woodwork, but in screws and in parts of machinery.

Shearing is a cutting or sliding stress; but in the case of cutting or sliding with the grain it is particularly known as detrusion.

*Designing Joints.*—In designing joints, the following principles, based upon those suggested by Professor Rankine, should be considered :—

1. To cut the joints and arrange the fastenings so as to weaken as little as possible the pieces of timber that they connect.

2. To place each abutting surface in a joint as nearly as possible perpendicular to the pressure which it has to transmit.

3. To form and fit every pair of surfaces accurately, in order to distribute the stress uniformly, and to proportion the area of each surface to the pressure it has to bear, so that each part may be safe against injury.

4. To proportion the fastenings so that they may be of equal strength with the pieces which they connect.

5. To place the fastenings in each piece of timber so that there shall be sufficient resistance to the giving way of the joints by the fastenings shearing or crushing their way through the timber.

*Classification of Joints.*—Joints may be arranged under the following heads :—

*a.* Lengthening Joints.—Lapping, fishing, tabling, scarfing, building up.

*b.* Bearing Joints.—Notching, cogging, halving, dovetailing, housing, mortice and tenon, chase mortising, joggle, tenon, housed and dovetailed tenon, tusk tenon, foxtail wedging, bridle.

*c.* Oblique Shouldered Joint.—Mitre, birdsmouth, halving and dovetailing, oblique tenon, single abutment, bridle.

*Lengthening oj Beams.*—Beams may be lengthened in various ways to suit the particular stress the timber has to resist.

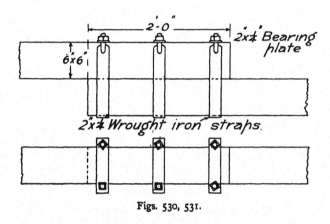

Figs. 530, 531.

This is rendered necessary if the desired length is not obtainable in the market.

*Lapping* consists in resting the end of one piece on the end of another piece, and securing the same with bolts if to resist tension; if compression, straps, as in figures 530 and 531, or as in scaffolding by ropes and wedges. The lap joint has a clumsy appearance, and should be used only for temporary structures.

Where the ends are required to be level, the length should be composed of an odd number of pieces.

*Fishing.*—Two pieces of timber having their ends abutting, with a plate of wood or iron secured on two parallel sides by means of bolts, are said to be fished, as in figures 532 and 533.

For the lengthening of beams, a plain fished joint is most economical for labour and material. In the event of the timbers shrinking, and thus causing great stress upon the bolts, coincident grooves are cut in the pieces joined, and the fish

plates and keys of hard wood or folding wedges inserted to assist
the bolts, as shown in figure 535; or sometimes tabling is resorted

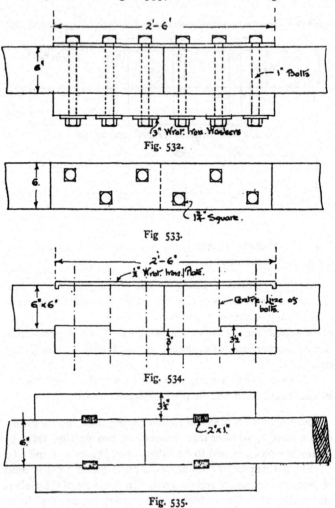

Fig. 532.

Fig 533.

Fig. 534.

Fig. 535.

Figs. 532–535.   Fished Joints,

to, as in figure 536, by bedding portions of one beam into another; but as this is costly, and reduces the effective section of the beam, it is not often employed. Where iron fish plates are employed, the ends are turned in, as shown in figure 534. In designing a fished joint, the following rules, *based* upon those given by Seddon, should be observed :—

1. The strength of fished joints depends on the effective sectional area of fish plates, being together equal in tensile strength to the effective sectional area of the tie.

2. On the sectional area of the bolts on either side of the joint being sufficient to resist shearing. Let the ultimate compressile, tensile, and shearing resistances of wrought iron be taken as 36,000, 49,000, and 49,000 lbs., and the ultimate compressile, tensile, and shearing detrusion resistances of good fir be taken as 5,000, 4,000, and 600 lbs., respectively per square inch of area opposed to force, then the shearing strength of wrought iron is to the tensile strength of fir as 49 is to 4, or, approximately, 12 to 1. Therefore the shearing area of the wrought iron should be one-twelfth the net tensile area of beams joined ; that is, in members connected by a fished joint, subjected to single shear with one plate only, the sum of the sectional area of the bolts on each side of the joints should equal one-twelfth the net tensile area of the timber joined; with fished joints with two plates (the usual practice), this being subjected to double shear, the sum of the areas of the sections of the bolts on each side of the joint should equal one-twenty-fourth the net tensile area of the timber joined. Generally, in any joint connected by bolts, the number required may be ascertained by making the sum of the minimum sectional areas opposed to the shearing force, which on shearing would cause failure of joint, to equal in area one-twelfth the net tensile area of the timber.

3. On the placing of the bolts in such a way, and at such distances from the ends of the timbers, as to prevent their drawing through them.

4. The bearing area of the bolts must be sufficient to

prevent their cutting their way through either the timbers or the fish plates.

*Scarfing* and *Fishing.*—The following difference may be noticed between a fished and scarfed joint.

In the fished joint the total length of pieces joined is not reduced, while in the scarfed joint the pieces are cut and fitted over each other so as to keep the same breadth and thickness

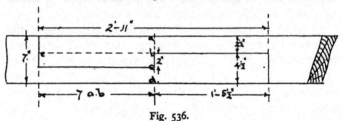

Fig. 536.

throughout, the total length of pieces joined only being less than the sum of the separate lengths, by the length of the scarf.

Scarfing should be used for the jointing of timber in the direction of its length, if neatness be necessary as well as strength.

Figure 536 shows a tabled joint, and if this be calculated as shown, dimensions for other scarfs may be reasoned out in a similar manner.

If area expressed by *b c* yields, the joint fails under compression; should *a b* or *c d* rupture, the joint fails under tension. By equating, it will be found for fir compressional area required is four, when the tensile area is five.

$\therefore$ *b c* = compressile area = $\frac{4}{14}$ depth of beam.

*a b* = *c d* = tensile area = $\frac{5}{14}$ depth of beam

*b e* is the area opposed to detrusion for fir; by equating, it will be seen that the necessary area should be $\frac{40}{8}$ of the tensile area, or approximately the detrusion area must be seven times the tensile area.

The lengths of scarfs may then bear the following proportion to depth :—

Length = 14 × $\frac{5}{14}$d. = 5 times depth $\begin{cases} \text{Without the assistance of} \\ \text{iron bolts.} \end{cases}$

When the scarf is assisted by bolts and fish plates the strength of the joints may be made equal to the strength of the net tensile area of the timber ; that is, if there are two lines of bolts in the plan, the sectional area will be reduced by the sum of the sectional area of fibres cut by two bolts. The length of the scarf may vary between three times the depth and zero times the depth, in which latter case it practically becomes an ordinary fished joint.

Figures 537 to 540 show two forms of joints to resist tensional stress. Figures 537 and 538 show a scarf constructed by tabling, tongued at ends, and secured by hard wood keys and wedges. Figures 539 and 540 show a simple form of scarf made by halving, tongueing, and tightening up by wedged keys. The shearing area along BF to be at least seven times BC, and BC should equal $\frac{5}{4}$ times the sum of the depths of the keys.

*Building Up.*—Curved ribs of domes, roofs, centres and standards of derrick crane towers, and the booms of trussed timber girders, may be economically built up as shown in the *Advanced Course* of this work.

Figures 545 to 550 illustrate a laminated arched rib, used as the supporting member of a timber bridge for a span of 60 feet.

The rib is constructed of six laminæ, each 3 inches thick, bent about a curve of 35 feet 9 inches radius. The ends are secured in cast-iron shoes, built into the abutments at each end. The laminæ are secured together by two rows of $\frac{3}{4}$-inch wrought-iron bolts spaced 2 feet apart. The joints in the laminæ are spaced at equal distances apart, say 8 feet, and not more than one joint in any cross section. The advantages of this method of building up over that by means of the

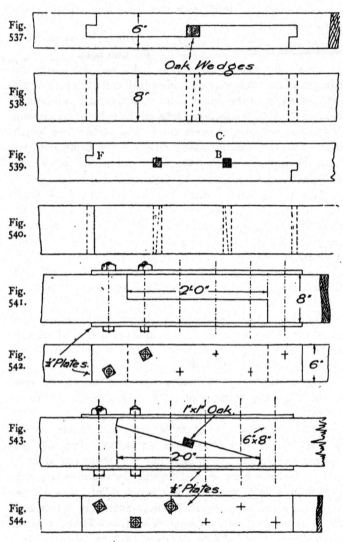

Figs. 537-544. Scarfed Joints.

vertical laminæ are: 1st, the comparatively small number of joints; and 2nd, the continuity of the grain is less interrupted.

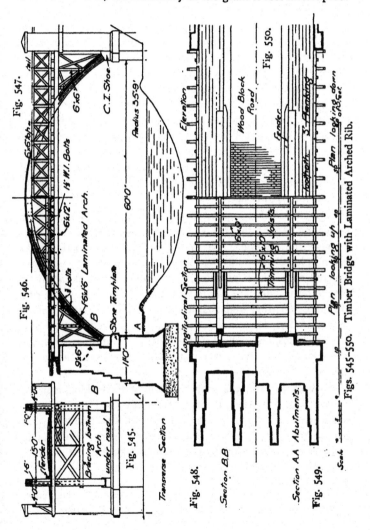

Figs. 545–550. Timber Bridge with Laminated Arched Rib.

The strength of such a rib may be considered equal to a solid rib with an equal depth, but of a thickness equal to the sum of the thicknesses of the separate pieces minus one.

To support the roadway of this bridge, 6″ × 12″ timbers are built into the arched ribs, and pass through the centres of the latter. The loads are transmitted to the arches at their haunches by means of upright struts from the underside of the beams to the top of the arches. The central portions of the beams are supported from the crowns of the arches by means of bolts, the spaces between the bolts and between the struts are cross braced, as shown in figure 546: this makes the ribs rigid, and together with the abutments prevents any tendency of the arched ribs to spring back to the straight. At the haunches the ribs are cross braced under the roadway, as shown in figure 545.

The roadway is supported by 6″ × 9″ cross beams, these projecting beyond the rib at each end to form the footpath. To give the necessary rise to the centre of the road, curved furring pieces are nailed on these timbers; the whole is then covered with 3-inch planking, and, upon this planking, wood blocks set in pitch are laid to form the road surface. The ribs are protected from vehicular traffic by means of stout timber fenders. The balustrade for the footway is fixed on the projecting beams, and is kept in the vertical position by inclined stays supported by every alternate beam, as shown in figure 545.

Figures 543 and 544 show the form of scarf which, assisted by bolts and fish plates, is most commonly used in practice to resist tensional stress: the simplicity, easiness of fitting and putting together by means of oak wedges render it a suitable and efficient joint for the purpose, the resistance to the shearing stress being offered by the wrought-iron bolts.

*Scarfs to Resist Compression.*—For joints to resist compression the abutting surfaces should be perpendicular to the pressure, consequently all shoulders should be square, and no splayed sallys or internal angles should be allowed. The following are suitable :—Scarfs similar to those shown in figures 537 to 540. The method shown in figures 541 and 542, formed by halving the timbers, with iron plates and bolts, is the simplest and most effective if plates and bolts are permissible.

*Transverse Stress.*—Wherever the span to be bridged is too great to admit of the use of timber in one piece, the beam employed should be trussed, so that every member may be under a direct length stress, as the joints that have been designed, although ingenious and costly, are not satisfactory.

Fig. 551.

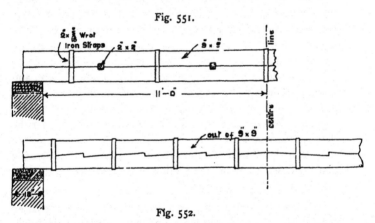

Fig. 552.

Figure 551 shows a method of building up, and is applicable to resist a transverse stress by placing one timber directly on the other, to obtain a deeper beam than could be readily obtainable in one piece. To resist the horizontal shearing stress, hard-wood keys, with the grain at right angles to the stress, are inserted. The timbers are held together by iron collars or straps, bolted as shown in figure 551.

Another method of obtaining the same result is by forming

the joint between the two members by a series of indents, and joining by iron collars or straps, bolted, taking care to arrange the shoulders of the indents, as shown in figure 552, to resist the compressional stress caused by the upper piece tending to slide over the lower.

*Notching.*—Joists resting on wall plates are commonly notched, as shown in figures 553 and 554, for the purpose of bringing all the upper surfaces to a required level, the shoulders assisting in forming a fixing and in keeping the work square. The plate should be housed in the case of joists, as shown in figure 557, to obtain a bearing for the whole depth of the joists.

*Cogging.*—Where wall plates, purlins, rafters, or other timbers have timbers crossing them, and it is desired to utilize the entire bearing depth of the timber, and to tie the lower other than by the fastenings, which often consist only of spikes, the pieces may be cut and fitted as in figure 556, two cheeks of rectangular section being taken out of the lower piece, and the upper member having a sinking to correspond to the projection left on the lower piece; this method is termed cogging. Where the end of the joist does not project over the plate it is advisable to form the cog nearer to the edge of the plate, as in figure 555, to obtain the greatest amount of shearing area behind the cog, as the section of the bearing surface has usually an excess of strength.

*Halving.*—Timbers that cross each other, and are required to be flush on one or two faces, and have a similar sinking in both to fit each other, are said to be halved together, and may be cut as ordinary bevelled or dovetailed halving.

1st. Ordinary halving, as in figures 558 to 560, is suitable for wall plates or braces of partitions crossing each other.

2nd. Bevelled halving, which assists in holding pieces together when spiked, as shown in figures 561 and 562.

3rd. Dovetailed halving is used where one member is to act as a tie for wall plates, as in figure 563 ; but as the dovetails

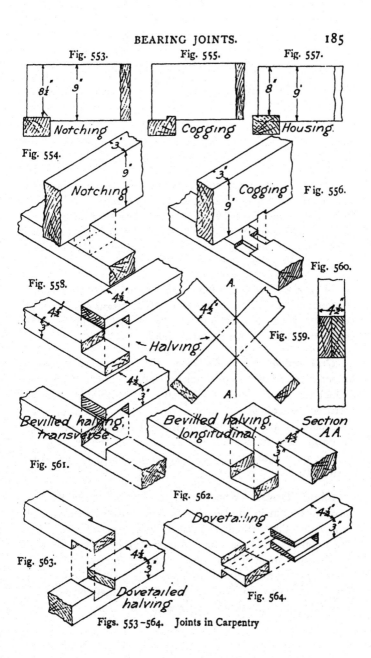

Fig. 553. Notching

Fig. 555. Cogging

Fig. 557. Housing

Fig. 554. Notching

Fig. 556. Cogging

Fig. 558. Halving

Fig. 559.

Fig. 560.

Section A.A.

Fig. 561. Bevilled halving, transverse.

Fig. 562. Bevilled halving, longitudinal.

Fig. 563. Dovetailed halving

Fig. 564. Dovetailing

Figs. 553–564. Joints in Carpentry

are usually made of great width in carpentry the pieces shrink, and the joint becomes loose, and fails for the purpose for which it was intended.

*Dovetailing.*—The joints of the angles of such timbers as curbs, pole plates, etc., which are usually deeper than the ordinary wall plate, are dovetailed as shown in figure 564.

*Housing.*—When the entire end or thickness of a piece of timber butts and fits into another piece, it is said to be housed. This joint is used for heavy work where other methods would be weak or expensive. The pieces joined would be secured by bolts, straps, tenons and wedges or dovetailing as in figures 571 to 574, or other fastenings.

*Mortice and Tenon.*—For simplicity of construction and efficiency for its purpose, the mortice and tenon joint is pre-eminently good; and is undoubtedly in some form or other used more than any other kind of joint.

*Ordinary Mortice and Tenon.*—The end of a piece of timber is cut so that a projection or tenon is formed which has a thickness $\frac{1}{3}$ that of the material; the adjoining piece is mortised to correspond to the tenon, and the latter may be secured either by wedges from the back, as in figure 565, by dowel pins from the faces as shown in the chapter on Joinery, or wedges and dowels may be combined for purposes of strength.

*Chase Mortices.*—Wherever it is required to fix timbers between the main timbers, which are already secured and immovable, the method of chase mortising may be adopted, which consists in fitting one end in the ordinary manner, and in the other main timber the mortice to receive the other extremity is cut as illustrated in figure 567, to allow the tenoned member to revolve into its ultimate position, that end which is fitted into the ordinary mortised main timber being used as the

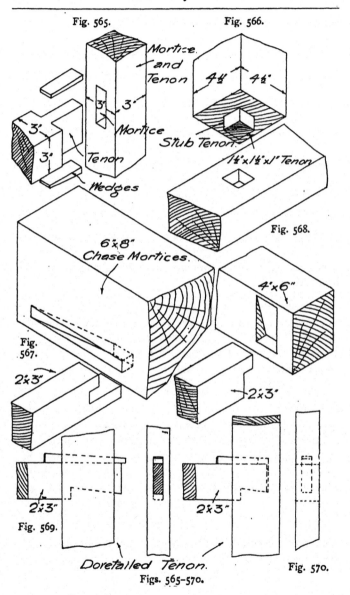

Fig. 565.

Fig. 566.

Mortice and Tenon

3″  3″

3″

3″

Mortice

Tenon

Wedges

4½″  4½″

Stub Tenon.

1½″×1½″×1″ Tenon

Fig. 568.

6″×8″
Chase Mortices.

4″×6″

Fig. 567.

2″×3″

2″×3″

2″×3″

2″×3″

Fig. 569.

Doretalled Tenon.

Fig. 570.

Figs. 565–570.

centre. The same purpose may be accomplished by cutting a vertical groove, as in figure 568, to allow the tenoned member to drop into its place; but by this method fibres are cut, which are not necessarily severed for the mortice, and so its strength is reduced. The better system is to make the horizontal chase, which has the advantage of not unnecessarily reducing its tensile strength. Ceiling joists in double and double framed floors are often fixed in this manner, and instead of cutting away a triangular piece for each joist, a rectangular groove in section, as shown in figure 626, is cut between mortices in every other space between the joists. To prevent any weakening of the binder a fillet is often fixed and the ceiling joists slotted. Both methods are shown in figure 626.

*Joggle*, *Stub*, or *Stump Tenon* are the names given to a short tenon. It is more particularly known as a joggle when cut, as shown in figure 566, which is used to prevent the foot of a post sliding away from its sill.

*Tenoned and Housed Joints.*—For purposes of bearing strength, where the ordinary tenon might not be sufficient, the rail is housed a short distance into the post or style, and tenoned in the usual manner. Figure 669 shows this joint, and the shoulders pulled closely together by means of a key.

*Dove-tailed Tenon.*—Where two pieces are required to be held together by a mortice and tenon, and to be taken apart occasionally, the tenon may be cut dovetail shape and the mortice made long enough to permit the wide part of the tenon to pass through, and when in this position it is secured by a hard-wood wedge, as in figure 569. This method may be adapted to secure pieces together, as shown in figure 570, when the back edge is not to be cut or is inaccessible.

*Tusk Tenon.*—Where floor timbers of equal depth are framed into each other it is desirable that the loss of strength should be as little as possible, and yet the members joined must be securely held together. The best bearing joint complying with these conditions is the tusk-tenoned joint. The tenon should be placed in the centre of the depth of the pieces joined, for two reasons. First, when it is in that position the stress caused by driving the wedge draws up both top and bottom shoulders equally tight. Secondly, the best position for the mortice is theoretically in the compressional fibres immediately adjoining the neutral layer, and in good specimens of northern pine timber beams, supported at both ends under loads, the ratio of the resistance of compressional to tensional stresses varies from 3 to 5 respectively in some specimens, and as given on page 177, from 5 to 4 respectively in others; the neutral layer would then vary as calculated by the principle of the moment of inertia (as stated in my *Advanced Course*) from ·45D to ·52D measured from the bottom layer of a rectangular beam, therefore it would be unwise to sink the mortice any nearer to the bottom than $\frac{10}{24}$D, as shown in figures 571 and 572. It would be slightly stronger if the underside of mortice were $\frac{11}{24}$D from the bottom layer, though the latter arrangement would make it much more difficult to obtain a tightly-fitting joint in timbers that have a tendency to be in winding. The depth of the tusk may be $\frac{8}{24}$D of the depth, that is, one half of the distance between the under surfaces of tenon and joist, and the thickness of the tenon $\frac{D}{8}$, as shown in figures 571 and 572.

In cases similar to binders tusk tenoned to large girders, especially wherever the tenons of two binders are arranged exactly opposite, the tenons of the former do not pass through the entire section of the latter, but only penetrate 3 or 4 inches, and are then spiked with large iron nails or secured with oak dowel pins or trenails, the aim being always to reduce the section of the main timber as little as possible.

Figures 573 and 574 show the method of trimming by the use of the stopped dovetail, a method which is largely employed in the North of England, and is especially applicable in positions where a member can only be conveniently dropped in from above.

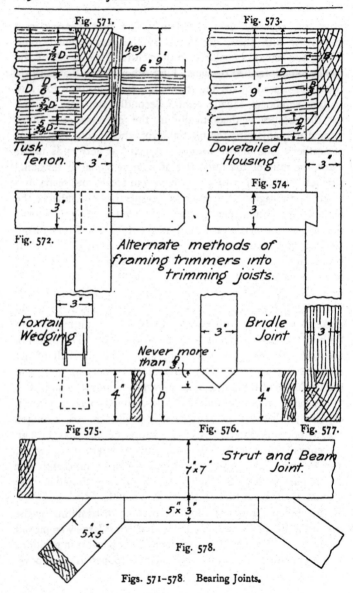

Fig. 571.

key

Tusk Tenon.

Fig. 573.

Dovetailed Housing

Fig. 572.

Alternate methods of framing trimmers into trimming joists.

Foxtail Wedging

Bridle Joint

Never more than 9

Fig 575.        Fig. 576.        Fig. 577.

Strut and Beam Joint.

Fig. 578.

Figs. 571–578.    Bearing Joints.

*Foxtail Wedging.*—A method of securing timbers, as shown in figure 575, where the back edge of the mortised member is inaccessible. It is mostly used in joinery work for securing the parts of framing, as in the rails and styles of doors in high-class work, in which the end grain of the tenon would be unsightly.

The mortice is sunk to from $\frac{1}{2}''$ to $1''$ of the back of the mortised member and is made slightly dovetailed in form. The tenon is neatly fitted, two saw cuts are made, one at each side of the tenon, about $\frac{3}{16}''$ from the side. Into each of these a wedge is inserted ; when the whole is ready, the tenon is glued and forced home. During this operation the wedges are urged into the saw cuts, causing the end of tenon to spread and fill the dovetailed mortice.

*Bridle Joint* is the name given to that joint, as shown in figures 576 and 577, that has its mortice or slot on the member that would be tenoned in the ordinary method, and the piece on which a sunk tenon is formed would in the usual joint be mortised. It is more expensive to make, but some prefer it to the mortice and tenon joint, as any bad workmanship can be easily detected.

*Strut and Beam Joints.*—Where struts are used to strengthen horizontal beams or shores, a straining piece is usually fixed to the beam, and the ends of the struts cut as shown in figure 578. Where the horizontal member is short, the straining piece is dispensed with and the ends of the struts are butted.

*Joints for Struts and Ties.*—These may be constructed as follows:—

(*a*). A cleat is bolted to the tie and the inclined strut is butted against the cleat, as shown in figure 579.

(*b*). By inserting an oak cog in notches cut in rafter and tie beam, to prevent sliding, as in figure 580.

(*c*). With single abutment and tenon, as shown in figures 581 and 582. This is simple, effective, and generally used when there is sufficient length of tie behind strut to resist shearing stress.

(*d*). *Bridle Joints.*—This is a method employed to increase the shearing area; it is especially applicable to such work where the length of tie in front of strut to resist the thrust is small, as shown in figures 583 and 584. In addition to having a large shearing area, it has the advantage that any defect in the fitting would be apparent.

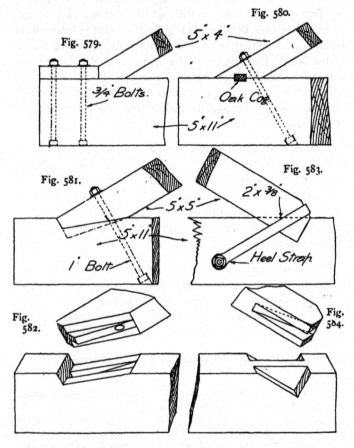

Fig. 580.

Fig. 579.

5˝ x 4˝

¾˝ Bolts.

Oak Cog

5˝ x 11˝

Fig. 581.

Fig. 583.

2˝ x ⅜˝

5˝ x 5˝

5˝ x 11˝

1˝ Bolt

Heel Strap

Fig. 582.

Fig. 584.

*Fastenings.*—Joints in carpentry should have white-lead applied with a brush over the surface forming the joint, and may be secured in their positions by the following fastenings :—

1. Pins { Wood Pins.
         Nails.
         Bolts.
2. Wedges.
3. Straps.
4. Sockets.

*Pins.*—These may be of wood or iron, and may be subdivided as dowel pins, trenails, spikes, dogs, screws, and bolts.

*Dowel Pins.*—When boards are required to be kept in the same plane surfaces, and yet not resist the shrinkage of the wood, dowels out of oak or hard wood of straight grain and split to the diameter required, and not cut with the saw, so that none of the fibres are cut across, are employed as in the example of flooring, figure 618. Dowels are used to fix mortice and tenon joints, and are especially applicable for large or triangular framing to pull the shoulders tightly up, which method is called drawboring, which consists of boring through the mortised piece, then marking the tenoned member and boring it, not exactly in the mark, but towards the outside corner of shoulder, and also towards the shoulder, which has the effect, when the dowel is driven with great force, of bringing the shoulder up closely, as shown in figure 833.

*Trenails.*—Hard-wood pins of large diameter are called trenails ; they are extensively used in shipbuilding or in positions where iron fastenings would be liable to rust. The force required to shear English oak across the grain is 4,000 lbs. per square inch.

*Nails.*—Nails are pieces of metal plate or wire, tapering or prismatic, usually pointed and headed, all of which are cast, cut, or stamped. They may be obtained in cast iron, wrought iron, steel, zinc, copper, brass, or in composition metal. Many

o

have one pair of sides parallel in length.  Such should always
be driven with those sides parallel to the length grain to reduce
the tendency of the nail to split the wood in its progress
through the latter after each blow.  Hard woods are bored to
receive the nails; but in the softer woods this is seldom
necessary.  There is a great variety of nails; those in common
use are manufactured from wrought iron or steel.

*Cast-iron Nails.*—These have been used for slating purposes, but are very brittle and are now discarded.

*Wrought-iron and Steel Nails.*—Cut clasps are cut by
machinery from wrought-iron and steel plates, to the form
shown in figure 585, and are the nails most commonly used.
Clasp nails 4 inches and above in length are known as spikes.
Nails cut from plates of wrought iron by manual labour, of the
form shown in figure 585, are particularly known as wrought
nails, and possess the property of bending sufficiently to clinch
without breaking.  Of late years, however, these have been
produced by machinery, which has been improved to cut
sufficiently and yet not to destroy the tenacity and pliability of
good wrought iron.

*Spikes* are iron nails above 3 inches in length, extensively used in fixing timbers together.  They require to be
hammered in with great force.  Spiking is a cheap method of
fixing.  The holding power is computed from experiments
made by Captain Fraser, R.E., to be in fir from 460 to 730 lbs.
per inch in length exclusive of thickness of cover plate.

*Brads.*—Tapering nails of parallel thickness with heads
projecting only on one side, as shown in figure 585, are used
to connect parts together where the nail hole is required to be
of the minimum size.

*Needle Points* are small steel pins, circular in section,
similar to the ordinary needle, but without eyes.  These are
used to fix hard wood moulding to joiners' work, and also in
the process of veneering.

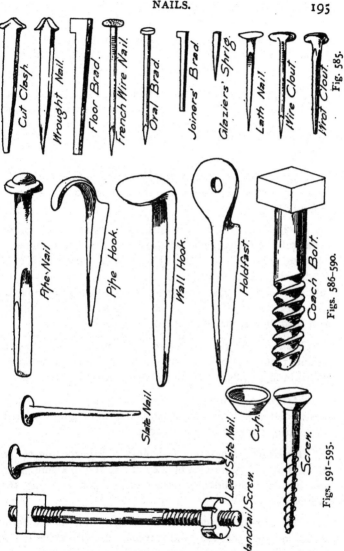

Cut Clasp. Wrought Nail. Floor Brad. French Wire Nail. Oval Brad. Joiners' Brad. Glaziers' Shrig. Lath Nail. Wire Clout. Wrot Clout. Fig. 585.

Pipe-Nail. Pipe Hook. Wall Hook. Holdfast. Coach Bolt. Figs. 586-590.

Slate Nail. Lead Slate Nail. Cup. Handrail Screw. Screw. Figs. 591-595.

Figs. 585-595. Nails and Miscellaneous Fastenings.

*Clout Nails.*—These are nails with shanks rectangular or circular in section, with large, flat, circular heads, as shown in figure 585. Small clout-headed nails are called tacks.

*Slating Nails.*—Zinc, copper, and composition nails of this type are largely used to secure slating, as they resist oxidation better than the iron. Slates are frequently fixed with lead nails (as shown in figure 592) where slates are laid upon iron lathing.

*Lath Nails.*—A form of iron clout with a tapering shank, square in section, the sides of the shank being rough, as shown in figure 585.

*Wire Nails.*—These are usually circular, elliptical, or rectangular in section, and are usually known as French nails. They are very tough and strong, and do not break so readily as the clasp, but whilst being driven in position are apt to split the wood. Those circular in section, as shown in figure 585, are often used in packing-case making, and the elliptical, with small heads, for securing mouldings to joiners' work.

*Glaziers' Sprigs.*—These are small iron nails square in section, reduced at one end to a point and without heads, as shown in figure 585, used to fix glass before puttying.

*Pipe Nails.*—These are of wrought iron, circular shank, chisel pointed with circular heads, as shown in figure 586, used for securing stack pipes to brickwork.

*Pipe Hooks.*—These are fastenings of wrought iron of the shape shown in figure 587, hooked at one end to secure pipes, and having a shoulder to permit of being driven with the hammer.

*Wall Hooks.*—These are of wrought iron of the form shown in figure 588, the head is turned at right angles to the shank. When used they are driven into the joints of the brickwork, and are employed for securing woodwork.

*Holdfasts.*—These have shanks similar to wall hooks, as shown in figure 589. The head is flattened out and a screw hole is drilled and a shoulder formed for the purpose of driving it. They are driven into the joints of brickwork, at the sides of timbers or framed woodwork, and the latter are secured by being screwed through the heads of the holdfasts.

*Coach-bolts.*—These are used for securing iron plates to timbers or timbers together. They have square heads to allow of turning by means of a spanner, and have wood screw threads, as shown in figure 590.

*Screws.*—These are used instead of nails where any vibration in fixing would be objectionable, or where pieces are required to be taken apart occasionally, as in figure 937. Figure 595 shows the form of screw used for fixing woodwork. They are usually made of wrought iron or of brass, the latter resisting oxidation better than the former metal. They are made with one end tapering to a point to permit of the screw entering more easily, the thread being of a coarse pitch. The shape of the head generally determines the name, such as round-headed, countersunk-headed, etc. Where it is necessary occasionally to withdraw these, brass cups, as shown in figure 594, are employed next to the countersunk head to form a better finish.

*Double-nutted Screws.*—These fastenings, as shown in figure 591, are used in joinery work for securing heading joints. They have a thread at each end, a square nut at one end, and at the other end a circular nut slotted on edge. The pieces to be joined are bored to receive the screw; a mortice is made to receive the nut in each member at the requisite distance from the joint and intersecting the screw hole; the square nut is dropped into one of these mortices, the screw is inserted and threaded on same, the latter being square cannot rotate. The circular nut is now placed into the mortice of the other member and the screw inserted into the hole and threaded on circular nut, the latter being turned by a screw-driver or a special punch till the joint is drawn very closely together.

The following formula from Hurst will give approximately the resistance of screws in wood :—

$f = d p l \times$ 42,000 for soft wood, and by 83,000 for hard wood, $d$ being the diameter of the screw, $p$ the pitch or distance between the threads, $l$ the length of the wood, all in inches, and $f$ the resistance in lbs.

*Bolts.*—Wrought-iron bolts and coach screws are used where spikes or screws would not offer sufficient tensile resistance. The bolts are screwed at one end for a tapped nut to be fixed on, and to prevent nut and head sinking into timber when tightened up are supplied with wrought-iron washers, which against oak should be two and a half times diameter of bolt, and for fir three and a half times.

The ratios of the dimensions of the heads of hexagonal-headed bolts and nuts and square-headed bolts and nuts, in terms of the diameter of the shank, are given in the chapter on Iron Roofs.

*Dogs* are pieces of flat or round wrought iron, bent at ends, which are pointed and hammered in to secure timbers that butt against each other, especially used in shoring, rough stagings, and temporary structures. The holding power in lbs. is computed by Capt. Fraser, R.E., to be 600 to 900 lbs. for each inch in length of spike.

*Wedges* are pieces of wood cut taper to secure joints together, as in the mortice and tenon joint; when used in pairs, which draw up more equally, as for scarfs, are called folding wedges.

*Keys.*—The name given to hard-wood wedges, especially used to pull joints closely together, as in the tusk-tenon joint, figure 571.

*Straps* are fastenings formed of metal bands, usually wrought iron, to enclose timbers together and keep them in the required positions, and may have ends forged and screwed which pass through bearing-plates, to which they are secured, or may be drawn up by gibs and cotters, as illustrated in figures 530, 531, 695 and 699, or may simply form a stirrup or heel-strap, and be secured by a bolt, as in figure 583, where it is used to prevent the end of the tie-beam being sheared.

The advantage of straps over bolts is that they need not cut into the timbers joined.

*Sockets* are made of wrought and cast iron to enclose the ends of timbers, to prevent splitting, and to retain the pieces in the required position, as in figure 711. When used at the feet of members they are called shoes, as in figure 713.

## CHAPTER VI.

# FLOORS.

*Definition.*—The tiers or levels· which divide a building into stages or stories are called floors. These may be of timber, or they may be constructed of fire-resisting materials ; for the latter, see *Advanced Course*.

*Classification.*—Floors for ordinary residential purposes are mostly made of timber, and may be divided under three heads :—

*Single-joisted Floors* include bridging joists.

*Double-joisted Floors* include bridging joists supported by binders.

*Framed Floors* or *Triple-joisted Floors* include bridging joists supported by binders, the latter usually being framed into girders which finally support the load.

*Single Floors.*—When the total weight upon a floor is carried by a single system of joists, which span or bridge an opening, it is termed a single floor, and the joists are known as bridging joists ; these are usually fixed 12 inches to 15 inches from centre to centre of joists ; although the simplest construction, with the same quantity of timber it forms the strongest floor. To these joists, laths may be nailed to receive the plaster, or ceiling joists are fixed, crossing the bridging joists at right angles, to which the lath and plaster is attached. There are two conditions governing the direction in which the joists are fixed : first, they are usually arranged to rest upon the walls which have the least span between; secondly, they are placed upon the walls which are weakened by the least number of openings ; these are usually party walls.

*Material.*—In England, timber joists, binders and girders are usually of northern pine, often known as Scotch fir or yellow deal. It is to be preferred to white fir or spruce, as it resists greater transverse bearing stresses, is more durable, and the difference of cost is slight. For works where great strength and large scantlings are required, as in wood floors of large spans for public buildings, pitch pine is commonly used.

*Loads upon Floors.*—In ordinary practice it is usual to construct floors sufficiently strong to support safely the following loads :—

Ordinary dwelling houses, including
    weight of floor ... ... ... $1\frac{1}{4}$ cwts. per ft. super.
Public buildings, lecture rooms, etc. $1\frac{1}{2}$ ,, ,,
Warehouses, factories, etc.... ... $2\frac{1}{2}$ to 4 ,, ,,

All bridging joists to be strutted at intervals, not greater than 6 ft., and the minimum breadth of all binders and girders to be ·6 of the depth.

*Calculation of Scantlings.*—The ultimate resistance of timber scantlings may be calculated from the following formula known as the empirical, as stated below :—

$$W = \frac{k \times b \times d^2}{l} \text{ for concentrated load at centre of beam}$$

$$W = \frac{2k \times b \times d^2}{l} \text{ for beams under distributed load.}$$

When $l$ = length in feet.
  ,,   $b$ = breadth in inches.
  ,,   $d$ = depth in inches.
  ,,   $W$ = breaking weight in cwts.
  ,,   $k$ = breaking weight in wood beams 1 foot long.
                 1 inch broad, 1 inch deep, loaded in centre
                 and supported at ends.

Value of *k*.

| Material. | | | | Central Breaking Weight in Cwts. |
|---|---|---|---|---|
| Spruce Fir | ... | ... | ... | ... | 3½ |
| Northern Pine | ... | ... | ... | ... | 4 |
| English Oak | ... | ... | ... | ... | 4½ |
| Canadian Oak | ... | ... | ... | .. | 5 |
| Pitch Pine... | ... | ... | ... | ... | 5 |
| Wrought-iron | ... | ... | ... | ... | 18 |

*Safe Working Load.*—It is usual to allow a factor of safety of one-fifth for dead loads ; thus the safe working load should not exceed one-fifth of the breaking weight, but where machinery is used, or where large numbers of people are moving in unison, as in a school, a factor of safety of one-tenth of the breaking weight should be taken.

Table giving the ultimate resistances of beams of northern pine, resting upon supports 1 foot apart, 1 inch broad, under transverse stress due to distributed loads, calculated by the formula

$$W = \frac{2k \times b \times d^2}{l},$$

also the safe working loads, ⅕ being taken as the factor of safety for northern pine, English oak, pitch pine and wrought iron.

| Dimensions. | Ultimate resistances in cwts. Northern pine. | Safe working loads in cwts. Northern pine | Safe working loads in cwts. on English oak. | Safe working loads in cwts. pitch pine. | Safe working loads in cwts. on wrought iron. |
|---|---|---|---|---|---|
| 1 × 4 | 128·0 | 25·6 | 28·8 | 32·0 | 115·2 |
| 1 × 5 | 200·0 | 40·0 | 45·0 | 50·0 | 180·0 |
| 1 × 6 | 288·0 | 57·6 | 64·8 | 72·0 | 259·2 |
| 1 × 7 | 392·0 | 78·4 | 88·2 | 98·0 | 352·8 |
| 1 × 8 | 512·0 | 102·4 | 115·2 | 128·0 | 460·8 |
| 1 × 9 | 648·0 | 129·6 | 145·8 | 162·0 | 583·2 |
| 1 × 10 | 800·0 | 160·0 | 150·0 | 200·0 | 720·0 |
| 1 × 11 | 968·0 | 193·6 | 217·8 | 242·0 | 871·2 |
| 1 × 12 | 1152·0 | 230·4 | 259·2 | 288·0 | 1036·8 |
| 1 × 13 | 1352·0 | 270·4 | 304·2 | 338·0 | 1216·8 |
| 1 × 14 | 1568·0 | 313·6 | 352·8 | 392·0 | 1411·2 |
| 1 × 15 | 1800·0 | 360·0 | 405·0 | 450·0 | 1620·0 |
| 1 × 16 | 2048·0 | 409·6 | 460·8 | 512·0 | 1843·2 |
| 1 × 17 | 2312·0 | 462·4 | 520·2 | 578·0 | 2080·8 |
| 1 × 18 | 2592·0 | 518·4 | 583·2 | 648·0 | 2332·8 |
| 1 × 19 | 2888·0 | 577·6 | 649·8 | 722·0 | 2599·2 |
| 1 × 20 | 3200·0 | 640·0 | 720·0 | 800·0 | 2880·0 |

To determine the safe resistance of a given beam under a distributed load, find the safe load of a beam 1 inch broad by the depth under consideration, multiply this by the breadth in inches, and divide by the length in feet of the given beam: the result is the strength of the beam.

*Example:* The following is an example showing the method of using the table :—

Let the double floor shown in figure 624 be taken. Determine the breadth of a beam of northern pine 7 inches deep to span an opening of 8 feet, and to have a safe resistance of 1½ cwts. per foot super. The distance from centre to centre of adjacent binders is 8 feet, and the bridging joists are fixed at 14 inches centre to centre. The load supported by each bridging joist will therefore equal the area multiplied by 1½ cwts. = 8 × 1½ × 1½ = 14 cwts. Let 7 inches equal the depth, then from the table given :—

7″ × 1″ × 1′ 0″ between supports will carry safely 78·4 cwts. distributed load.

7″ × 1″ × 8′ 0″ ,, ,, 9·8 ,, ,,

$$\therefore b = \frac{14}{9\cdot8} = 1\cdot43 \text{ inches nearly.}$$

But as it is not wise to have floor joists less than 2 inches in thickness, as they tend to split while nailing the floor boards, let 2″ × 7″ be selected.

Next determine the dimensions of binders of northern pine, 20 feet between walls, and spaced 8 feet centre to centre, to carry safely a distributed load of 1½ cwts. per foot super. The distributed load in cwts. carried by each binder will be therefore, 20 × 8 × 1½ = 240 cwts. Let the depth be given at 15 inches, then from the table given :—

15″ × 1″ × 1′ 0″ between supports will carry safely 360 cwts. distributed load.

15″ × 1″ × 20′ 0″ ,, ,, 18 ,,

$$\therefore b = \frac{240}{18} = 13\cdot3 \text{ inches ; to this add } 1\cdot66 \text{ inches to compensate for any}$$

loss of strength due to the cogging; this will therefore necessitate a binder 15″ × 15″.

This binder might be substituted by an iron flitch girder to reduce the scantling. Let a depth of 13 inches be taken, and the beam to be of pitch pine, then from the table given :—

13″ × 1″ × 1′ 0″ wrought-iron between supports will carry safely 1216·8 cwts. distributed load.

13″ × 1″ × 20′ 0″ wrought-iron between supports will carry safely 60·84 cwts. distributed load.

$$\therefore 240 - 60·84 = 179·16 \text{ cwts.}$$

13″ × 1″ × 1′ 0″ pitch pine between supports will carry safely 338 cwts. distributed load.

13″ × 1″ × 20′ 0″ pitch pine between supports will carry safely 16·9 cwts. distributed load.

$$\therefore \frac{179·16}{16·9} = 10·6 \text{ inches nearly, and to compensate for any loss of strength}$$

due to the cogging, say 12 inches; this will therefore necessitate a pitch pine binder 12 × 13 inches, and a 1 × 13 inches wrought-iron flitch.

*Basement Floors.*—To comply with the model bye-laws, the site enclosed by the external walls of a house is required to be covered with cement concrete at least 6 inches in depth, the object being to prevent the rising of ground gas, and also to sterilize the soil, thus preventing any vegetable growth, and between the concrete and the under-side of floor timbers a distance of at least 3 inches to allow of the timbers being ventilated. Under the basement floors of houses half-brick walls, as shown in figure 596, are built generally about 6 feet apart to carry sleeper plates, to which the joists are fixed. The nearer the sleeper walls are constructed, the less will be the scantlings of joists required.

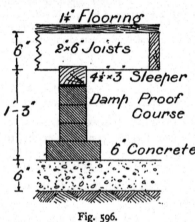

Fig. 596.

*Upper Floors.*—Floors above the basement are constructed with trimmed openings to form well holes for staircases, lifts, trapdoors and to clear flues; in order to satisfy the model

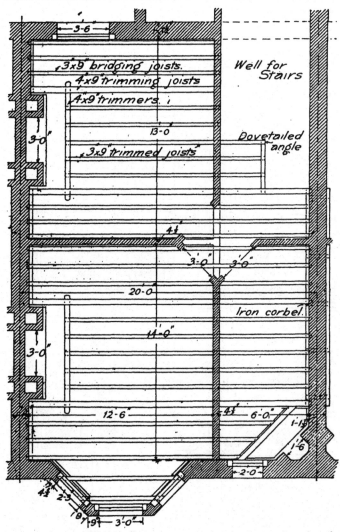

Fig. 597. Plan of Upper Floor, showing the trimming to clear flues in Chimney Breasts, &c.

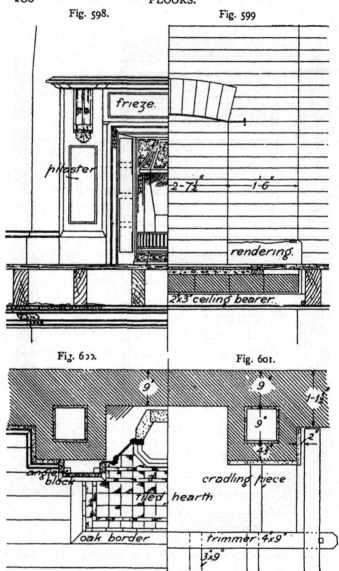

Fig. 598.　　　　　　　　Fig. 599

*frieze.*

*pilaster*

2-7½"　　　　1-6"

*rendering.*

2x3" ceiling bearer.

Fig. 600.　　　　　　　　Fig. 601.

9"　　　9"　　1-1½

9"

2"

*angle block*

*cradling piece*

*tiled hearth*

*oak border*　　　*trimmer 4x9"*

3"x9"

Figs. 598–601.　Example of Fireplace Construction.

Fig. 602.

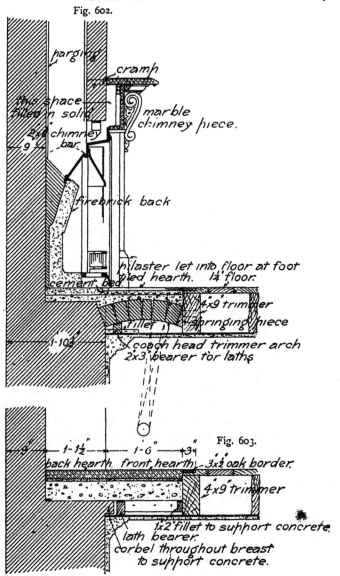

harging

cramp

this space
filled in solid

marble
chimney piece.

2½ chimney
bar.

9"

firebrick back

pilaster let into floor at foot
tiled hearth. 1¼ floor.

cement bed

4"x9" trimmer

fillet

springing piece

1-10½"

coach head trimmer arch
2x3 bearer for laths

Fig. 603.

9"      1-1½"          1-6"        3"

back hearth. front hearth. 3x¾ oak border.

4"x9" trimmer

1x2 fillet to support concrete.
lath bearer.
corbel throughout breast
to support concrete.

Figs. 602, 603   Section through Fireplace.

bye-laws, which require that no wood-work may penetrate walls nearer than 9 inches from the flue, and no wall plate or joist may be within a distance of 2 inches from a flue wall less than 9 inches thick without it is rendered in cement.

Figure 597 shows the trimming to clear flues in chimney breasts, the joists being arranged in the shortest direction; this necessitates two trimming joists and one trimmer, which is carrying the ends of five trimmed joists. All the joints between, trimmed, trimmer, and trimming joists are tusk tenoned, as described in the chapter on Joints in Carpentry to obtain the maximum bearing strength. To comply with the model bye-laws, 18 inches of fireproof material, 7 inches thick, must be provided in front of the fireplace, a coach-head trimmer arch is turned, as shown in figure 602, or a slab of concrete supported by a fillet nailed on the trimmer and resting on a corbel course on the breast, as shown in figure 603, which may extend from trimming joist to trimming joist, and never less than 9 inches on each breast. The coach-head trimmer arch is brought to a level surface by means of concrete, which forms a bedding for the hearth. Hearths are usually formed by slabs of stone, slabs of marble, tiles, or floated in cement. Where a stone is bedded it usually is 2 inches thick, and an extra ½ inch must be allowed for bedding. Marble hearths are usually 1 inch in thickness, and ½ inch allowed for bedding. White marble hearths and mantelpieces should be coated at back with a slip of plaster of paris, otherwise the lime mortar would cause dirty brown stains to appear on the front surface. Three-quarters of an inch must be allowed for the thickness of tiles and bedding, but for hearths finished by floating in cement, 1 inch is the thickness allowed. The floor is stiffened with herring-bone strutting, the rows of which should not be greater than 6 feet apart. Figure 604 shows an arrangement of herring-bone strutting.

*Section through Fireplace.*—Figure 602 is a section showing

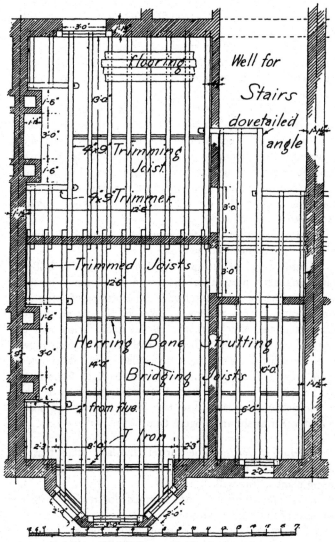

Fig. 604.   Plan of Floor, with Herring-bone Strutting.

the method of forming a rough coach-head trimmer arch, with wood springing piece nailed to trimmer. The ceiling underneath is upheld by small joists resting on fillets 1½ inch by ¾ inch, nailed to trimming joists; or the ribs used as the temporary centre of the arch are allowed to remain to answer that purpose.

In wide chimney breasts the arch is sometimes stopped before it reaches the trimming joists, and a shaped piece of joist timber may be bedded on the back of the arch, provided it is 9 inches away from the opening to which the floor boards are secured. Sometimes a short joist is placed against the end of the arch, and tusk tenoned into trimmer, the other end resting on an iron bar or corbel, which is fixed into the brickwork.

Figure 604 shows a trimming joist instead of trimmer forming the abutment of the arch.

Figure 604 gives the outline plan of floor with bridging joists arranged, parallel to the width of chimney opening, and extending from front to back. The bearing of joists over the opening for bay window is provided for by an inverted rolled steel tee, which permits of flat soffit below and the joists to be continuous. The external angle of joists in well hole for stairs is made as a dovetailed joint, as shown in figure 564. Where the distance from back to front is too great to permit of a continuous joist, advantage may be taken of the partition walls to use two joists lapping and spiked to each other as shown in figure 604.

*Bearing of Joists.*—The model bye-laws state that no timber must be built in party walls nearer than 4½″ to the centre plane of the wall.

There are various methods adopted for resting or fixing the ends of the joists upon walls.

1st. The joists are often built into walls, as shown in figure 604, and bear directly upon the brickwork, but this is objectionable, as the weight is not thoroughly distributed over the wall, and the ends of the timber are subjected to dry rot.

2nd. A better method is to build in a wrought-iron bearing bar (2″ by ¼″) upon which the joists rest, as shown in figure 605. This distributes the pressure, and is not open to the disadvantages of a wood plate. Pockets are sometimes formed in the brickwork allowing the timber at least ½ an inch clearance about the ends of joists, to allow for the circulation of air.

3rd. The joists may rest upon and be spiked to a wall plate which receives its bearing from a set-off or projection of the wall, as is usual at the lowest floor and at other heights of walls where the thickness is diminished by half a brick. This is a very good arrangement, as the wall plate then fulfils its function of distributing the pressure of the floor without reducing the section of the brickwork, as shown in figure 607.

4th. The method of corbelling is adopted for thin walls to take the bearing of joists. Continuous corbels are formed of two or more courses of bricks set off 1¼″ each course, as shown in figures 134 and 135; party walls are often diminished in thickness 4½″; a projection of 2¼″ is thus obtained on each side. This is insufficient to form the bearing for the joists, but may be made wide enough by corbelling out, 1st, by one corbel course of 2¼″ projection or by wrought-iron corbels placed 3 feet apart as shown in figure 606. The best corbels to support the wall plate are of wrought iron (as shown in figure 606), and are made 4 inches wide and ½ or ¾ inch thick, bent at the ends as shown and built into the walls about 3 feet apart. They should be let in the wall 9 inches with sufficient weight above, which should always be noted, and care must be taken that they rest upon good hard bricks or stone to prevent crushing at the edge.

*Bearing Joints of Joists upon Wall Plates.*—Joists may simply rest upon the wall plates or binders, the lower surface

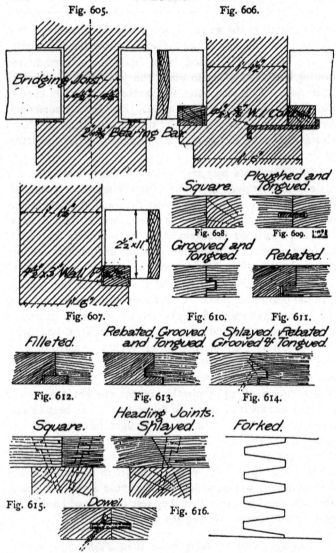

Fig. 605.

Fig. 606.

Bridging Joist.

2¼ Bearing Bar.

Fig. 607.

Square.
Fig. 608.

Ploughed and Tongued.
Fig. 609.

Grooved and Tongued.
Fig. 610.

Rebated.
Fig. 611.

Filleted.
Fig. 612.

Rebated, Grooved, and Tongued.
Fig. 613.

Shlayed, Rebated Grooved & Tongued.
Fig. 614.

Heading Joints.
Square.
Fig. 615.

Shlayed.
Fig. 616.

Forked.
Fig. 617.

Dowel.
Fig. 618.

Figs. 605–618. Bearing of Joists, and Joints of Flooring.

of joist resting upon the upper surface of binder or plate, and fixed by being nailed ; by this method the full depth of the joist is obtained, which is the chief aim, with the least labour.

They are often notched, but this is not good, as if there be a shake or defect in the timber below the notch it is liable to rupture by a load, which if the depth were not cut would be safely carried by the joist, as the effective depth of the joist is reduced by the depth of the notch. If, however, the plate be housed, as shown in figure 607, to take the bearing of the whole depth of the joist, loads tend to force together any shakes in the timber, and the full value of its depth is obtained.

The wall plate often has a fillet about 1″ by $\frac{3}{4}$″ nailed or worked on its upper surface, the joist being notched over ; this is known as a cogged joint. The cog should always be placed on the inside edge of the wall plate so as to give the latter as great an area as possible to resist the detrusion or sliding of the projecting part.

*Joints in Bridging Joists.*—If it is necessary to make a joint in the length of a bridging joist, it should always be arranged to come over a binder or the head of a partition. If there be no ceiling underneath, they should be bevelled halved together, care being taken that each joist has a bearing for its whole depth in the notching, as shown in figure 627. If, however, the continuity of the centre line of the joists is of no great importance, the joists should be laid lapping side by side and spiked together, as shown in figure 604. This is a stronger method, giving much greater bearing, and the joists are easily nailed together.

*Trimming.*—This is the method of arranging timbers to carry the ends of those bridging joists which are cut short to form openings or well holes and to clear flues.

Those timbers which are cut short are termed trimmed joists, those which çarry their ends and distribute the weight on the walls or other timbers are known as trimmers. The joists which carry the trimmers are called trimming joists, and these are generally parallel to the ordinary bridging joists, as shown in figure 597.

In practice trimmers are made of the same scantling as the trimming joists, the latter usually being one inch broader than the bridging joists.

*Strutting.*—Joists laid with boards on top, and lath and plaster underneath, are found to be insufficient to prevent lateral bending and vibration under pressure. To remedy this, which is injurious to the walls and to the timber members strutting is placed in straight continuous rows at intervals of not more than 6 feet apart, which reduces to a great extent this defect and adds greatly to the stiffness and strength of a floor by distributing the pressure from any concentrated load.

The following are two methods of strutting :—

*Herring-bone Strutting* consists of pieces about 2 inches by 1½ inches, fixed as shown in figure 620. It is usual, instead of boring for nails, to make saw cuts at ends of struts, and through these channels to drive the nails. By doing this any danger of splitting is avoided. This method has the recommendation that it acts where its leverage to prevent bending is greatest, viz., at the edges of joists, and is used largely in dwelling-houses, although the labour in fixing is comparatively costly.

*Solid Strutting.*—One inch or 1¼ inch boards, nearly the same depth as the joists, are cut in tightly, and nailed between the joists, as in figure 619. This method is greatly used for dwelling houses, floors of factories and heavy buildings ; and if cut, so that its entire shoulder buts against the joists, is undoubtedly the firmer method; but without care is exercised the shoulders are apt to bite on the centre only. Sometimes long bolts of round bar iron are passed through all the joists in the centre of their depth, and the strutting crushed together by bolting. By this method no vibration is transmitted to the walls through the strutting. This undoubtedly is the idealistic and practical method.

*Joints of Flooring.*—The following are the general methods of arranging the longitudinal joints of flooring (which may be drawn together by means of special cramps, called dogs for fixing), and may be divided into two classes as follows :—

(*a*) Joints with visible fixing, as plain, ploughed and tongued, rebated, and rebated and filleted joints, as in figures 608 to 612.

(*b*) Joints with secret fixing, as dowel, rebated, grooved and tongued, and splayed, rebated, grooved and tongued, as shown in figures 613 to 618 and 614, and these may be again subdivided into joints which do and do not admit of dust passing through when the boards have shrunk.

*Plain or Square Joint Laid Folding.*—Square joints are forced together, if iron cramps are not obtainable, in the following manner :—About six battens are taken and laid into position ; one outside is fixed and the others are pushed together, and the boundary marked. The loose outside batten is now fixed ⅜ inch nearer to the board already fixed, and the others are placed with a plank across them, on which two or three men jump. This crushes the boards together, and closely-fitting joints are obtained, and while kept in this position they are nailed to the joists.

*Ploughed and Tongued.*—A groove is made on each edge of the board lower than the centre, so as to obtain as much wearing thickness as possible, and a wood or iron tongue is inserted which prevents the edges curling up and any dust passing through, as shown in figure 609.

*Grooved and Tongued.*—Figure 610 shows this method of forming a dust-proof joint.

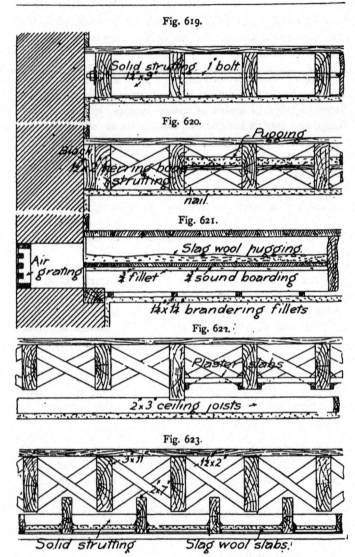

Fig. 619.

Solid strutting   1" bolt
1½"x9"

Fig. 620.

Pugging

Slag

1½"x2" herring bone
strutting

nail.

Fig. 621.

Slag wool pugging.

¾" fillet      ⅞" sound boarding

1¼"x1¼" brandering fillets

Fig. 622.

Plaster slabs

2"x3" ceiling joists →

Fig. 623.

3"x11"     1½"x2"

2"x7"

Solid strutting          Slag wool slabs.

Figs. 619–623.  Sound-Resisting Floors.

*Rebated Joint.*—For floors this joint, as shown in figure 611, has no special advantages to compensate for its extra expense, and is a construction not much used. It prevents dust passing through.

*Rebated and Filleted.*—The advantage of this method is that the greater part of the depth can be worn before the fillet is exposed, and is used in barracks and similar places where floors undergo heavy wear. Figure 612 shows this method.

*Secret Joints.*—Important or polished floors are often required to be secured in such a manner as not to show their fixing, nor unsightly nail holes, as would occur in ordinarily fixed flooring; they should be free to shrink or swell without splitting, which is accomplished by fixing on one edge only. Figure 618 shows an effective method, the boards being secured by dowel pins, and screwed to joists through one edge only. It was a method greatly used before the introduction of machinery, and should only be used upon a counterfloor.

Figures 613 and 614 show the methods by rebate, groove and tongue, secured by nails or screws on one edge only, and are dust proof. The expense of these joints is the great objection to their general use. The method shown in figure 614 is to be preferred, as the projecting portion for fixing is stronger than in the preceding.

*Heading Joints.*—The joints between the ends of floor-boards are called heading joints, and may be square, splayed, or forked.

*Square Heading Joints.*—These are usually secured by two nails on each side of joint, as in figure 615.

*Splayed Heading.*—When the ends of boards are bevelled, only one board need be nailed, as in figure 616; the splayed heading will effectually secure the end of adjoining board. This is the method usually adopted.

*Forked Headings.*—These headings are made with a series of truncated triangular forks, which fit into each other, inclined to the length about 10°, as in figure 617. They are very expensive to make, and are only used in very good work.

*Headings for Secretly Fixed Boards.*—These may be formed in a manner similar to the joint shown in cross section, figure 609.

*Straight Joints.*—This is the name given where the heading joints of adjoining boards are arranged on the same joist; such work looks unworkmanlike, as the plan of all heading joints should present a bonding appearance.

*Prevention of Sound.*—To prevent the passage of sound between the rooms of lower and upper floors, the following methods may be adopted :—

First.—By reducing the number of through timbers the continuity of transmission is interrupted, which lessens the sound, every fourth or fifth joist being made deeper than the others. To these the ceiling joists are secured, as shown in figure 622.

Secondly.—The most effective consists of a double system of joists, each system being entirely separate, not connected in any part. With this system no sound can be carried by conduction through the floor timbers, and if all the members are bedded on felt, and the bridging joists strutted, as shown in figure 623, the passage of sound will be reduced to a minimum.

Thirdly.—Layers of list or felt are placed under every bearing of girders, binders, bridging joists, and floor boards to reduce vibration and sound.

Fourthly.—Various methods of pugging are used to prevent the passage of sound waves through floors, the object being to absorb the sound waves. The methods of pugging alone are never quite successful in preventing the passage of sound, and are objectionable as they induce dry rot in the timber, and the circulation of air is often impeded, if not entirely stopped, if the pugging is not near the top or bottom of the joists. The pugging employed may consist of coarse plaster, slabs of coke breeze concrete, layers of slag wool, or slabs of fibrous plaster.

The fibrous plaster slabs are sometimes placed flush with the tops of the joists, the entire floor space being covered with felt immediately beneath the boarding in order to deaden the sound. This is effective in reducing the sound, but does not protect the lower portion of the joists from fire. The better plan is to place the plaster slabs as close to the ceiling as possible, as shown in figure 623, and have the tops of the joists only covered with felt. The flames would have a much smaller space to play upon, and the floor could be strutted and ventilated in the usual way.

*Ceiling Joists.*—These are the small joists that carry the lath and plaster ceiling only, and are suspended to other timbers. The usual dimensions are 2 inches × 3 to 5 inches, as smaller pieces might be split by the nails. They are fixed about 12 inches apart, and may be secured to the upper timbers by simply spiking, chase mortising or notching over fillets planted on sides to receive them, as in figures 622 and 635.

If ½-inch is multiplied by the number of feet spanned by the ceiling joists, it will give the necessary depth, when the thickness is 2 inches.

*Lathing.*—Laths are fixed about ⅜ inch apart, to the under side of joists, the opening and space in rear allowing the plaster to pass through and protrude, thus affording a key for the plaster, which key should not be interrupted at any part for more than 2 inches continuously. Joists above 2 inches in thickness should be counterlathed, which consists in nailing one or more rows of laths on the under side of thick joists. Laths should be fitted with a butted and not a lapped joint, and the heading joints should be broken every 18 inches.

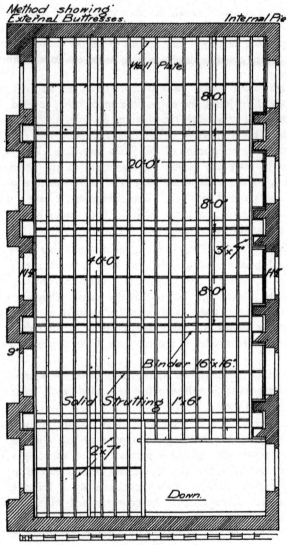

Fig. 624. Plan of Double Floor.

*Brandering.*—Fillets 1″ × 1″ are sometimes nailed at right angles to the under side of joists to receive the lathing, as shown in figure 621, to prevent the plaster ceiling cracking, through any slackness or slight movement of any separate joist, and also to allow of more complete circulation of air about the joists. These fillets are fixed at the most economical distances for the laths. This method is called brandering, the fillets being practically small ceiling joists.

Fig. 625.        Fig. 626.        Fig. 627.

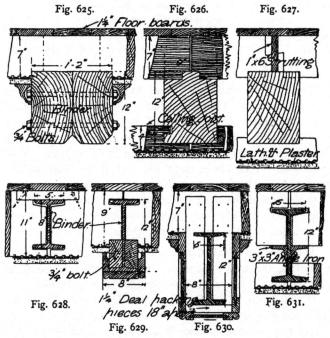

Fig. 628.        Fig. 631.

Fig. 629.        Fig. 630.

*Double Floors.*—These floors consist of bridging joists and binders, with or without ceiling joists, and are used where the supporting walls are perforated with openings. Columns, piers, or the uninterrupted parts of walls are arranged to support the binders. The binders stiffen the floors, and if ceiling joists fixed to the binders are used the passage of sound is minimised, as previously explained in this chapter.

Double floors are generally used now for all large floors, as it is usually possible to obtain steel girders of convenient section to carry the usual loads over any span.

Figure 624 gives the plan of a double floor, calculated to support safely a load of 1½ cwts. per foot super; the binders rest on stone templates built into walls. Ceiling joists, if any, are usually fixed exactly under the bridging joists.

Scantling for binding joists of wood may be calculated from the table given on page 202, as shown in the example worked on page 203. They are generally placed at distances, from centre to centre, not exceeding 10 feet.

*Double Floor Sections.*—Figures 625 to 631 show ordinary methods of arranging binders to carry floor and ceiling joists. Where the extra height obtained is considered a greater advantage than efficiency of sound resistance, the lath and plaster ceiling may be attached to the bridging joists, as in figure 628. The binder may be worked as illustrated in figure 625, which gives the ceiling a panelled appearance. If the projection of the binder in lower room is an objection, the bridging joists are often housed a great portion of their depth and notched over the binders, as in figures 626 and 627.

*Steel Binders.*—Rolled steel joists are now extensively used as binders, the upper flange of which may support the floor joists, and the lower flange the ceiling joists, as in figure 631.

If the depth of the floor is required to be as shallow as possible, such joists for binders are useful, as the floor may be constructed without increasing the depth required by the bridging joists, as shown in figure 628.

Figures 629 and 630 show two methods of rolled steel joists used as binders, to give the ceiling a panelled appearance.

The methods of calculating the necessary sections for steel joists are given in the *Advanced Course.*

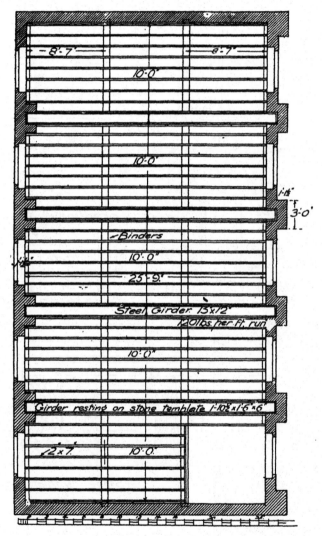

Fig. 632. Plan of Framed or Triple-joisted Floor.

Fig. 633.

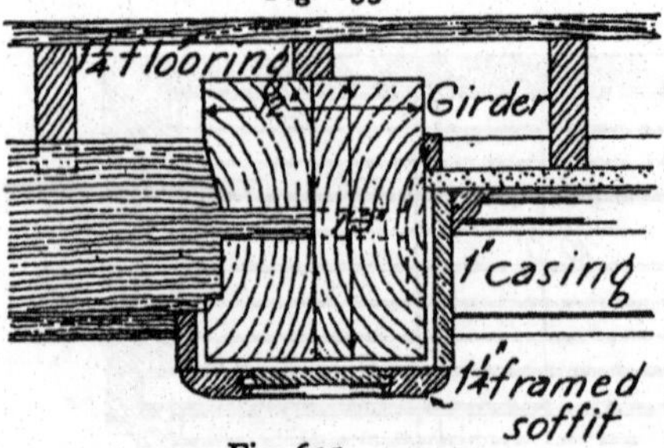


Fig. 634.

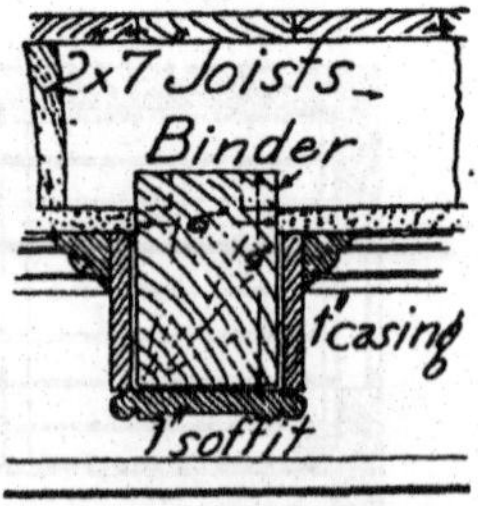


Fig. 635.

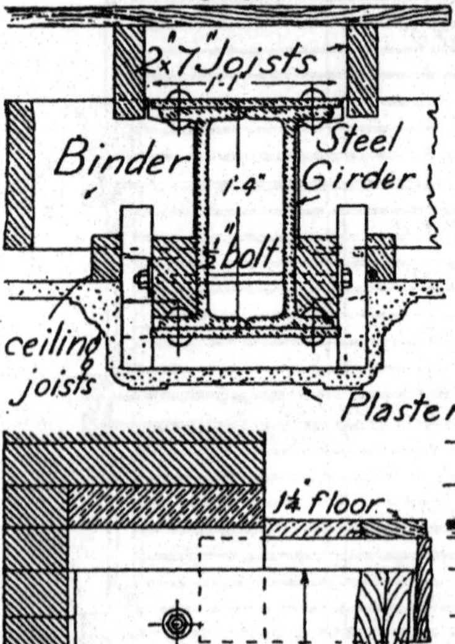


Fig. 636.

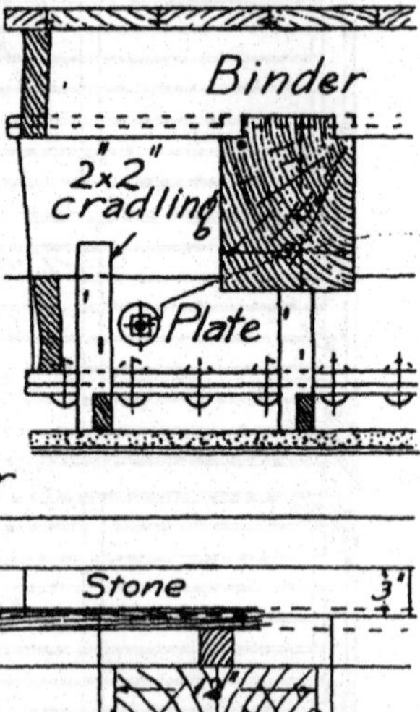


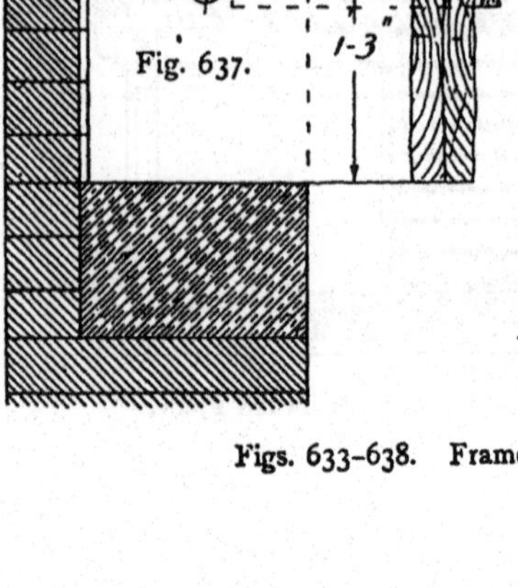


Fig. 637.

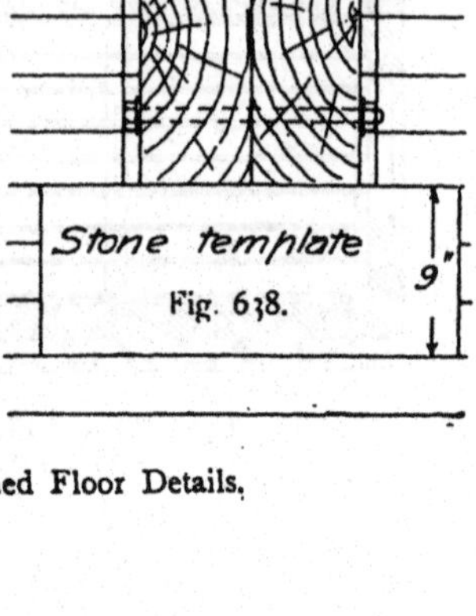


Fig. 638.

Figs. 633–638. Framed Floor Details.

*Framed Floors*, or triple-jointed floors, are composed of bridging joists, binders and girders, with or without ceiling joists, and are used in large factories, warehouses, and public buildings, where the span to be covered is great, or the number

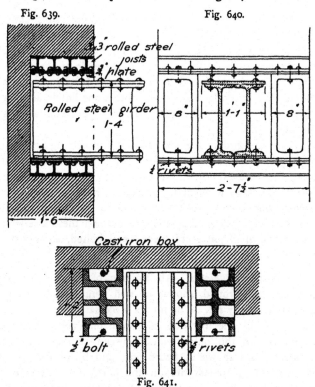

Fig. 639.   Fig. 640.

Fig. 641.

of supports comparatively few, and have all the advantages of a double floor.

It is not usual to place wood girders from centre to centre greater distances than 10 feet apart.

Figure 632 shows a framed floor, calculated to support a load of 1½ cwt. per foot super.

Q

Figures 633 to 636 give methods of applying wood and rolled steel joists as girders.

The binders in figures 633 and 634 are fixed by tusk tenoning and spiking into the girder.

*Pockets in Walls.*—All timbers penetrating walls should rest on stone templates, with a clear space of ½ inch at least, to allow of a free circulation of air around the ends of timbers, and to prevent dry rot being induced, the ends of the timbers being in no case allowed to touch the brickwork. Small cavities may be bridged at the top with stone lintels, as shown in figures 637 and 638, or the larger ones with a rough relieving arch.

Where the girders are supported by projecting piers care should be taken that the rebatement is not so large as to reduce the sectional dimensions of the pier to an unsafe area. If this cannot be avoided a template formed by small steel joists is bedded, and two cast-iron blocks are bolted to it to form the sides of the rebatement, and upon these a series of small rolled steel joists are placed and bolted, as shown in figures 639 to 641.

*Ventilation of Floors.*—All timbers of floors, especially those near the ground, should be well ventilated to prevent destruction by dry rot. The proper necessary ventilation may be obtained by the insertion of iron or terra-cotta air bricks at intervals in the outer walls between ceiling and floor board levels, as shown in figure 621, and by boring a number of horizontal holes through the joists at the centre of their depth, or by brandering, so that the air may circulate through all.

The ventilation is more complete when, instead of a few iron air bricks, a course of perforated bricks is inserted at two or more sides of the floor.

# CHAPTER VII.

# PARTITIONS.

———

*Definition.*—Partitions are screens used to divide flats into rooms, in addition to which they also often act as deep trusses to support floors.

The general tendency of modern building is to design the structure as a steel skeleton, and then to clothe it with brick or masonry work. The floors, under these conditions, are supported directly by the walls, and the flats are divided up into rooms by partitions, each of which constitutes a load upon the floor. This simplifies the construction and renders the division of any flat independent of any other. The essentials for such partitions are lightness, facility for fixing joinery work, simplicity and rapidity of construction, and, in some instances, fire resistance. To comply with these conditions a great number of partitions have been devised of terra-cotta, brick, concrete, ferro concrete, concrete slabs, plaster slabs, corrugated iron and plaster, timber, and brick and timber.

Figure 642 is a good type of the terra-cotta class. These partitions consist of hollow blocks $12'' \times 2\frac{1}{2}'' \times 6''$ with joggled joints, with a smooth egg-glazed surface, or with dovetailed sinkings formed to receive the plaster, and are made by Messrs. Picking & Co., London, N.; they have been extensively used. They are fire-resisting, sound-resisting, strong and light compared with bricks, and are only about half the thickness of brick partitions

The weight of 100 square feet of this partition is 1,232 lbs. exclusive of plaster.

*Brick.*—Brick walls from half brick and upwards in thickness are often used for partitions, which are built and bonded in the usual manner.

Several varieties of bricks are specially made into which nails or screws may be driven to fix joinery work. They are also much lighter than the ordinary brick.

*Coke Breeze Concrete.*—Partitions are now largely built of coke breeze concrete; 4 parts of breeze to 1 of Portland cement. As the concrete has to be cast *in situ* between vertical boarding, it is not economical unless it is required to be repeated several times. The thickness of these partitions is usually about 4 inches. Coke breeze concrete will receive nails and screws for the fixing of joinery work.

*Ferro Concrete.*—Where coke breeze concrete partitions of the usual thickness are above 10′ in height, they must either have an increase in thickness or they must be supplemented with light steel rods, placed at intervals of about 3′ apart, preferably in a vertical position, being fixed at top and bottom with light bar iron heads and sills. Such partitions are light, fire-resisting, and very strong.

*Breeze Concrete Slabs.*—Breeze concrete slabs from two to four inches in thickness, and up to six feet in length by one foot in width, are cast, built and bonded together in the usual manner. They invariably are provided with some form of joggle joint, which endows them with stability, and simplifies the process of fitting them together. They are usually set in plaster of Paris; this material sets in a few minutes, and facilitates the rapid construction of the wall. The coke breeze slabs are easily cut, rendering their adaptation to any position easy. For their construction temporary vertical posts about 3″ × 3″ are strutted between floor and ceiling about 3′ apart and perfectly plumb ; the face of the slabs are kept against the uprights when bedding, rendering the plumbing of each individual block unnecessary. These partitions have all the advantages of the concrete partitions, but are not so costly to erect, as they do not require the expensive cradling necessary for concrete partitions cast *in situ*.

*Plaster Slab Partitions.*—A large number of these partitions are now manufactured, all of which are similar in character to the Mack partition, figure 643, or those of the London Fire-proof Plate Wall Company. Most of these partitions are the subject of patents, varying only in the dimensions of the slabs, the form of the joggles, or the aggregate that is mixed with the plaster of which they are formed.

In some instances the partitions are supplemented by steel rods of small diameter passing vertically through the centre of the slabs.

The following descriptions of the Mack, London Plate Wall, and Kulm partitions are fairly representative of all.

*The Mack Partitions* consist of plaster slabs with reeds embedded (this constitutes the peculiarity of this patent), with a joggled joint and set in plaster, as shown in figure 643. They are used for partitions and ceilings.

Extreme lightness renders it possible to construct them in large slabs; this results in great economy in construction, owing to the fewness of the joints and the rapidity with which partitions can be erected. They are made in thicknesses of $\frac{3}{8}$ inch, $\frac{1}{2}$ inch, 1 inch, $1\frac{1}{4}$ inch for ceilings, and 2 inches, $2\frac{3}{4}$ inches, 4 inches for partitions ; in slabs from 26″ × 20″ to 6′ × 1′. They have been largely used in Germany, America, France, Scotland and England. They are fire-resisting, sound-resisting, light, and can be easily cut with a saw. They work out cheaper than the ordinary lath-and-plaster partitions. Special slabs with asphalte backing may be obtained to form

damp-proof courses.   The weight of 100 square feet of these
partitions 2¼ inches thick is 890 lbs.

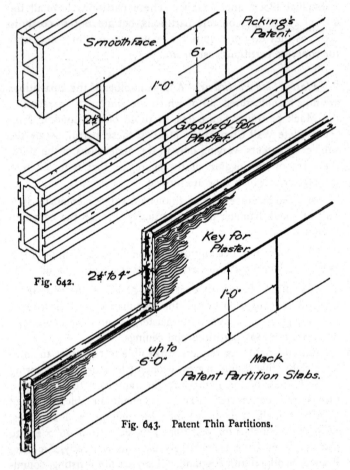

Fig. 642.

Fig. 643.   Patent Thin Partitions.

*The London Plate Wall Co. Partitions.*—These consist
of plaster slabs 20″ × 26″ and 17″ × 22″, and of thicknesses
varying from 2″ to 4″.   These slabs as shown in figures

644 to 653, have a semicircular joggle on their top edges and a corresponding groove on their bottom edges; there is also a semicircular channel on their two vertical edges. Each slab is formed with ten vertical circular-stopped sinkings and one vertical circular perforation in the thickness. The slabs are erected with a proper bond; the through perforations coincide with the vertical joints of the adjacent slabs. As each course is built these are filled up with cement grout, which passes along the bed joints and up a short distance into the vertical stopped sinkings, thus forming a series of cement joggles which effectively hold the plates together. In large walls steel rods are placed through these vertical perforations to add to the transverse strength. The steel rods may be suspended from girders above, and thus relieve the floor from the weight of the partition. This arrangement practically becomes a ferro-plaster construction. Although great supporting power has not been claimed for these partitions, they have in light buildings been used 4″ in thickness as internal walls to support floors.

The slabs are made with both smooth and rough faces.

*The Kulm Partition Slab* is a recent patent. The aggregate is formed of crushed pummice, the matrix is a fire-resisting cement. The slabs are bedded in Portland cement mortar. The advantages claimed are as follows : (*a*) lightness, a square of 100 feet super, two inches thick, weighing 7·4 cwts. ; (*b*) rapidity of construction ; (*c*) very fire-resisting, tests showing a resistance of over 2,000° Fahrenheit ; (*d*) joinery is easily fixed to these. These are also largely used.

*Dovetail Corrugated Steel.* — This is a patent of the Fireproof Co., Ltd., and consists of sheets of thin steel corrugated with a dovetailed corrugation, the depth of which is from ½ to ¾ of an inch. The sheets are about 3 feet in width, and are placed in small channel or ⊢ steel uprights, which latter are riveted to channels on floor and ceiling. The

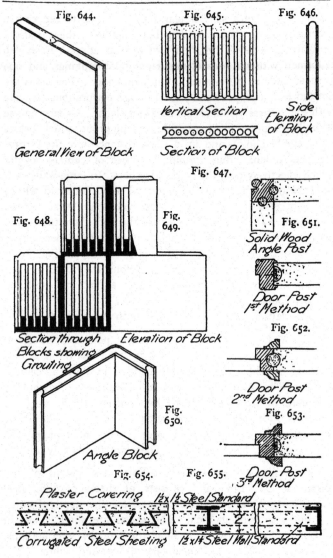

Fig. 644.

General View of Block

Fig. 645.

Vertical Section

Section of Block

Fig. 646.

Side Elevation of Block

Fig. 647.

Fig. 648.

Fig. 649.

Section through Blocks showing Grouting

Elevation of Block

Fig. 650.

Angle Block

Fig. 651.

Solid Wood Angle Post

Door Post 1st Method

Fig. 652.

Door Post 2nd Method

Fig. 653.

Door Post 3rd Method

Fig. 654.

Plaster Covering

Corrugated Steel Sheeting

Fig. 655.

1½ x 1½ Steel Standard

1½ x 1½ Steel Wall Standard

Figs. 644 to 655.

corrugations form a key for the plaster, and the whole thickness of the partition when complete is from $1\frac{1}{8}$ to 2 inches in thickness. For sound-resisting purposes, a double partition is erected, having a void of about 2 inches between. Figures 654 and 655 give sections of this partition.

*Timber Partitions* may be divided into two classes, common and trussed. The common partitions are employed where the floors are sufficiently strong to take the weight of the partition, the trussed where the partitions can only receive support at their extremities from the walls. Trussed partitions are sometimes required to assist in supporting floors.

*Characteristics.*—The advantages of timber partitions are, first, lightness, an important consideration when the partition cannot have a solid bearing along its whole length, and can only be supported by a wall at either end. Secondly, rigidity, which is obtained by the triangulation of the framing ; this also gives the walls a substantial support and aid in resisting external pressures. Thirdly, the facility of construction in any position, and the convenience with which loads may be transmitted and concentrated in any part or parts of the walls.

Timber partitions lack the fire-resisting properties of brick or concrete walls, they do not prevent the passage of sound to any great extent, and should not be erected in basements or floors next to the ground unless they are thoroughly protected from damp, brick or stone walls being preferable in this position.

Common partitions consist of a sill and a head with a number of uprights called studs tenoned to them, the whole fitting tightly between the floors below and above, as shown in figure 656.

Care should be taken when placing the sill in position that it lies either at right angles to the floor joists below, or directly over one of the joists if it be parallel to them, so that a solid and uniform support may be obtained along its entire length.

Care must be taken that common partitions are not placed
on the floor boards between the joists, as in the event of
the floor being taken up to be relaid, the partition would
fail.

The studs, which consist of 4″ × 2″ quarters, placed 1′ 0″ to
1′ 3″ from centre to centre, are stump tenoned into the head
and sill, and are stiffened in the direction of their length, by
nogging pieces out of 4″ × 2″, cut tightly between the studs and
nailed, or else out of strips 2″ × ¾″ let into the face of the studs
and nailed, thus binding the whole together.  The door studs,
head and sill are formed of  4″ × 3″ or 4″ × 4″.

*Bricknogged Partitions.*—These are used for the prevention
of fire and the passage of sound, and are a form of, common
partition, the spaces between the studs being filled in with
brickwork 4½″ thick; the studs are made from 4″ × 3″ quarters,
placed a distance of some multiple of a brick apart, usually
either 2′ 3″ or 3′ 0″, and about every 2′ 0″ in height a wood
bonding strip out of  4″ × ⅛″ should be placed horizontally
between the studs and nailed to them, as shown in figures 657
and 658.  Bricknogged partitions must always have a solid
support along their entire length, otherwise they are liable to
settle and crack the plaster and cornice.  Half-brick partition
walls should always be built in cement mortar.

If the partition does not exceed 10′ 0″ in height, it might
with advantage be made entirely of brick,  half-brick in thickness built in cement.  Partitions above 10 feet in height supporting floors should have wood studs and bonding strips
introduced to avoid any injury from vibration or shocks.

*Theory of Trussing.*—Trusses are deep-braced girders.  The
following conditions require to be satisfied in a perfect truss :
First, rigidity ; secondly, each member to be subjected to a
direct stress ; thirdly, the pressures to be transmitted in a
vertical direction upon the walls.  In order to comply with
the first condition, the truss must be triangulated, the triangle

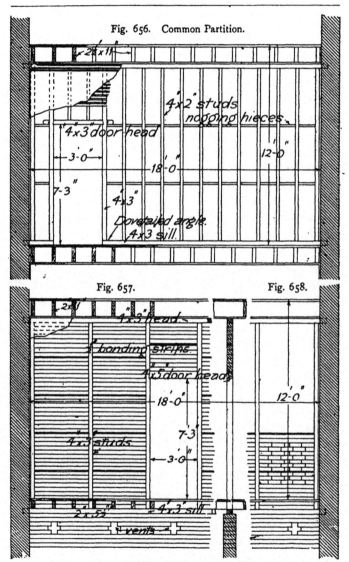

Fig. 656. Common Partition.

4x2" studs
nogging hieces.

4"x3" door-head

3'-0"

12'-0"

18'-0"

7-3"

4x3"

Dovetailed angle.
4x3 sill.

Fig. 657.      Fig. 658.

4x3" head.

3" bonding strips

4x3 door head

18'-0"

12'-0"

7'-3"

4x3" studs

3'-0"

2x5½"    4x3 sill.

vents

Figs. 656-658. Bricknogged Partition.

being the only polygon that is unalterable in form when stressed within the limits of its resistance. To comply with the second condition, the loads must be applied at the apices of the triangles; and to satisfy the third condition, the truss must be efficiently tied at the points of support. This last applies to dead or vertical loads only; inclined stresses necessitate inclined reactions.

In setting out trussed frames the centre lines of the members of the proposed truss should always be drawn first, to avoid transverse stresses.

The partitions are constructed of timber or of timber and steel. Timber alone is unsuitable for forming tensional joints, and these latter always require to be supplemented by steel straps. It often becomes more economical to substitute a steel bolt for the entire tensional member.

Trussed partitions may be classified as simple or compound. Simple partitions may be constructed where the doorways are convenient or the sill of the partition is parallel to the joists below, as shown in figure 659. If the door openings are placed near to the walls, or if the direction of the length of the partition traverses the joists below, as shown in figure 660, in which case the continuity of the sill is broken unless the sill is placed beneath the joists and protrudes into the room below, as shown in figures 661-665. In either case a compound partition would be formed. A perfect truss, generally of the N girder type, is used, as shown in figures 660 and 661, the lowest boom of which would be placed above the height of the doors, the irregular framing between the door openings being suspended from this truss.

If the partitions and the doorways are convenient, the lower portion of the partition may be trussed, as shown in figure 666; this reduces the stresses in the upper part, rendering it possible to use slighter timbers, and distributes the pressures over more parts of the walls.

Fig. 659.

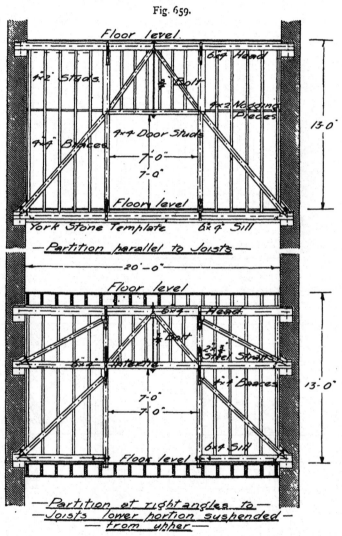

*Partition parallel to Joists*

*Partition at right angles to — Joists lower portion suspended — from upper*

Fig. 660.

Fig. 661.

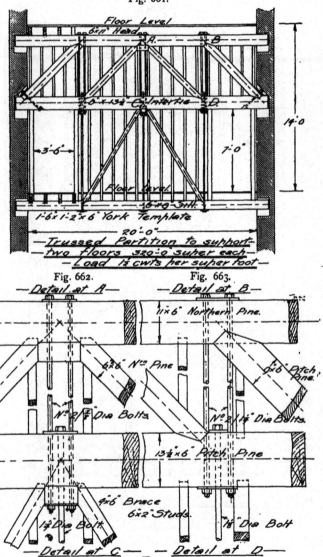

Floor Level.
6·11" Header.
A.    B
6" x 13¾" C. Intertie    D.
3'-6"    7'-0"
14'-0
Floor Level.
6"x9-3ill.
1·6 x 1·2 x 6" York Template
20'-0"
—Trussed Partition to support—
—two floors 320·0 super each—
—Load 1½ cwts per super foot—

Fig. 662.    Fig. 663.
—Detail at A—    —Detail at B—
11"x6" Northern Pine.
6x6 N⁰ⁿ Pine    6x6 Pitch Pine.
N⁰ 2½" Dia Bolts.    N⁰ 2 1½" Dia Bolts.
13½"x6" Pitch Pine
4"x6" Brace
6"x2" Studs.
1½" Dia Bolt.    ½" Dia Bolt
—Detail at C—    —Detail at D—
Fig. 664.    Fig. 665.

Fig. 666.

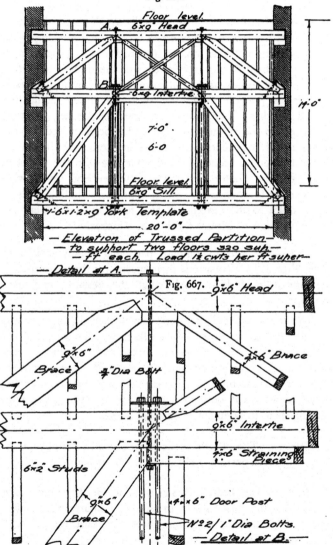

*Floor level.*

A. 6×9" Head

B. 6×9" Interfie

7'-0"

6'-0"

*Floor level.*

6×9" Sill.

1'-6"×1'-2"×9" York Template

20'-0"

14'-0"

— *Elevation of Trussed Partition* —
— to support two floors 320 sup-
— ft each. Load 1¼ cwts per ft super—

— *Detail at A.* —

Fig. 667.

9×6" Head

9×6" Brace

9×6" Brace

¾"Dia Bolt

9×6" Interfie

6×2" Studs

7×6" Straining Piece

9×6" Brace

4×6" Door Post

Nº 2/1"Dia Bolts.

— *Detail at B.* —

Fig. 668.

*Studs.*—The studs, or vertical members, are stump tenoned into the sill head and intertie, similarly to door stud shown in figure 671, but where they intersect a brace they are simply cut to the bevel and nailed on, not tenoned into it, which would weaken the brace. The special duty of the studs is to form a fixing for the lathing, and to interfere with the key of the plaster as little as possible; their thickness, therefore, should not exceed 2 inches. The door studs should be stouter, generally 4″ by 4″, to resist the vibration and shocks caused by the slamming of the doors; the angles are sometimes taken off these members to reduce the width on face with the object of preserving the plaster key, but a better method, on all wide members, is to counterlath, which consists in nailing laths at right angles to the covering lathing to all the main members of the partition of greater width than 2 inches.

*Sills.*—The sills in common partitions usually consist of 4″ by 3″ or 4″ by 4″ quarters, their special use being to receive the ends of the studs and to distribute the pressure evenly over the supports.

The sills in trussed partitions vary in their depth according to the weight to be carried. The sills receive the weight of the ordinary studs, which causes the former to be under a cross stress, and are supported at their two ends by walls, and at intervals by being hung to the uprights by steel straps or bolts; they also receive at both extremities the thrust of the braces, causing them to be under a tensional stress.

*Intertie.*—The intertie is the lower boom in the upper truss of a compound partition, as shown in figures 660 and 666.

*Heads.*—The head serves as a fixing for the studs; as a support for the joists above, as shown in figures 659 to 661; and in some cases as a straining beam to resist the thrust of the braces at their upper ends. It should always be connected at its extremities to the lower horizontal members by bolts, tensional bars, or straps.

*Braces.*—The object of the braces, or inclined members, is to give rigidity, and to transmit the weight on to the walls, and care should be taken that these should always be arranged to be in compression, to avoid the expense of tensional connections.

When wooden upright members are employed, the ends of the braces are usually framed into the upright, with the exception of the lower ends when they occur at the extremities of the sill or intertie, at which part they should be framed into the latter members by a bridle or an abutment joint, as shown in figures 667 and 573. Where steel bolts are used as tensional members, the joint at both head and sill should be a bridle, as in figure 666, except where the extremities of two struts meet, and then they should be mitred and butted against each other and the horizontal member, and stump tenoned into the latter, as at the head shown in figure 667.

*Joint for Door Head.*—Figure 669 shows a housed and tenoned joint for door head. The housing is to obtain bearing strength, and the tenon is keyed to draw the shoulders up tightly.

*Fixing of Partitions.*—The sill, head and intertie should take their bearing on stone templates. If the partitions are arranged to act as cross walls they should be fixed, but if only to support themselves the ends of the timbers should rest in pockets so as to be thoroughly ventilated at all points.

*Trussed Partitions.*—The scantlings of the principal members of trussed partitions vary with the span and the load to be carried, and should be calculated for each case.

*Weight of Partitions and Floors.*—The following table will

R

be found useful in determining the weight of partitions and floors :—

| | | | |
|---|---|---|---|
| Spruce Fir ... | ... | ... | 32 lbs. per foot cube. |
| Northern Pine | ... | ... | 36 lbs. ,, ,, ,, |
| Brickwork ... | ... | ... | 112 lbs. ,, ,, ,, |
| Coke Breeze Concrete (4 to 1) | ... | ... | 72 lbs. ,, ,, ,, |
| Wrought·Iron | ... | ... | 480 lbs. ,, ,, ,, |
| Lath and Plaster | ... | ... | 10 lbs. per foot super. |

The weight of a square of partitioning may be taken at 1,480 to 2,000 lb. per square of 100 feet superficial.

The weight of a square of single-joisted flooring without counter flooring, 1,260 to 2,000 lb. per square.

The weight of a square of framed flooring with counter flooring, 2,000 to 4,000 lb. per square.

If computing a partition to carry a floor the load on the floors must be added to the above given weights of the flooring.

*Fixing for Plaster.*—Wire netting fixed to the timber studs and expanded metal lathing about iron girders are often used as the fixing for the plaster instead of wood laths. Fibrous plaster slabs are also largely used, being nailed or screwed direct to the partitions.

*Regulations.*—The model bye-laws require that the open space inside any partition wall of a new building or between the posts in any wall of such building be stopped with brickwork, concrete, pugging, or other incombustible material at every floor and ceiling. With this exception, the model bye-laws do not control the construction of partition or division walls which simply separate one room from another in the same house.

In the first schedule of the London Building Acts Amendment Act, 1905, defining fire-resisting materials and

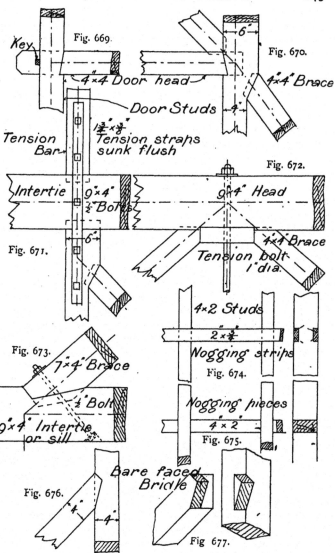

Fig. 669.

Key

4"×4" Door head

Door Studs

Tension Bar

Tension straps sunk flush

1¾"×⅜"

Fig. 670.

6"

4"×4" Brace

4"

Fig. 672.

Intertie

9"×4"

½" Bolts

9"×4" Head

4"×4" Brace

Tension bolt 1" dia.

Fig. 671.

6"

4×2 Studs

2"×¾"

Nogging strips

Fig. 674.

Fig. 673.

7"×4" Brace

½" Bolt

9"×4" Intertie or sill

Nogging pieces

4"×2"

Fig. 675.

Bare faced Bridle

Fig. 676.

4"

4"

Fig 677.

Figs. 669-677. Details of Joint and Connections of Partitions.

R 2

referring to partitions, clause 6 says :—" In the case of internal partitions enclosing staircases and passages, terra-cotta brickwork, concrete or other incombustible material not less than three inches thick."

*Joints of Partitions.*—Figures 669 to 677 show the details for the various joints and connections of partitions.

The principles upon which they are constructed are explained in the chapter on Joints in Carpentry.

# CHAPTER VIII.

# WOOD ROOFS.

_____

*Definition.*—The coverings or upper parts of buildings constructed over the enclosed space to keep rain and wind out and to preserve the interior from exposure to the weather, are called roofs. These should incidentally tie the walls and give strength and firmness to the structure.

For utilitarian purposes, the inclination of the roof is made as flat as possible for the purpose of economizing the timber and covering material. The pitch of a roof is governed, first by climatic conditions, secondly by the covering material used. First for any given covering the milder the climate the flatter the pitch that may be given to the roof, secondly the joints in the covering are the vulnerable points in the roof, the larger and thinner the material, the flatter may be the pitch; thus for lead, which may be used in sheets of about 7 feet by 3 feet and of about ⅛th of an inch in thickness, the pitch need not be more than about 5°. With plain tiles of a dimension of 10″ × 6½″ × ⅜″ a pitch of 45° is necessary for efficiency. The inclination or pitch of the roof is indicated in two ways, first the ratio of the rise to the span, secondly in degrees. The latter is the method to be preferred.

*Materials.*—The following materials may be used as coverings for roofs. The following are the minimum inclinations :—

| Covering. | | | | Angle with Horizon. | | Ratio of Rise to Span. |
|---|---|---|---|---|---|---|
| Asphalte | ... | ... | ... | 1° | 4 | $\frac{1}{8}$ |
| Lead ... | ... | ... | ... | ,, | ,, | ,, |
| Zinc ... | ... | ... | ... | ,, | ,, | ,, |
| Copper | ... | ... | ... | ,, | ,, | ,, |
| Corrugated Iron | ... | ... | 11 | 18 | $\frac{1}{10}$ |
| Asphalted Felt | ... | ... | 18 | 26 | $\frac{1}{6}$ |
| Slates, Large ... | ... | ... | 22 | 0 | $\frac{1}{5}$ |
| ,,   Ordinary | ... | ... | 26 | 33 | $\frac{1}{4}$ |
| Thin Slabs of Stone ... | ... | ,, | ,, | ,, |
| Pantiles | ... | ... | ... | 33 | 40 | $\frac{1}{3}$ |
| Plain Tiles | ... | ... | ... | 45 | 0 | $\frac{1}{2}$ |
| Thatch | ... | ... | ... | ,, | ,, | ,, |

In localities exposed to much wind and rain the pitch of slates should not be less than one-third of the span. The angle of 45° is the best for slates and plain tiles, and admits of the use of the interior of the roof for the construction of garrets.

The timber used for members of roofs is usually northern pine, but for roofs of the class known as open timbered, pitch pine, oak, and chestnut.

*Classification.*—These may be classified in a similar manner to floors :—

(*a*) Single, consisting of one system of timbering, such as flats, lean-to and couple roofs, with common rafters only.

(*b*) Double, consisting of two systems of timbering, such as roofs formed by common rafters, resting on purlins, supported by walls.

(*c*) Triple - membered or trussed, consisting of common rafters, purlins, and trusses, such as mediæval, hammer-beam, and bracket roofs, arched rib trusses, couple-close, collar-beam, and all triangulated trusses.

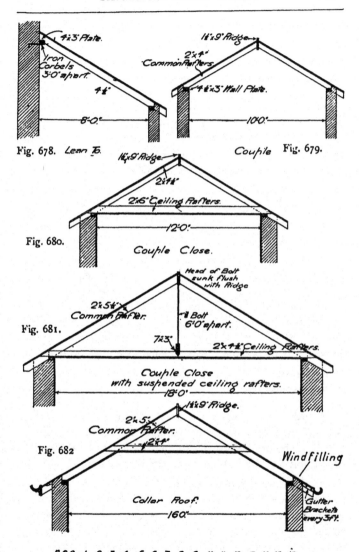

Fig. 678. Lean To.

Couple Fig. 679.

Fig. 680. Couple Close.

Fig. 681. Couple Close with suspended ceiling rafters.

Fig. 682 Collar Roof.

Figs. 678-682. Single-Raftered Roofs.

Amongst the numerous forms of wooden roofs, the following are considered in the elementary stage :—

Flat, lean-to or shed, couple, couple close, collar, double rafter or purlin, king-post and queen-post roofs.

*Flat Roofs*.—Are similar to floors, but having a water-proof top, the upper surface being constructed to drain water off, they may be made as a lead flat, as shown in figure 1063, or as an asphalted concrete roof, or in other ways, but as water or snow is not thrown off, any defect will be severely tested.

*Lean-to* or *Shed Roofs*.—As shown in figure 678, are roofs ormed with one slope only, and used for outhouses against main buildings and for sheds.

*Couple Roofs*.—Are roofs composed of rafters with their feet fixed to wall plates, with their heads butting against a ridge piece ; there is no tie, they depend for their stability upon the abutment afforded by the walls, as in figure 679. Such construction should only be used in permanent structures where the mass of the wall is sufficient to safely resist the outward thrust. For this reason this is an expensive form of roof.

The scantlings given for couple close roofs are applicable for this method.

*Couple Close Roofs*.—For roofs about 12 feet in span, ties are used to prevent the walls being thrust out by the rafters. The ties are usually formed by fixing the ends of the ceiling joists to the feet of rafters, as in figure 680.

As the span increases, the tendency of the ceiling joists is to sag ; to prevent this a plate is placed in the centre and on the upper surface of the ceiling rafters, and the latter are spiked to the plate, which is supported by a king bolt at intervals of about 6 feet to the ridge piece, as shown in figure 681. The space in the roof is not altogether a waste, as it acts as a screen and a non-conductor of heat' to the rooms immediately underneath.

Scantlings for couple close roofs, of northern pine, rise ¼ span slated with Countess slates on boarding.

| Span from centre to centre of Wall Plate. | Rafters. | Ridge Board. | Ceiling Joists. | REMARKS. |
|---|---|---|---|---|
| 8 feet | 3 × 2 | 7 × 1½ | 4 × 2 | |
| 10 ,, | 3½ × 2 | 7 × 1½ | 5 × 2 | |
| 12 ,, | 4 × 2 | 7 × 1½ | 6 × 2 | |
| 14 ,, | 4½ × 2 | 7 × 1½ | 7 × 2 | If these spans have a king bolt as shown in figure 681, the depth of the ceiling joists may be reduced to one half of these dimensions. |
| 16 ,, | 5 × 2 | 8 × 1½ | 8 × 2 | |
| 18 ,, | 5½ × 2 | 8 × 1½ | 9 × 2 | |

*Collar Roofs.*—Roofs of this description are constructed by fixing wood ties halfway up to the rafters, to support them, and carry ceilings which are attached to the under-side, and the lower part of rafters, as shown in figure 682. If the wall plates are securely fixed in the walls, and the latter are of an ordinary thickness, this method answers without ties, and has the advantage of adding extra height to the room below; but directly the wall plates give, the collars become stretched owing to the feet of rafters spreading, and the result is that the walls have the tendency to fall outwards, besides which the non-conductivity of heat is reduced, as compared with the preceding method.

Scantlings for collar beam roofs of northern pine, collar halfway up without a tie, rise ¼ span slated with Countess slates on boarding. If the pitch is 45°, add 1 inch to the depth of the rafters.

| Span from centre to centre of Wall Plate. | Rafter. | Ridge. | Collar. | REMARKS. |
|---|---|---|---|---|
| 8 feet | 3 × 2 | 7 × 1½ | 3 × 2 | |
| 10 ,, | 3½ × 2 | 7 × 1½ | 3½ × 2 | |
| 12 ,, | 4 × 2 | 7 × 1½ | 4 × 2 | |
| 14 ,, | 4½ × 2 | 7 × 1½ | 4½ × 2 | If there is no ceiling the collars may be reduced in depth, but they should be made thicker to prevent side buckling, as they will no longer be stiffened by the lathing. |
| 16 ,, | 5 × 2 | 7 × 1½ | 5 × 2 | |
| 18 ,, | 5½ × 2 | 7 × 1½ | 5½ × 2 | |

*Double Rafter* or *Purlin Roofs.*—Above 18 feet span the roofs already described will not answer unless the timbers are abnormally large, the expense and unsightliness of which prohibit their use. Above that length of span either purlin or framed roofs should be used. If an intermediate support can be obtained, as in the majority of dwelling-houses, purlin roofs, as shown in figures 683 to 688, may be used with advantage. Figure 683 shows a type of purlin roof in which the space inside the roof is not intended for use; the ceiling joists should be in one length or connected over a partition by lapping and nailing, to form a tie for each pair of common rafters; at the extremities of ceiling joists plates for fixing the feet of rafters are nailed. Halfway up the length of the rafters purlins are fixed, having a support on the gable wall at each end, and obtaining intermediate supports at intervals not greater than 6 feet apart by collars placed under the purlins and nailed to the rafters, and also by braces nailed to the collar and butting at the lower extremities on a plate nailed to the ceiling joists.

Where no intermediate support can be obtained, trussed purlins should be employed, as shown in figures 684 to 687; these are supported by the gable walls, and afford a support for the roof rafters and also for the ceiling joists.

Figure 687 shows a type of purlin roof in which the space inside the roof is utilised; the floor joists extend from wall to wall, being lapped and nailed over the partition wall about the centre. The construction is similar to that previously described. Beneath the upper purlins a collar is nailed to every pair of rafters and acts as a ceiling joist. The lower purlin is trussed as an N girder, the spaces between the triangulated members being filled in as studs to receive the plaster. The lower boom should be kept clear of the

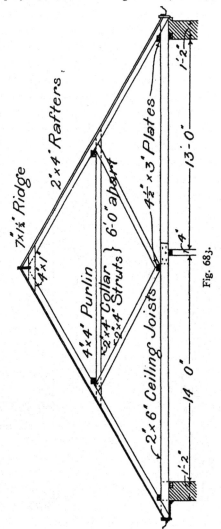

7"x1½" Ridge

2"x4" Rafters

4½"x3" Plates

1'-2"

13'-0"

4"

Fig. 685.

4"x1"

4"x4" Purlin

2"x4" Collar } 6'-0" apart
2"x4" Struts }

2"x6" Ceiling Joists

14'-0"

1'-2"

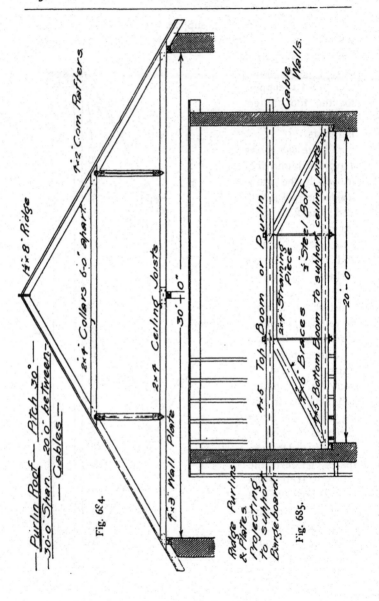

*Purlin Roof — Pitch 30° —*
*30'-0" Span — 20'-0" between*
*— Gables —*

14"×8" Ridge

7"×2" Com. Rafters

2"×4" Collars 6'-0" apart.

2"×4" Ceiling Joists

30'-0"

4"×3" Wall Plate

Fig. 64.

Gable Walls.

2"×4" Straining Piece

4"×5" Top Boom or Purlin

¾" Steel Bolt

5"×6" Braces

4"×5" Bottom Boom to support ceiling joists

20'-0"

Ridge Purlins
& Plates
Projecting
to support
Bargeboard.

Fig. 685.

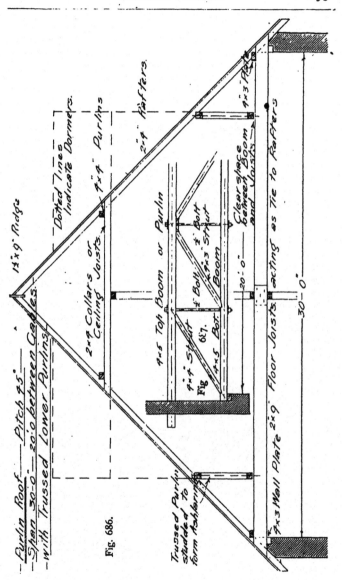

Fig. 686.

Fig. 687.

Purlin Roof.—Pitch 45°.—Span 30·0—20·0 between Gables.—with Trussed Lower Purlins.

14"×9" Ridge

Pitch 45°

Dotted Lines indicate Dormers.

2"×9" Rafters.

4"×4" Purlins

1"×3" Piece

2"×9" Collars or Ceiling Joists.

Top Boom or Purlin

2" Bolt

2"×3" Strut

Clear space between Boom and Joists

4"×5" Top Boom or Purlin

4"×4" Strut

2" Bolt

4"×5" Bot. Boom

20'—0"

30'—0"

2"×9" Floor Joists acting as Tie to Rafters

4"×3" Wall Plate

Trussed Purlin studded to form Ashlering

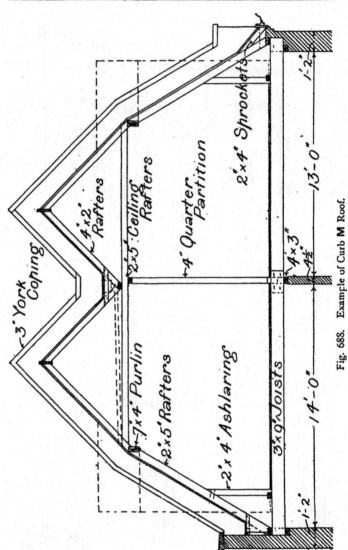

Fig. 688. Example of Curb M Roof.

floor joists, so that the roof is independent of the floor joists for their support. The upright studs of the purlin frame will form what is known as the ashlaring. Dotted lines indicate the position in which dormers would be fixed.

Figure 688 is what is known as a curb M roof. Its construction follows the principles already described in the two immediately preceding examples. The valley gutter is drained by a trough or box gutter or lead pipe as indicated by dotted lines.

The upper portion is often omitted and a flat constructed. Where this is done there is less height and consequently less brickwork required in gables, party walls, and chimney stacks, but the rooms in such roofs are more susceptible to the variations of the external temperature.

*Construction of Trusses.*—The theory of trussing has already been given in the chapter on Partitions.

Trusses are usually arranged at a distance of not more than 10 feet apart. Where it is desirable to have the trusses a greater distance than 10 feet apart, trussed purlins of the types shown in figures 451 and 687 would be used to support the roof coverings, and distribute the weight on the trusses.

The trusses are timbers framed together, and consist of principal rafters, tie beam, posts, and struts. These frames carry purlins and sometimes pole plates, which in their turn support the common rafters, and should be so constructed that every member is in direct tension or compression (with the exception of the tie beams carrying ceilings), the reason being that timber can resist very much greater loads when in direct tension or compression than when subject to transverse stress. This advantage is gained by designing triangulated frames, and arranging that the loads are supported at the angular points of the triangles. In setting out these frames it is wise to draw first the centre line of all the principal members, and to arrange the intersection of the centre lines of principal rafter

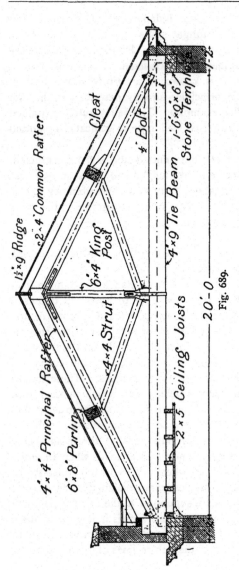

Cleat
Bolt
1½×9 Ridge
2×4 Common Rafter
4×9 Tie Beam
1-6×0×6 Stone Templ
6×4 King Post
4×4 Strut
4×4 Principal Rafter
6×8 Purlin
2×5 Ceiling Joists
20'-0
Fig. 689.

and tie beam immediately over the centre of bearing, or within the middle third of the supporting wall.

The truss should be so arranged that the distances between the supporting points of the principals, or the unsupported lengths of common rafters, are not more than 8 feet. Experience has shown 2″ × 4″ is the most economical scantling for common rafters under the ordinary loads; these require a support every 8 feet. The kingpost form would thus be applicable for a roof with a principal rafter 16 feet long. The correct position for purlins is directly over the joints of principal rafters and their struts, and for the feet of principal rafters directly over their supports, and the struts should be inclined to the horizontal as much as possible, and their feet on king-posts, kept as low down as is practicable, as shown in figure 689; the ends of tie-beams should not be built into walls, so as to come in direct contact with the brickwork, which is more or less damp, according to the hygrometric state of the atmosphere, but should be so arranged where the tie beam supports the ceiling that the timbers of roofs are just sufficiently ventilated to prevent the dry rot without destroying the valuable non-conductive properties of the comparatively still dry air chamber.

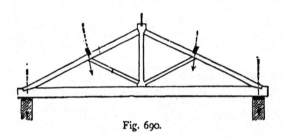

Fig. 690.

Figure 689 gives an illustration of a king-post roof-truss for 20-feet span, showing methods of arranging timbers to form taper-shaped parapet and eaves gutters. In London and surrounding districts all external walls, if adjoining gutter, to have parapet walls, at least one brick thick, 12 inches high, the height being measured from highest point of gutter, and party walls to have parapets 15 inches high above roof. The sole piece of taper-shaped gutters should never be less than 9 inches at lowest point, so as to be wide enough for a man to stand upon it.

*Feet of Principal Rafters.*—These should rest immediately over the walls, as in figures 689 and 691, the best position being the centre line of rafter meeting the centre line of tie

s

Fig. 691.

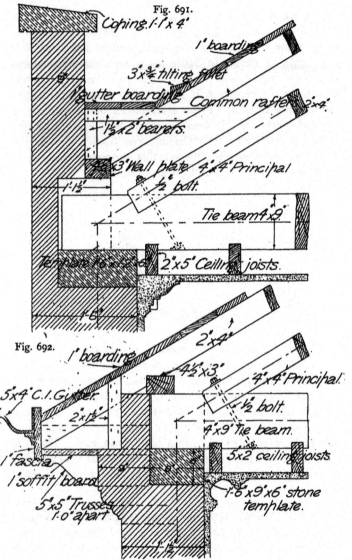

Coping. 1'·1" x 4'

1' boarding

3" x ¾" tilting fillet

8"

1" gutter boarding   Common rafter 2"x4"

1½"x 2" bearers.

3" Wall plate   4"x 4" Principal

1'·1½"

½" bolt.

Tie beam 4"x 9"

Template to stone.   2"x 5' Ceiling joists.

1'·6"

Fig. 692.

1" boarding

2"x 4"

4½"x 3"

5"x 4" C.I. Gutter.

4"x 4" Principal

½" bolt.

2"x 1½"

4"x 9" Tie beam.

5"x 2" ceiling joists.

1' fascia

1' soffit board

1'·6"x 9"x 6" stone template.

5"x 5" Trusses
1'·0" apart.

1½"

Figs. 691, 692.   Wall Plates, Gutter Bearers and Eaves Gutters.

beam perpendicularly over centre of bearing, as in figures 691 and 692.

*Tie-beams.*—It is advisable to let the ends of tie-beams rest on stone templates, over thickness of walls, or even project somewhat over where possible, as in eaves roofs, so that the load may be taken directly by the walls; but whichever method is adopted, care should be taken to ensure a sufficient circulation of air, to prevent timbers being attacked by the dry rot. This may be accomplished by having dormers with louvre boards, or by purposely formed ventilating fleches with louvre boards fixed at the highest parts of roofs, as shown in the chapter on Roofs in my *Advanced Course*, or, in roofs for stations, workshops, etc., where plenty of ventilation is no objection, by having open soffits to the eaves.

*Pole Plates* are timbers acting as girders supported only by the trusses, carrying the feet of common rafters, and are necessary where parallel gutters are to be formed, as in figure 1056, or where all the load must be carried by the piers or portions of walls supporting the tie-beams.

*Wall Plates.*—If taper-shaped or eaves gutters are to be constructed, the brickwork may be built up to carry wall plates, as in figure 691, instead of using pole plates, thus receiving and distributing equally a portion of the load of the roof directly on to the walls.

*Gutter Bearers.*—The gutter boards, usually 1 inch in thickness, may be supported by 1½-inch by 2-inch, or 3-inch by 2-inch gutter bearers, which are halved and nailed to every rafter, as shown in figure 691.

*Eaves Gutters.*—One method of forming eaves gutters is shown in figure 692, and shows a wall plate directly supported by the wall.

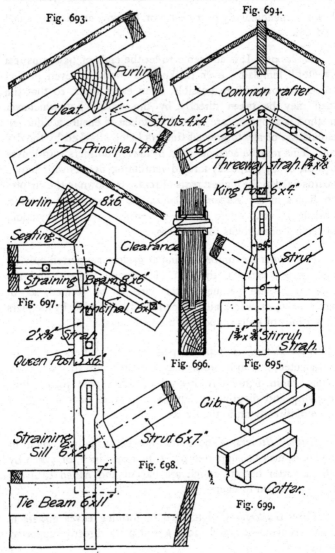

Fig. 693.

Fig. 694.

Purlin

Cleat

Struts 4"x4"

Principal 4"x6"

Common rafter

Threeway strap 1¼"x¾"

King Post 6"x4"

Purlin 8"x6"

Seating

Clearance

Straining Beam 8"x6"

Fig. 697.

Principal 6"x6"

Strut

6"

1¼"x¾" Stirrup Strap

2"x¾" Strap

Queen Post 5"x6"

Fig. 696.

Fig. 695.

Straining Sill 6"x2"

Strut 6"x7"

Gib

Fig. 698.

Cotter

Tie Beam 6"x11"

Fig. 699.

Figs. 693–699.  Details of Wood Roofs.

*Head of Principal Rafter.*—Figure 694 shows a method of forming the junction of principal rafter with the king-post by a single shoulder, and which may be secured by three-way straps, as shown in figure 694.

*Ridge Boards* are the highest timbers of roofs ; they form the abutments for the heads of common rafters, which are usually spiked to them, but in good work are housed in. The ridge board may be held in position by slotting the head of the king-post, as in figure 694.

*Struts and King-post Joint.*—This is made similar to the joint at the head of principal rafter and king-post ; the width of the foot of the king-post is usually sufficient to admit of the joint shown in figure 695, which, being simple and easy to make, should always be used.

*King-post and Tie-beam Joint.*—These are sometimes held together by bolts only, or by straps secured with bolts, the tie-beam being brought to the required camber by ropes suspended from the ridge (1 inch in 20 feet generally being considered sufficient to prevent any unsightly settlement), the shoulders of king-post having been cut to the required length ; but the best arrangement is that of gibs and cotters, which allow the joint to be drawn up at any time an open shoulder is observable, by leaving clearances, as shown in figures 695 and 696.

*Purlins.*—The horizontal members of the roof, supported by the trusses, and which carry the common rafters between the pole or wall plates and the ridge piece. If $2'' \times 4''$ common rafters are used, they should not be more than 8 feet from the next support of common rafters, and should be cogged to the principal rafters directly over the struts. In some forms of roofs, where common rafters are dispensed with, they carry the boarding to receive slating, as in figure 705.

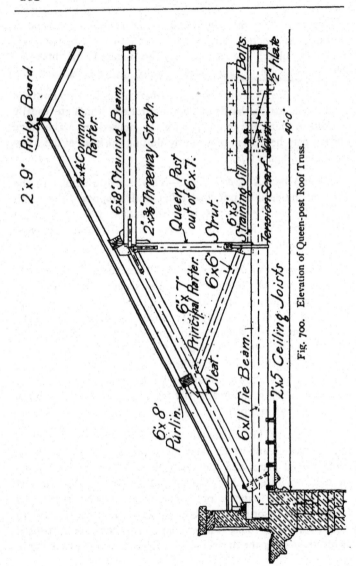

Fig. 700. Elevation of Queen-post Roof Truss.

*Junction of Struts and Principal Rafters.*—These may be arranged as in figure 693, and are sometimes held together by straps, and should be placed directly under the stress transmitted by the purlin above, as in figure 690; the latter distributing the load from roof above when acted on by the wind on that side.

*Common Rafters.*—The sloping timbers to which roof coverings are attached. In framed roofs they are supported by pole or wall plates, purlins and ridge pieces, and should not have an unsupported length of more than 8 feet, they should not be in plan more than 15 inches apart, and are generally arranged to clear the heads of king-posts. Common rafters should have a minimum thickness of 2 inches and a depth of 4 inches. If the span is greater than 8 feet they should be increased in depth $\frac{1}{2}$ inch for each foot in excess.

*Cleats.*—Pieces fixed on principal rafters, straining beams, etc., to prevent purlins tilting; they are usually simply spiked on, as in figure 693, but in very good work may be housed.

*Sprocket Pieces.*—The upper surface of the common rafter is often made concave by planting pieces of timber of the same breadth as the common rafter of curved shape in elevation, termed sprocket pieces, at the feet of rafters, as shown in figure 701. This allows of slate or tiled coverings to fit with closer joints on their tail ends. Sometimes sprocket pieces are used to improve the appearance of roofs.

*Wind Filling.*—The brickwork that is carried up between rafters to the underside of the roof boarding, as shown in figure 682.

*Queen-post Roof Truss.*—Figure 700 shows an elevation of a truss, suitable for spans from 30 to 45 feet, the parts being easily understood from what has been already stated about the king-post truss. The tie-beam, if of northern pine, is usually in two lengths, and scarfed. A straining sill is sometimes fixed between queen-posts to act as an abutment for the queen-posts, and as a fish-plate to the scarf joint.

Figures 697 and 698 show the joints at the head and foot of a queen-post truss. The joints are similar to those already explained for the king-post truss.

Scantlings for wood roofs, of northern pine—trusses 10 feet apart, ¼ pitch, slate covering *(Hurst's Handbook)*:—

| Span in feet. | Tie-beam. | King-post. | Queen-post. | Principal Rafters. | Straining Beam. | Struts. | Purlins. | Common Rafters. |
|---|---|---|---|---|---|---|---|---|
| 20 | 9½×4 | 4×3 | — | 4×4 | — | 3½×2 | 8×4¾ | 3½×2 |
| 22 | 9½×5 | 5×3 | — | 5×3½ | — | 3¾×2¼ | 8¼×5 | 3¾×2 |
| 24 | 10½×5 | 5×3½ | — | 5×4 | — | 4×2½ | 8½×5 | 4×2 |
| 26 | 11½×5 | 5×4 | — | 5×4¼ | — | 4¼×2½ | 8¾×5 | 4¼×2 |
| 28 | 11½×6 | 6×4 | — | 6×3½ | — | 4½×2¾ | 8¾×5¼ | 4½×2 |
| 30 | 12×6 | 6×4½ | — | 6×4 | — | 4¾×3 | 9×5½ | 4¾×2 |
| 32 | 10×4½ | — | 4½×4 | 4½×6¾ | 6¾×4½ | 3¾×2¼ | 8×4¾ | 4½×2 |
| 34 | 10×5 | — | 5×3½ | 5×6½ | 6¾×5 | 4×2½ | 8¼×5 | 3¾×2 |
| 36 | 10½×5 | — | 5×4 | 5×6¾ | 7×5 | 4¼×2½ | 8½×5 | 4×2 |
| 38 | 10×6 | — | 6×3¾ | 6×6 | 7¼×6 | 4½×2½ | 8½×5 | 4×2 |
| 40 | 11×6 | — | 6×4 | 6×6½ | 8×6 | 4½×2½ | 8½×5 | 4¼×2 |
| 42 | 11½×6 | — | 6×4½ | 6×6¾ | 8¼×6 | 4½×2¾ | 8¾×5¼ | 4½×2 |
| 44 | 12×6 | — | 6×5 | 6×7 | 8½×6 | 4½×3 | 9×5 | 4¾×2 |
| 46 | 12½×6 | — | 6×5½ | 6×7¼ | 9×6 | 4¾×3 | 9×5½ | 5×2 |

The following gives the nature of the stresses that the members of king and queen-post roofs are calculated to resist :—

*Compression.*—Principal rafters, straining beams, struts.

*Tension.*—King-posts, queen-posts, tie-beams without ceilings attached.

*Transverse or Cross.*—Purlins, common rafters.

*Transverse and Tension.*—Tie-beam with ceiling attached.

*Mansard.*—The outline of the Mansard truss is given by

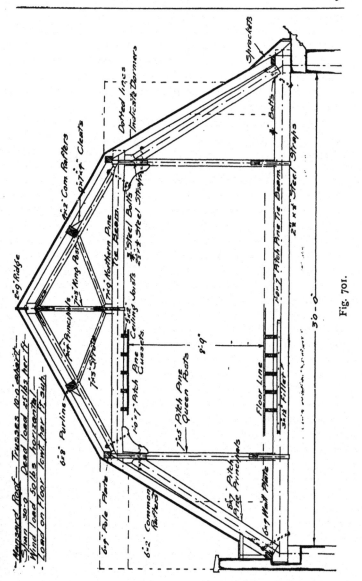

Fig. 701.

placing the outline of a king-post truss immediately over a queen-post truss.

Figure 701 gives all the necessary details of a Mansard roof truss. They are often used in large cities to obtain the maximum amount of cube space in buildings with the minimum obstruction of light and air.

To reduce the stresses upon the fastenings to a minimum two gusset pieces are often introduced in the place of perfectly triangulating the truss, which would interfere with the clear space in the roof.

# CHAPTER IX.

# COMPOSITE ROOFS.

*Definition.*—Roof trusses built up of timber and steel members are known as composite trusses.

*Composite Trusses.*—The tensional members of trusses in composite roofs are often constructed of steel, because the latter has much greater tensile strength than timber; moreover, the rectangular section in beams is wasteful, as the mortices or housings made at the joints in cutting the fibres considerably reduce the effective sections, consequently large pieces of timber have to be used to do the required work; this adds great weight to be carried by the supports, and interrupts the passage of light in open roofs.

The objection hitherto to the use of composite trusses has been the necessity that was considered of using cast-iron shoes and heads for the junction of the principal rafters with their ties. Where a large number of similar trusses are required the cost of the patterns for casting would be relatively small, but where only a few trusses are wanted or they are variable in design the expense of these members is prohibitive. Since the great development of the steel industry has taken place several sections are now on the market from which these junctions can easily be prepared.

Composite roofs of wood and iron are subject to this objection, that the timber principals neither expand nor contract appreciably with the changes of temperature, while the steel members do so considerably. This can easily be imagined to have the effect of lengthening the tie until it ceases to act, or until the feet of the principal rafters spread; or, again, the contraction of the steel tie-rod may have the effect of bending the wood principals or overstraining the tie-rod; but this may be provided for by arranging the joint at the head with a small amount of movement.

In designing the joints simplicity in construction should be aimed at, and ordinary rolled or easily forged sections should be used, and expensive forked connections should be avoided.

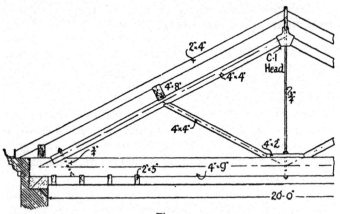

Fig. 702.

In open roofs, iron ties may be advantageously used, but ceiling joists are more easily fixed to timber tie-beams. The principal rafters in composite roofs are always of timber, owing to the facility with which purlins may be fixed, and to the high resistance of rectangular sections of timber to compression.

*Typical Trusses.*—The following are typical composite trusses :—For spans of from 20 to 30 feet

Figure 702 shows a wrought-iron king bolt and cast-iron head, the joint at the foot of the struts being made by introducing a straining piece. Figure 703 illustrates a truss for a 20-feet span, the tensional members being of mild steel, while the struts and principal rafters are of timber. All the connections in this truss are such as can be made by an ordinary smith.

The German truss, as shown in figure 704, may be used, where it is desirable to gain increased height in the chamber below, introducing a king bolt and tie-rods, those being the members in tension.

This form follows the lines of the timber collar beam truss, but with tie-rods to overcome the horizontal thrust instead of the wood gusset pieces, shown in the chapter on Roofs in my *Advanced Course*.

There is a cast-iron shoe at the connection between the principal and the tie-rod, also a cast-iron head to receive the upper ends of principals.

Figure 705 shows steel tension and tie-rods with cast-iron struts, the principals alone being of timber. There is a cast-iron shoe at the joint between principal and tie-rod. The junction at heads of principal rafters is secured by two $\frac{3}{8}''$ steel gusset plates bolted through : these form a fork to receive the upper ends of tie-rods. The ridge board is supported by cleats nailed on to principal rafter, as shown in figure 706.

Figures 706 to 708 show details of the joints.

For spans from 30 to 40 feet, the king bolt and queen bolt forms of truss may be used, those members being of steel and constructed as illustrated in figure 710. The tie-beams being of timber and well tied up by the king and queen bolts, render these trusses suitable for supporting a ceiling.

*Feet of Principal Rafters.*—Steel ties may be secured to wood principals by the aid of a cast-iron shoe, which receives foot of principal, and is drilled to receive tie as illustrated in figure 713. Also the feet of principal rafters may be received in a steel shoe, as shown in figures 719 and 720. The feet of

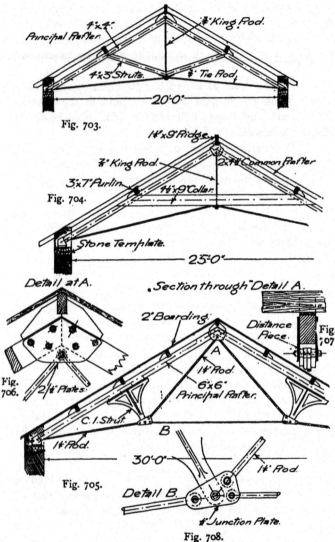

Fig. 703.

Fig. 704.

Detail at A.

Section through Detail A.

Fig. 706.

Fig. 705.

Detail B.

Fig. 708.

Figs. 703-708.　Composite Trusses.

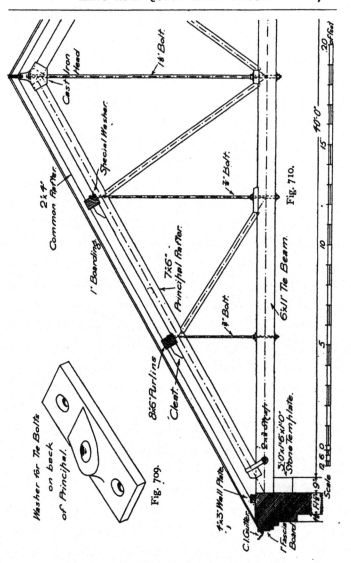

Figs. 709, 710. King and Queen-Bolt Truss.

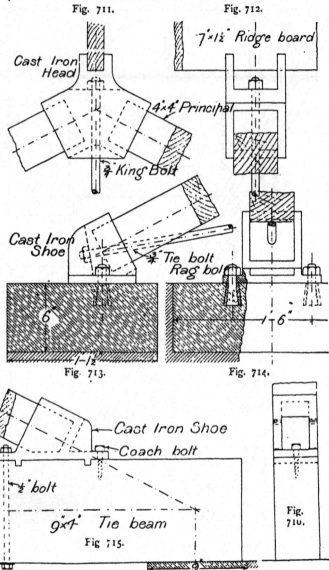

Fig. 711.

Cast Iron Head

4"x4" Principal

¾" King Bolt

Fig. 712.

7"x1½" Ridge board

Cast Iron Shoe

¾" Tie bolt

Rag bolt

6"

1-1½"

Fig. 713.

1-6"

Fig. 714.

Cast Iron Shoe

Coach bolt

½" bolt

9"x7" Tie beam

Fig. 715.

9"

Fig. 710.

Figs. 711-716. Cast-iron Heads and Shoes.

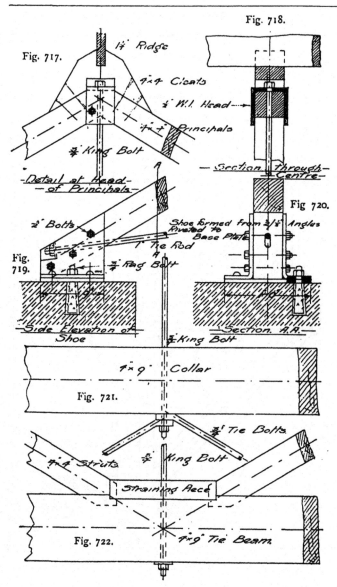

Fig. 718.

Fig. 717.

1¼" Ridge

4×4 Cleats

½" W.I. Head

4×4 Principals

¾" King Bolt

—Detail at Head—
—of Principals—

—Section through—
Centre

Fig 720.

2" Bolts

Shoe formed from 3/¼" Angles
Riveted to
Base Plate

1" Tie Rod

Fig.
719.

¾" King Bolt

Side Elevation of
Shoe

Section. A.A.

¾" King Bolt

4×9" Collar

Fig. 721.

¾" Tie Bolts

4×4 Struts

¾" King Bolt

Straining Piece

Fig. 722.

4×9" Tie Beam.

T

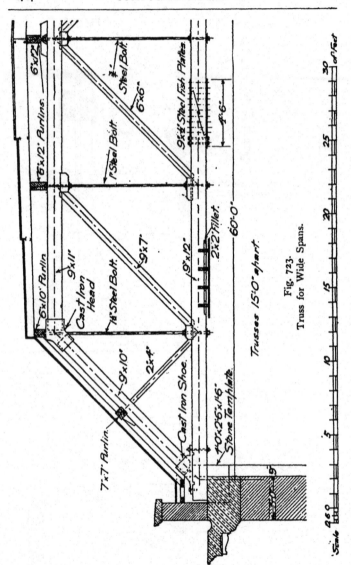

Fig. 723.
Truss for Wide Spans.

principal rafters are at times received by a cast-iron shoe fixed to the wooden tie-beam, as shown in figures 715 and 716.

*Head of Principal Rafter.*—Cast-iron heads are made to form the abutment for the upper ends of principal rafters, with a hole cast for the king bolt to pass through, as shown in figures 711 and 712. An alternative arrangement is by abutting the heads of the principal rafters together, and enclosing them by a U-shaped piece of steel plate inverted and drilled to receive the king bolt as shown in figures 717 and 718.

*Timber Strut, Principal and Queen Bolt.*—The usual joint between the strut and principal is a single abutment with queen bolt passing through a hole bored through the principal, and fixed by a nut screwed on bevelled washer to back of principal, as shown in figure 709.

*Feet of Struts.*—This joint may be made by fixing a straining piece, as shown in figure 722, timber struts and tie-beam being used.

*Feet of King or Queen Bolts.*—These may pass through the straining pieces, which are indented into the tie-beam and receive the wood struts. The tie-beam is secured by screwed nuts and washers on king bolt over straining piece and under tie-beam, as illustrated in figure 710.

N *Truss.*—Wide spans may be effectively bridged by composite trusses built up in the form of an N girder, in which the long members are in compression and the short members in tension. Figure 723 shows a truss on this principle suitable for an effective span of 60 feet. The upper covering has been arranged for a lead flat. It may be noted that in such type of structures in steel, the long members are usually arranged as ties, and the short as struts.

# CHAPTER X.

# IRON AND STEEL ROOFS.

———

*Steel Roofs.*—The great superiority of mild steel over timber in stress-resisting properties, the great varieties of economical and suitable sections in the market, the ease with which they can be packed and transported, and their resistance to fire and insects, have caused them to gradually displace timber as materials for roofs of large or even moderate dimensions, and great attention has been given to the most economical methods of designing them.

*Stresses on Trusses.*—Every member of a truss should be in either direct tension or compression ; but in smaller roofs the purlins are often distributed along the length of a principal rafter, as shown in figure 750, causing the member to be in transverse stress ; in that case the distance between the struts should not be great, otherwise the principal rafter will have to be wastefully large. Struts should be arranged as short as possible, and as many braces as possible be in tension.

*Cast Iron* has been largely employed for shoes and short struts, but the modern tendency is to eliminate all cast iron in these structures. Mild steel is generally employed for every part of a truss, as illustrated in figures 732 to 739.

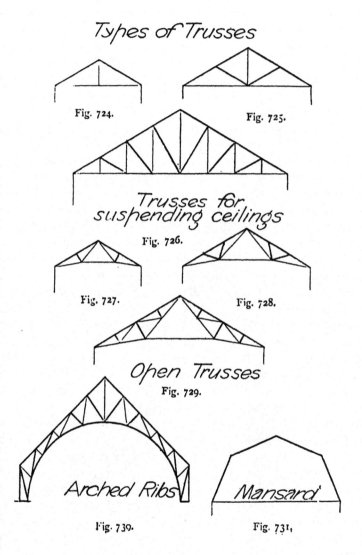

Types of Trusses

Fig. 724.

Fig. 725.

Fig. 726.

Trusses for suspending ceilings

Fig. 727.

Fig. 728.

Fig. 729.

Open Trusses

Arched Ribs

Fig. 730.

Mansard

Fig. 731.

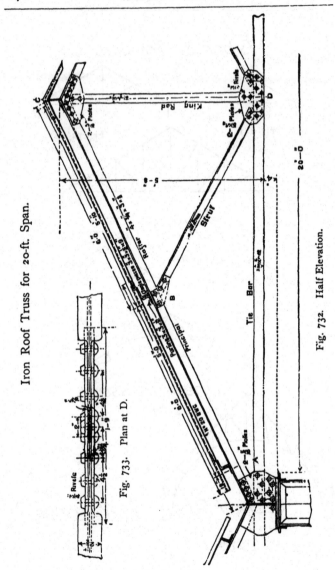

Iron Roof Truss for 20-ft. Span.

Fig. 733.    Plan at D.

Fig. 732.    Half Elevation.

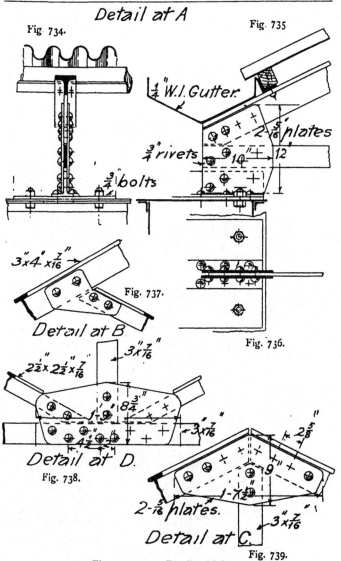

Detail at A

Fig. 734.

Fig. 735

$\frac{1}{4}$" W.I. Gutter.

$2-\frac{5}{16}$" plates

$\frac{3}{4}$" rivets

10"    12

$\frac{3}{4}$" bolts

Fig. 736.

$3"x4"x\frac{7}{16}"$

Fig. 737.

Detail at B

$3"x\frac{7}{16}"$

$2\frac{1}{2}"x2\frac{1}{2}"x\frac{7}{16}"$

$8\frac{3}{4}"$

$3"x\frac{7}{16}"$

$4\frac{1}{2}"$

Detail at D.

Fig. 738.

$2\frac{5}{8}"$

9"

$1-7\frac{1}{2}"$

$2-\frac{5}{16}"$ plates.

$3"x\frac{7}{16}"$

Detail at C.

Fig. 739.

Figs. 734-739.   Details of Joints.

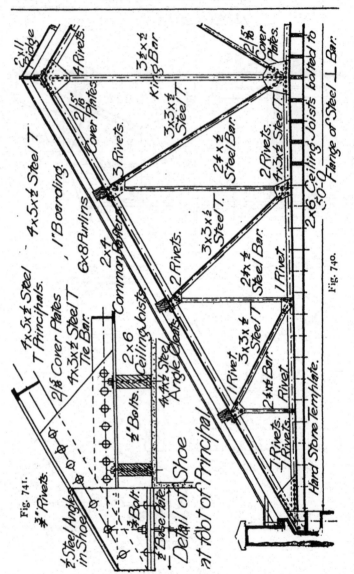

2"x11" Ridge

4 Rivets.

3½"x½" Bar
King Bolt.

2/16" Cover Plates

2/5" Cover Plates

4x5x½ Steel T

3x3x½ Steel T.

3 Rivets.

2#x½ Steel Bar.

2 Rivets.
4x3x½ Steel T.

4x5x½ Steel - T Principals.

1" Boarding.

6x8 Purlins

3x3x½ Steel T.

2 Rivets.

2#x½ Steel Bar.

2x4 Common Rafters.

1 Rivet.

Fig. 740.

2x6 Ceiling Joists bolted to Flange of Steel ⊥ Bar.

20'-0"

Hard Stone Template.

1 Rivet.

3x3x½ Steel T.

2#x½ Bar

1 Rivet.

1 Rivets. Rivets.

4x5x½ Steel T Principals.

2/16 Cover Plates

4x3x½ Steel T Tie Bar.

2x6 Ceiling Joists.

2x6 Ceiling Joists.

4x4x½ Steel Angle Cleats.

½" Bolts.

Fig. 741.

¾" Rivets.

½ Steel Angles in Shoe.

½" Rivets.

½" Bolt.

¾" Base Plate.

Detail of Shoe at Foot of Principal

*Typical trusses.*—There are four general types of steel trusses for roofs :—

1. Those for supporting ceilings, as shown in figures 724 to 726.
2. Open trusses, as shown in figures 727 to 729.
3. Arched rib truss, as shown in figure 730.
4. Mansard truss, as shown in figure 731.

*Trusses supporting Ceilings.*—Figures 732 to 741 illustrate this type of truss. In these it is necessary to have a flanged member for the tie such as a tee, to facilitate the fixing of the ceiling joists, which are secured by being bolted to the flange of the tie, as shown in figures 740 and 741. The outline of trusses with vertical members are suitable for hip-ended roofs.

*Open Steel Trusses.*—In this type, for appearance, the ties are often made of steel rods, and are provided with a large camber to reduce the length of the struts. They are usually constructed of two trussed beams, the compressional booms or principal rafters being secured together by gusset plates which also secure the upper ends of the tensional members. The central points of the two trussed beams are held together by a horizontal tie, the combination thus forming the truss, as shown in figures 742 to 749. This type is suitable for gable-ended roofs, but does not lend itself for hipped ends.

The third type of roof truss is the arched rib, and is illustrated in figures 780 to 787. These are suitable for large halls and open galleries. The trusses may be exposed to view or they may be ceiled about the line of the arched rib, the ceiling being attached to expanded metal lathing fixed to light steel joists, which latter are riveted at their extremities to the arched ribs. The principals and the arched ribs are constructed of pairs of steel angles or channels, this arrangement being most convenient for connections to be made. The ties are constructed of steel bars, the struts of steel angles. To

form the connections, bearing plates are necessary to afford
sufficient bearing area for the required number of rivets.   The
greatest stresses occur about one-third the way up the principals,
where the distance between the arched rib and the principals
is the minimum.   At these parts it becomes more economical

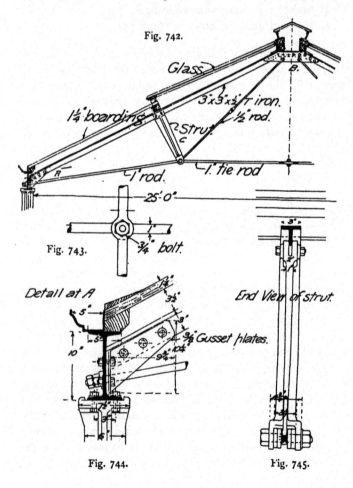

Fig. 742.

Fig. 743.

Detail at A

Fig. 744.

End View of strut.

Fig. 745.

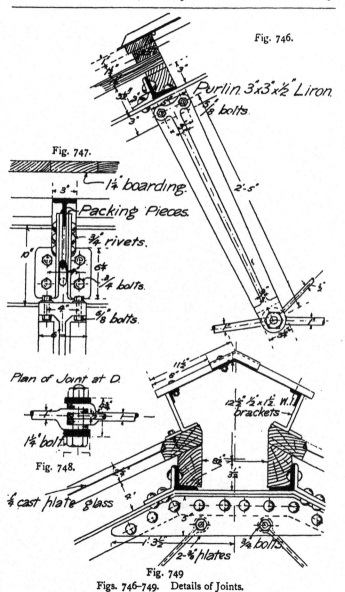

Fig. 746.

Purlin 3"x3"x½" L iron.
⅝" bolts.
2'-5"

Fig. 747.

1¼" boarding.
Packing Pieces.
¾" rivets.
¾" bolts.
⅝" bolts.

Plan of Joint at D.
1¼" bolt.
Fig. 748.

11½"
6"
12½"½"x1½" W.I. brackets→

¼" cast plate glass

2-⅝" plates
¾" bolts
1'-3½"

Fig. 749

Figs. 746–749.   Details of Joints.

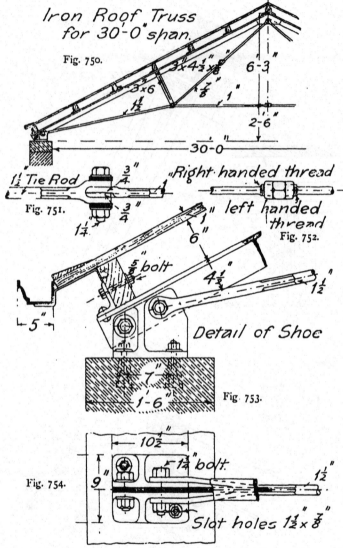

Iron Roof Truss
for 30'-0" span.

Fig. 750.

3×4½"×⅜"

6'-3"

3×6

½"

1"

1½"

2'-6"

30'-0"

1½" Tie Rod    ¾"    1"

Fig. 751.    ¾"

1¼"

Right handed thread

left handed thread

Fig. 752.

1"

6"

⅝" bolt

4½"

1½"

Detail of Shoe

5"

7"

1'-6"

Fig. 753.

Fig. 754.

10½"

1½" bolt

1½"

9"

Slot holes 1½" × ⅞"

Figs. 750-754.   Iron Roof Truss for 30-feet Span.

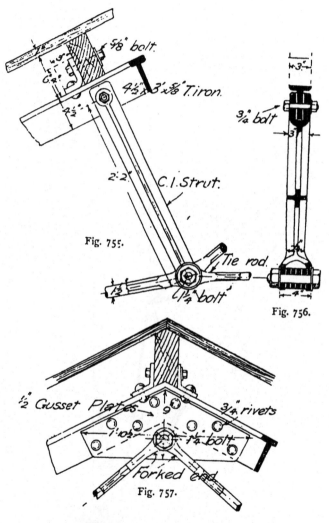

Fig. 755.

Fig. 756.

Fig. 757.

Figs. 755-757.   Details of Joints.

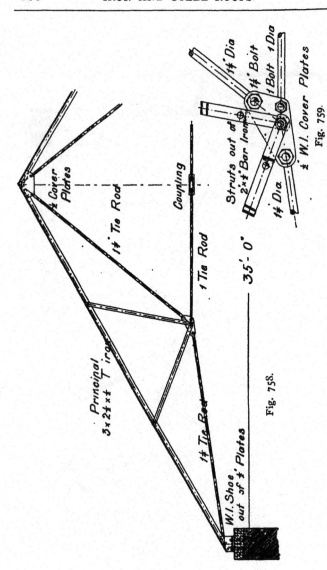

Fig. 758.

Fig. 759.

Figs. 758, 759. Method of Arranging the Junction of the five Members of a Truss.

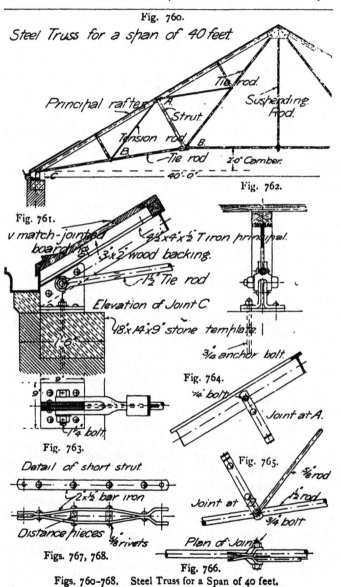

Fig. 760.

Steel Truss for a span of 40 feet.

Tie rod.

Principal rafter.      A.

Suspending Rod.

Strut.

Tension rod.

B.      B.

Tie rod.      2'.0" Camber.

C.      40'.0"

Fig. 761.

v match-jointed boarding.

3½'x4'x½' T iron principal.

3'x2' wood backing.

1½' Tie rod.

Elevation of Joint C.

48'x14'x9' stone template.

1'.6"

9

9

1¼' bolt.

Fig. 763.

Detail of short strut.

Fig. 762.

¾' anchor bolt.

Fig. 764.

¾' bolt.

Joint at A.

Fig. 765.

⅝' rod.

Joint at B.

½' rod.

¾' bolt.

2'x½' bar iron

Distance pieces      ⅜' rivets

Figs. 767, 768.

Plan of Joint.

Fig. 766.

Figs. 760–768.    Steel Truss for a Span of 40 feet.

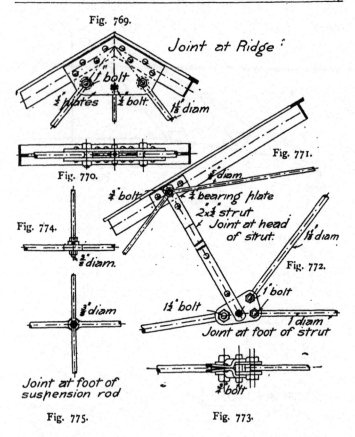

Fig. 769.

*Joint at Ridge*

1" bolt

½" plates    ½" bolt.    1⅛" diam

Fig. 770.

Fig. 771.

⅞" diam

¾" bolt    ⅞" bearing plate
2×½" strut
*Joint at head of strut.*

Fig. 774.

1⅛" diam

¾" diam.

Fig. 772.

¾" diam

1" bolt

1½" bolt    1 diam

*Joint at foot of strut*

*Joint at foot of suspension rod*

¾" bolt

Fig. 775.        Fig. 773.

to insert solid webs instead of separate bars and bearing plates, the webs being stiffened along the compressional lines by plates or angles riveted to them. These trusses should always be made sufficiently rigid, so that the deflection is inappreciable, to prevent outward thrusts on the walls. Where the spans exceed forty feet, provision should be made for freedom of expansion and contraction, usually by fixing one end only and placing the other end on rollers. The stresses on a truss of this type for a 40-feet span are determined graphically in the *Advanced Course*.

The fourth type, as illustrated in figures 788 to 793, has the outline of the Mansard truss; the internal bracing in this

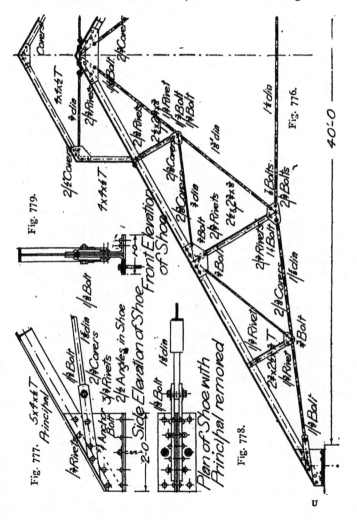

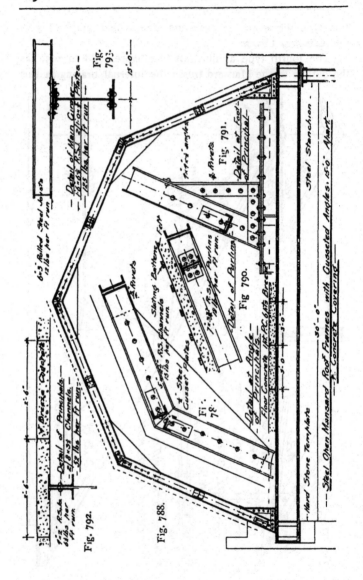

Fig. 793.

Detail of Main Gusset Plate—
1/2×6 R.S.A. 37 cwt. Plate—
123 lbs. per ft. run.

6×3 Rolled Steel Joists
12 lbs. per ft. run

Detail of Principals—
2/8×3½ Channels—
52 lbs. per ft. run

½ Bridge Cover Plate

Fig. 792.

4×3 R.S.A. per
44 lbs. per
ft. run

Tripei angles

½ Rivets

Fig. 791.

Detail at Foot
of Principal

Steel Stanchion

Steel Gusset Plates

½ Rivets

Sliding Battens—Felt

7×3¼ R.S. Purlins
19 lbs. per ft. run

Detail of Purlins

Fig. 790.

5×3½ R.S. Channels
13 lbs. per ft. run

½ Rivets

Fig. 789.

Detail at Angle
of Principals

18 P.C. 6 Fts. Breeze
Floor Concrete

Hard Stone Templets

Fig. 788.

Steel Open Mansard Roof Frames with Gusseted Angles, 15′-0″ Apart—
4″ Concrete Covering

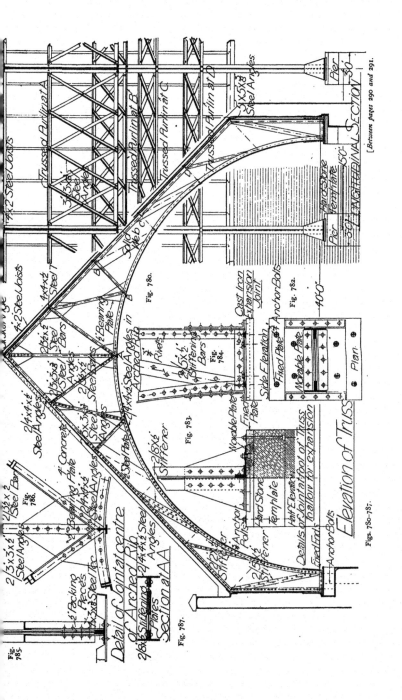

Fig. 785.

2½x3x½" Steel Angles
Fig. 786.
2½x2½x½" Steel Bar
2" Packing Pieces
½" Bearing Plate
2½x4x½ Steel Angles
4x3x½ Steel Angles
3x3x½ Steel Angles
1x5x Steel Tee

Detail of Joint at centre of Arched Rib

2/8x½ Stiffening Plates
2/4x4x½ Steel Angles
Section at AA
Fig. 787.

4½x2 Steel Joists
4x4x½ Steel
2½x2 Bearing Plate
4x4x½ Steel Angles in Arched Rib
2½x5x½ Steel Bars
2½x3x½ Steel Angles
5x3x½ Steel Angles
4" Concrete Plate
½" Concrete Plate

2/4x4x½ Steel Angles in Arched Rib
8" Plate
3x½" Rivets
Fig. 783.

Trussed Purlin at A
Trussed Purlin at B
Trussed Purlin at C
Trussed Purlin at D

3-3½x3 Angles

4x2 Steel Joists

Solid Web C

B

Fig. 780.

2½x5x½" Stiffening Bars
Fig. 784.

Cast Iron Expansion Joint
Fig. 782.

Side Elevation
Fixed Plate
Movable Plate

Movable Plate
Fixed Plate

2/8x½ Stiffener

Details of Joint at foot of Truss to allow for expansion

Movable Plate
Fixed Plate
Anchor Bolts
Hard Stone Template
2½x2½ Stiffener
Fixed Ends
Anchor Bolts

Half Elevation

Fixed Plate Anchor Bolts

40'-0"

Plan.

Elevation of Truss

3x5x8 Steel Angles

Pier 36'

Face Stone Template
Pier Template
50'

LONGITUDINAL SECTION

Pier 36'

[Between pages 290 and 291.]

Figs. 780-787.

type is usually omitted to give clear space in the chamber. These are suitable for large business premises of fire-resisting construction. The main floor girders are used to form the ties; the principal rafters consist of double channels connected at their angles by gusset plates, which take the place and do the work of the internal bracing. Each of the principals becomes a fixed beam under a distributed load. The purlins consist of steel channels and steel joists at the extremities and centres of the ribs respectively. To these ribs are attached light 2″ by 4″ steel joints at about 2′ 6″ centres, the space between these being filled by cement breeze concrete. The stresses are, shown in the *Advanced Course*, determined by graphic statics for a mansard truss of 30-feet span with internal bracing, the gusset plates being considered subjected to the same stresses as their respective displaced internal bracings, and the principal rafters to a bending stress.

Figures 724 to 731 show outlines of steel trusses for spans varying from 15 to 50 feet. Thin lines indicate the members in tension, thick lines in compression.

*Riveting and Bolting.*—In making the connections with members rivets may be employed up to one inch in diameter. Three-quarter inch diameter rivets are the most usual. Where a larger diameter is required bolts are used, and also for the convenience in fixing when *in situ*; bolts are more expensive than rivets.

The following table gives the dimensions of heads and nuts square and hexagon for bolts in the ratio of the diameter of the bolt. Figures 525 to 529 illustrate these :—

| | Diameter of Head and Nut between Faces. | Diameter of Head and Nut across Angles. | Thickness of Head. | Thickness of Nut. |
|---|---|---|---|---|
| Hexagon ... ... | 1·75 | 2 | ·75 to 1 | 1 |
| Square ... ... | 1·5 | 2·12 | ·75 to 1 | 1 |

The following table gives the dimensions of the various rivet heads in the ratio of the diameter of the shank, and are shown in figures 521 to 524 :—

|  | Diameter of Head. | Thickness of Head. |
|---|---|---|
| Cup ...　...　...　... | 1·7 | ·6 |
| Countersunk...　...　... | 1·6 | ·4 |
| Pan head　...　...　... | 1·6 | ·7 |
| Conical　...　...　... | 2 | ·75 |
| Conoidal　...　...　... | 2 | ·7 |

*Hipped Roofs.*—Trusses with perpendicular members are better for the fixing of the half trusses of hipped roofs.

*Sections to resist Tension.*—These may be round, or flat for preference, but any section that lends itself for simple connections may be employed ; if one or more flat bars are used their greater dimension should be vertical, to reduce sagging or deflecting under its own weight.

*Sections to resist Compression.*—Angle, tee, and channel sections are generally used. Double channel, kept apart by pipe distance pieces with small bolts passing through and riveted at ends, forms a stiff section, but is more costly than tees. Angles arranged apart in a similar manner render the connections easy to make. Flat-bars, kept apart by distance pieces so arranged as to bulge them out in the centre and secured by rivets, form a good and simple joint, as shown in figure 768. Short struts may be made of cast iron of cross section, with forked ends, as in figure 745, for a bolt to pass through and form the connection.

*Sections under Transverse Stress.*—Rolled steel joists, angles, or tees are used for purlins, as in figure 732.

*Pitch of Roofs.*—The pitch is governed by the nature of the covering and the architectural requirements.

*Camber of Tie Rods.*—Tie rods are usually cambered $\frac{1}{30}$ to $\frac{1}{40}$ the span, but in some cases more. This is done to reduce the length of the struts and to gain greater height above the floor level.

*Tightening up Joints of Trusses.*—Tension or tie rods are drawn tightly up or slackened by means of gibs and cotters, or by adjustable screw couplings or union joints, as in figure 752. In the latter case the tension rod is in two pieces, the adjoining ends of which are thickened and screwed with right and left handed threads; the rotation of the coupling piece then draws the ends together or urges them farther apart. In modern practice it is usual to make each member the exact length, obviating the necessity of such joints.

*Plus and Minus Threads.*—Threads worked upon thickened ends of rods are known as plus threads, and those formed on rods without thickened ends are called minus threads, as the latter reduce the diameter of the solid rod.

*Design of Joints.*—Joints of trusses should be designed on the fork, eye, and pin principle. All welds should be avoided in the members or their joints, but the expense in forming forged forks and the uncertainty of the proper welding of the joint, render it preferable to eliminate forks and accomplish the purpose by cover plates. The centre lines of members should be drawn about each joint, all meeting in a point.

There are two general methods of construction:—1st. The interior member is constructed as an eye, taking the common intersecting point as the centre, and the outer member or members forked, as in figure 751. A bolt, calculated as described in Part 2, is inserted, and acts as the common axis for all the members about the joint to rotate, as shown in figure 755. A better method is shown in figures 772 and 773. This represents the junction of three ties and a strut; the three ties have forged eyes and are connected by a pair of cover plates; the strut is formed of two pieces of steel bar riveted together, forming at the ends a natural fork.

2nd. All members included between the cover plates are forged flat, and widened at extremities to one common thickness, or packed up to one common thickness, and eyes are drilled; the two cover plates enclose the whole of the members, having drilled holes coinciding with the eye holes of the members, the centres of which are in the centre lines at a distance sufficiently far from the common point of intersection to allow for the necessary width of metal about the eye hole, and a small clearance between the parts. Nutted bolts are then placed through cover plates and members. It may be noticed that a slight adjustment may cause the centre lines not to meet all in a common point, but the error is usually so small that it may be neglected. The cover plates practically act as a common fork, and the enclosed struts and ties as a number of eyes, as shown in figure 738.

*Junction at Feet of Principal Rafters.*—These may be constructed by riveting cover plates over web of principal rafter and flat tie rod, as shown in figure 732, by forking the tie rod over the web of the principal rafter and shoe, as in figure 761, or by arranging cover plates to fork over the web of the principal and bolting or riveting to steel shoes.

*Expansion Joints.*—The shoes are bolted on one side of the building to stone templates, and on the other by passing bolts through elliptical holes formed in the shoes, or by means of two cast-iron plates, the upper fitting in the lower with a dovetail joint to allow of movement, the lower plate being bolted to the templates, as shown in figures 782 to 784, to allow the truss, by being free at one end, to move in a horizontal direction, and thus to expand and contract under changes of temperature without hindrance.

*Junction of Members at Foot of King Bar.*—These may be arranged, as shown in figure 740, by riveting to cover plates where all the members are of flat iron. All trusses, to prevent them bulging laterally, should be tied together by rods at the feet of king bars, or may be secured by ceiling rafters bolted to the tie, as shown in figure 740.

*Junction at Foot of Strut with Tension and Tie Rod.*—This may be constructed with unequal forks to tie rod when in two pieces, the strut forked over both, and the whole bolted together, as in figure 748, or the horizontal portion of tie rods and lower jaw of tension rods may be forged slightly eccentric and eyes drilled, the strut and lower portion of the tie rod forked, and all bolted together, as shown in figure 751; or, as in the example of truss for 50-feet span, by the use of cover plates, figure 740.

*Junctions with two Struts and Tie Rod.*—This joint often occurs in small trusses about 35-feet span, and may be constructed as in figures 758 and 759.

*Junctions at Heads of Principal Rafters.*—These may be formed by means of gusset plates, as shown in figure 740, or with tension rods forged flat at end, and eyes drilled, and thickness of web made out, if necessary, with packing pieces, and the whole riveted or bolted together according to its size, as shown in figure 758.

*Junction of Struts with Principals.*—Where cast-iron struts are used, forks are cast on ends of struts for bolting to principal rafters and tie rods, as shown in figures 755 and 756; if flat bars or steel tees are employed for struts, straps are riveted to each side of the web of the struts to enclose web of principal rafter, to which it is riveted or bolted.

*Trusses.*—Trusses up to 20-feet span are usually made without struts, and consist of steel tees, principal rafters, and flat or round steel ties and king bolts, as shown in figure 724; the principal rafter may be bent to any curve, and the feet prevented from springing by the tie forming the chord of the arch.

Up to 30-feet span, trusses may be formed by steel T principals and flat bar for tensional members, the joints being held together by riveting to cover plates.

Figures 750 to 757 show a method of forming trusses suitable for spans of 30 feet, using steel rods for tensional members, steel tees for principals, and cast iron of cross section for struts.

Figures 742 to 749 give all necessary details and methods of securing shoes to cast-iron stanchions or to stone templates, and method of forming ventilators at apex. The truss for 30 to 40-feet spans is sometimes economically arranged by using two cast-iron struts, which are supported at the junction of tension and tie rods, and in their turn uphold the principal rafter. Figures 758 and 759 show method of arranging the junction of the five members.

40 feet to 50-feet spans may be roofed, as illustrated in figures 760 to 779.

*Coverings of Iron Roofs.*—Corrugated iron, as in figure 734, is especially suitable for places where it is desirable to collect the rain water. Slates, zinc, and glass, as illustrated in figures 797 and 749, are usually the coverings employed for iron roofs.

*Fixings for Coverings.*—There are five general methods for preparing the ground for fixing the roof coverings. 1st. Wood purlins resting against and being bolted to a steel angle cleat, as shown in figure 794, with common rafters and

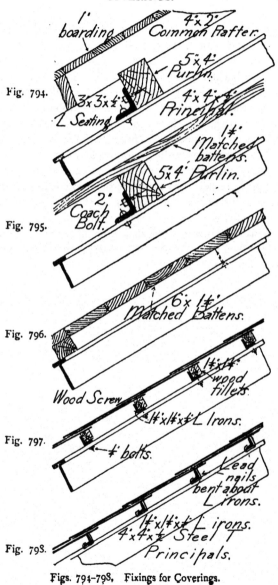

Fig. 794.

Fig. 795.

Fig. 796.

Fig. 797.

Fig. 798.

Figs. 794-798, Fixings for Coverings.

boarding in the usual manner. 2nd. Purlins fixed at more frequent intervals than in the preceding case, and stout boarding fixed directly thereto, as shown in figure 795. 3rd. The principals are backed with timber 2 inches in thickness for their whole length, the steel principal being drilled at intervals of 2 feet for the wood screws to secure the backing, and on this, stout boarding, the thickness of which varies with the distance between the principals, is fixed, as shown in figure 796. 4th. Small steel angles are bolted to principals at intervals equal to the gauge of the slate used; to the small steel angles, wood fillets $1\frac{1}{4}$ by $1\frac{1}{4}$ are fixed; to these the slates are nailed. This dispenses with the use of roof boarding, as shown in figure 797. 5th. Occasionally, in order to dispense with timber, small steel angles are fixed as before, the slating being secured to the same by long lead nails which are bent about the steel angles, as shown in figure 798.

# CHAPTER XI.

# JOINERY.

*Definition.*—Joinery is the art of preparing and framing pieces of timber together to form the finishings of houses, and comprises all wood-work fixed for decorative purposes, as well as for convenience or utility, apart from structural work.

*Design of Joinery.*—Joinery is not required to resist great stresses as in the case of carpentry works, appearance, not strength, being the primary consideration; all the skill of the joiner is therefore expended in constructing the work in such a manner that it shall resist deterioration.

There are three ways in which joinery becomes defective. First, by the warping or splitting of the material; secondly, by the joints coming apart owing to the expansion and contraction of the material; and, thirdly, by the decay of the material.

The following conditions should therefore be strictly adhered to in designing and constructing joinery :—

1st. Frame all parts in such manner as to allow for shrinkage or free expansion, at the same time proportioning the thickness to prevent warping, this being best determined by experience.

2nd. In order to render the amount of shrinkage inappreciable, all fixed parts, as the members in framing, should be made as narrow as possible.

3rd. All stuff should be thoroughly seasoned, and chosen from the heart wood of a tree, all sapwood being rejected.

All the exposed surfaces are usually prepared to take a covering of paint or polish for decorative purposes, or to preserve them from the effects of the atmosphere. The wood employed is called stuff, and comprises battens from 2 to 7 inches in width and $\frac{3}{4}$ to 3 inches thick, boards 7 to 9 inches wide and 2 to $4\frac{1}{2}$ inches thick, and planks any width above.

*Operations.*—The operations of joinery include sawing, planing, shooting, chamfering, rebating, grooving, scribing, moulding, mitreing, and the construction of joints and frames in such a manner as will allow the wood to shrink and swell without hindrance.

*Sawing.*—The cutting of wood by means of saws. On machines these would be rack, band or circular, but for hand use may be rip, hand, tenon, dovetail, key and fret saws.

*Planing.*—The taking shavings off wood by means of planes. This is described as dressed or wrought, and when brought to a level surface, is said to be out of winding.

*Shooting.*—The planing of the edges of boards straight and square with face.

*Rebating.**—This consists in forming a rectangular sinking in the edge of a piece of stuff, as shown in figure 799. Rebates are usually worked by one of the following four methods :— First, up to about $1\frac{1}{2}$ inch in width by $\frac{3}{4}$ inch in depth, by means of a side fillister, as shown in figure 801. This tool has an adjustable fence and stop to regulate the width and depth respectively ; also a vertical cutter, that works in advance of the cutting-iron to ensure a sharp arris.

Similar size rebates may be worked with a rebate plane as shown in figure 802. A temporary fence is nailed on the

* The descriptions of Rebating, Housing, Bending and Veneering are extracted from the author's "Carpentry Workshop Practice" (Cassell & Co., Ltd., price 1s. 6d.), to which the reader is referred for fuller information on the subject.

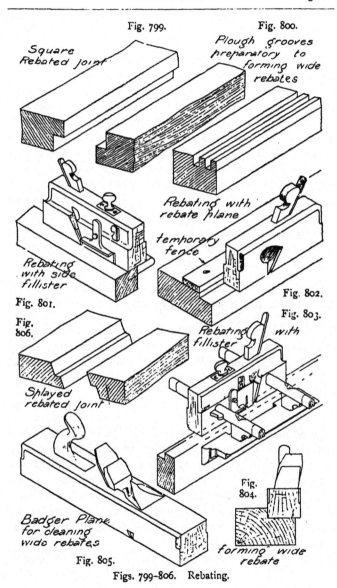

Fig. 799.

Square Rebated Joint

Fig. 800.

Plough grooves preparatory to forming wide rebates

Rebating with rebate plane

Rebating with side fillister

Fig. 801.

temporary fence

Fig. 802.

Fig. 806.

Fig. 803.

Rebating with fillister

Splayed rebated joint

Badger Plane for cleaning wide rebates

Fig. 805.

Fig. 804.

forming wide rebate

Figs. 799–806.  Rebating.

surface of the stuff to regulate the width; the depth is marked by a gauge.

Rebates up to 1 inch by 1 inch on the back edges of stuff are usually worked by a fillister, as shown in figure 803. The adjustable fence of this fillister differs from that of the side-fillister, but in other respects it is similar. Rebates wider than 1½ inch are worked as follows:—A series of grooves are worked in the stuff by means of a plough to the required depth of the rebate, as shown in figure 800; the stuff not removed by a plough is cleared out with a chisel, and the rebate is cleaned with a badger plane, as shown in figure 804. A badger plane is shown in figure 805; this is in many respects similar to a jack plane, but has the iron inserted on the skew, so that a shaving may be taken out up to the side of the plane. Rebated joints in doors or sashes that are required to fit very closely are made with splayed joints, as shown in figure 806.

*Plough Grooving.*—Grooves made parallel to edges of boards, with a special tool termed a plough.

*Cross Grooving.*—The operation of forming a sinking across the grain, and is often done with a saw and router, as in the case of housings for risers and treads into strings, or a groove is sunk entirely across a board by a special plane with cutting teeth, called a grooving plane.

*Housing.*—This is the term given to the sinking of the edge of one piece of stuff into another. There are three kinds. Figure 809 is known as the plain-housing; it has the whole of the edge sunk into the side of the second member. Figure 808 shows the shouldered housing; this is employed where the two sunk members are required to be kept an exact distance apart. Figure 807 shows the dovetailed housing; this is employed where neither nails nor screws can be used. There are two methods of forming housings.

First, where the housing is required across the entire width of a member. A temporary fence is nailed, and the sinking formed with a grooving plane, as shown in figure 811. These planes are made in various widths and each provided with a stop to regulate the depth of the sinking, and with two cutters in advance of the plane-iron to form sharp arrises.

Secondly, where the housing is stopped. A small portion of the housing at the stopped end is sunk with a chisel, the

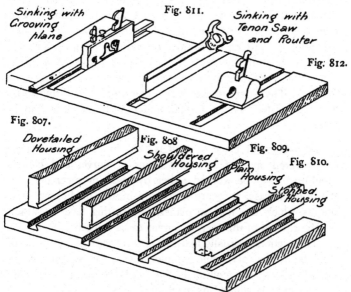

Sinking with Grooving plane

Fig. 811.

Sinking with Tenon Saw and Router

Fig. 812.

Fig. 807.

Dovetailed Housing

Fig. 808

Shouldered Housing

Fig. 809.

Plain Housing

Fig. 810.

Stopped Housing

Figs. 807–812. Methods of forming Housed Joints.

sides of the housing are cut with a tenon saw, the bulk of the material is removed with the chisel, and the sinking taken to the exact depth by means of the router, as shown in figure 812.

*Mitreing.*—This is the process usually applied to the junction between two pieces of stuff that are jointed at an

angle one with the other. If the two pieces to be joined are
of an equal width, the mitre cut bisects the angle.

*Scribing.*—The process of fitting skirtings, linings or
framings to irregular surfaces, also of cutting one moulding
to fit a similar moulding in an internal angle instead of
mitreing.

*Tongues or Feathers.*—Thin pieces of wood, usually pine
veneer, cut $\frac{7}{8}$ inch wide, about $\frac{1}{8}$ to $\frac{1}{4}$ inch thick, inserted in
plough grooves, to prevent dust, light or air passing through
the joint of battens or boards, should the latter shrink. If
cut with the grain they are called straight tongues or feathers.
When cut across the grain, they are known as cross tongues or
feathers. The latter are used where considerable shearing stress
is to be resisted, as the former under such conditions would
split.

*Moulding.*—A piece of stuff is said to be moulded when
it is worked on its face into a series of curves and bands;
mouldings may be worked either by hand or machine. In
the former case, they are stuck by means of hollows and
rounds, which are narrow planes, the cutting faces of which
are the reverse of the varying curves to be worked.

The section is usually marked on the ends of the piece of
stuff by means of a zinc templet cut to the section of the
moulding. Where mouldings are produced on a large scale
they are usually worked by machinery.

To ensure mouldings being cleanly worked, select the wood
slightly cross-grained, as shown in figure 821, and work that
way of the edge running in the direction with the grain. This
will prevent the tearing up of the surface, which is bound to
occur if the grain is either curly or in the wrong direction, as
all moulding planes have single irons only.

*Chamfering.*—Taking the arrises or edges off boards;
forming a bevel, generally by means of a plane. When two

chamfered edges are placed together, as in figure 852, it makes what is known as a V-joint.

Chamfers are usually worked to an angle of 45° to the face. If the chamfer be taken from end to end of a piece of stuff it is usually worked with the ordinary bench planes; but if it be stopped it is more convenient to use the chamfer plane shown in figures 813 and 814, which allows the chamfer to be worked to within ⅛ inch of the stop. There are several forms of stops, three of which are shown in figures 815, 817, and 818—the plain, the moulded, and the broached stops. The plain and moulded stops are expeditiously formed by cutting to a templet worked to the shape of the stop, as shown in figure 816. The broach stop, of which there are several kinds, must be carved.

*Bending.*—The bending of wooden members is usually performed in one of five ways—(1) kerfing, (2) building up, (3) stave and veneer method, (4) cooper-jointed method, (5) blocking and bending.

(1) The method of kerfing consists in making a series of transverse saw-cuts through the piece to be bent to within $\frac{1}{16}$ inch from the finished face. The face should be well damped; the piece may then be bent to a curve, the quickness of which depends upon the thickness and frequency of the saw-cuts. If the required surface is concave the saw-cuts open when the piece is bent, feather-edge slips are fitted in the saw-cuts, and glued as shown in figure 822. If a convex surface is wanted the face is damped, the saw-cuts are filled with glue, and the piece is bent to fit a shaped bearer, which is screwed to the back, as shown in figure 823. Kerfing is an inferior method of bending, the position of the saw-cuts being in most cases discernible on the finished surfaces.

(2) The surfaces forming the method of building up are constructed of a series of narrow members cut to the curve required, and built and bonded similar to masonry, as shown in

x

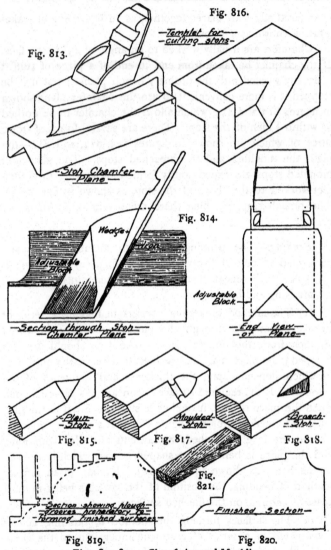

Fig. 813.

Fig. 816.

Templet for cutting stops

Stop Chamfer Plane

Fig. 814.

Wedge

Iron

Adjustable Block

Adjustable Block

Section through Stop Chamfer Plane

End View of Plane

Plain Stop

Moulded Stop

Broach Stop

Fig. 815.

Fig. 817.

Fig. 818.

Section showing plough grooves preparatory to forming finished surfaces

Fig. 821.

Finished Section

Fig. 819.

Fig. 820.

Figs. 813-821.　Chamfering and Moulding.

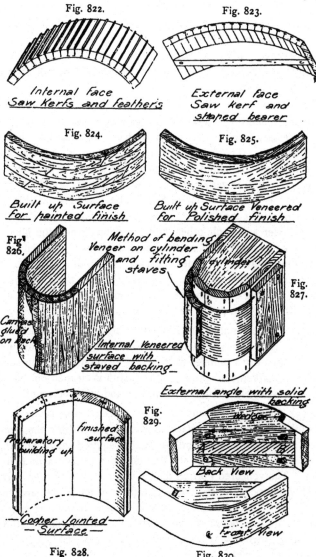

Fig. 822.

*Internal face*
*Saw Kerfs and feathers*

Fig. 823.

*External face*
*Saw kerf and*
*shaped bearer*

Fig. 824.

*Built up Surface*
*for painted finish*

Fig. 825.

*Built up Surface Veneered*
*for Polished finish*

Fig. 826.

*Method of bending*
*Veneer on cylinder*
*and fitting*
*staves*

*cylinder*

Fig. 827.

*Carriage*
*glued*
*on back*

*Internal Veneered*
*surface with*
*staved backing*

*External angle with solid*
*backing*

Fig. 829.

*Wedges*

*Back View*

*Preparatory*
*building up*

*finished*
*surface*

*Cooper Jointed*
*Surface*

Fig. 828.

*Front View*

Fig. 830.

Figs. 822–830. Methods of forming Curved Wood Surfaces.

figure 824, and glued together. These surfaces when cleaned off form a sufficiently smooth ground for a painted surface. If a polished surface is required it will be necessary to veneer the surfaces in order to hide the numerous joints, and show a continuous grain.

(3) Sharp, concave curves are usually built up by the stave and veneer method. For this a cylinder is required to give the necessary form to the veneer, and is constructed as shown in figure 827. The veneer, which is well damped or steamed, is bent round the cylindrical surface, and is kept in position by pieces of stuff covering its extremities and screwed to the cylinder, as shown in figure 827. The staves, consisting of pieces of stuff 1 inch wide, $1\frac{1}{2}$ inches in thickness, are fitted to the cylinder and each other, and glued up; when the glue is dry they are cleaned off at the back, and canvas is glued over the whole surface to generally assist in binding the staves together, as shown in figure 826. The screws are then withdrawn from the cylinder, the work removed and cleaned up, as shown in figure 826.

(4) Cooper-jointing, as shown in figure 821, consists in glueing up a series of narrow boards usually with a tongued joint; when dry a templet to the curve required is laid on each end and marked, and the superfluous stuff is cleaned off. This method is largely employed for panels and wood columns.

(5) Blocking and bending.—The piece of stuff to be bent is sunk from the back surface to within $\frac{1}{16}$ inch from the front face, a block is built up and cut to the required shape, and the extremity A is screwed to the first member, as shown in figure 829. The veneered portion of the first member is then thoroughly damped, and the contact surfaces of the veneer and the block are then glued, the veneer is bent round the block, wedges are inserted as shown to draw the veneer tightly about the block, then screws are inserted as shown at B. Figure 830 shows a front view.

*Veneering.*—Stuff—that is, wood used by the joiner—is termed veneer when it is cut into thin sheets, usually not exceeding $\frac{1}{18}$ inch in thickness; it is generally prepared from rare or costly woods possessing a beautiful colour or grain. It is used to enhance the appearance of ornamental woodwork by covering some or all of its visible parts with a layer of a more strikingly marked piece of similar or other wood. Veneer may be laid on work in large sheets, or it may be cut into shaped pieces and laid to various designs. For the latter purpose veneers of different woods are often used in combination, and veneers may be dyed several different colours to obtain any desired effect.

To veneer a piece of stuff, the following processes in the order here given are usually carried out. First, the board to be veneered, after having been planed up, is traversed with a toothing plane, to remove any inequalities and to render the surface in a better condition to receive the glue.

The toothing plane has a stock similar to that of a smoothing plane ; it has a single-iron pitched vertically. The face of the iron is grooved with a number of small grooves parallel to its side, so that when the iron is sharpened from the back, as in the ordinary plane iron, the edge produced has a number of teeth.

If the board to be veneered is cut tangentially, the heart side should be chosen to be covered, in order that the natural tendency for the annual rings to straighten may always tend to keep the veneer taut. After having been toothed, the surface should be sized, to render it less porous, and at the same time the reverse side should be damped, to prevent the board from casting. The next process is to match and joint the pieces of the veneer on a board. They are temporarily secured in position to the surface with needle points. Strips of paper are then taken (damped on their upper sides and glued on the other) and laid on the veneer over the joints. The paper, on drying, draws the pieces close together and makes neat joints. In

fixing the veneer, the board should be covered plentifully with
glue of a medium consistency.    There are two methods of
securing the veneer—(1) with a caul hammer, (2) with a caul.
The caul hammer is used on flat plane surfaces, where the
veneer is in one piece and no tendency to crumble.    After the
board has been glued, the veneer is laid in position, its upper
surface wetted, a hot flat iron is rapidly passed over the wetted
upper surface to soften the glue, which is squeezed out by
dragging the caul hammer over the upper surface, commencing
at the centre and working diagonally towards the edges to
ensure all air blisters and superfluous glue being worked out.

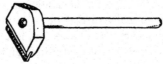

Fig. 831.

The caul hammer, as shown in figure 831, is a  tool similar
to a rake, but having a horizontal  piece of iron about 4 inches
in length and $\frac{1}{4}$ inch thick, with an edge straight in length and
rounded transversely.

If the veneer is jointed to any pattern, or is of a crumbling
nature, or is glued on a curved surface, it is usual to press it
into position with a caul, which is removed when the glue has
dried.   The caul consists of a piece of stuff cut to the reverse
of the surface to be veneered, or if it is a flat plane surface,
the caul, as shown in figure 832, is formed of boards, jointed
to the dimension of the work to be veneered, so that when
fixed the whole of the veneer is subjected to pressure.   In
the latter case cambered stretchers are employed to ensure
the pressure being applied at the centre of the surface before
the edges.   Before applying the caul, it is heated to keep
the glue in a melted condition while the caul is being fixed.
Paper should be placed between the upper surface of the

veneer and the caul, which latter should be well rubbed with soap to prevent the glue coming through the veneer and adhering to the caul. On wide surfaces in important work it is usual to veneer both sides of the work, because the veneer, in drying, contracts, and causes the surface on which it is glued to cast; but if the stuff be veneered on both sides the opposing tendencies preserve a true surface.

The veneer is brought to a finished surface ready for polishing, first with a fine toothing plane, and secondly with a scraper, and rubbed with fine glass paper in the direction of the grain.

*Framing.*—Large areas requiring to be covered with wood-work should comply with the conditions previously stated in this chapter.

A frame is usually constructed by joining together narrow pieces of stuff, in width about 3″, with a mortice and tenon joint, as shown in figures 893 to 896. The pieces enclose a number of rectangular or other shaped spaces, and usually have grooves on their inside edges, into which the edges of pieces of stuff termed panels are fitted. The framing completely encloses the panel, preventing it from twisting, but allowing it to expand and contract freely. The great bulk of the area should be taken up by the panels, only a relatively small portion being taken up by the framing.

The vertical members that are mortised, forming part of the boundary of the frame, are termed styles·; all other vertical members between the styles are termed muntins, and the horizontal members are termed rails.

*Method of Glueing up Framing.*—When all the members have been prepared and fitted, the framing is laid upon bearers (the upper surfaces of which are fixed "out of winding"), and all the tenons and shoulders are coated with hot glue; the frame is then compressed by an iron cramp; it is then secured by wedges, dowel pins, or wedges and dowel pins combined,

which have been dipped in hot glue. In frames for exterior
work white lead is used instead of glue, and dowels are often
employed, especially where mortices or slots are cut with a
saw, as in figure 833.

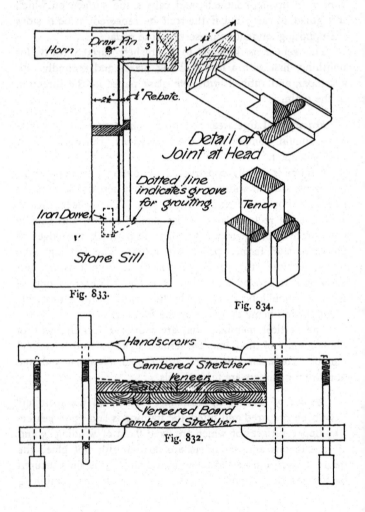

Horn

Draw Pin

3"

Detail of
Joint at Head

2½"      ⅛" Rebate.

Dotted line
indicates groove
for grouting.

Tenon

Iron Dowel

Stone Sill

Fig. 833.

Fig. 834.

Handscrews

Cambered Stretcher

Veneer

Veneered Board

Cambered Stretcher

Fig. 832.

*Panelling and Moulding.*—Specifications of framing should include the method in which the panels are treated, and the manner and extent of the moulding of the frame.

The panels may be either flat, raised, raised and moulded, bead flush, bead and butt or flush only. The framing may be square or moulded.

There are three methods of moulding frames: first, by

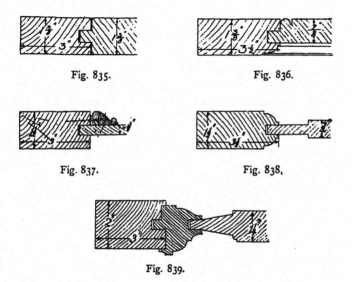

Fig. 835.

Fig. 836.

Fig. 837.

Fig. 838.

Fig. 839.

planting; secondly, by working the moulding on the frame; thirdly, by framing the mouldings, and grooving them into the frame.

Figure 837 shows a section of square framing and flat panel, with a planted moulding. Figure 838 shows a section with the moulding stuck on the solid, and a raised panel. Figure 839 shows a framed bolection moulding on one side and a sunk moulding on the other, with a bevelled raised and moulded panel. The object of bevelling is to add strength to the panel by having the latter of greater thickness than the circumscribing groove. Care must be taken to make the thickness of the panel that enters the framed bolection moulding parallel, otherwise, if any shrinkage takes place, the panel becomes loose if the bevelling be continued to the edge of the panel. In order to gain strength the panels are often made flush with the framing, as shown in figures 835 and 836.

The method of planting mouldings is bad, and should never be used for good work ; but where adopted, care should be taken that the mouldings are nailed to the frame and not to the panel.

Small mouldings are best stuck on the solid. These mouldings should be scribed together at the angles where the section and position render it possible, otherwise they are mitred.

Mouldings exceeding 1 inch in width should be framed. The framed mouldings are usually mitred, tongued, glued, and screwed at angles.

Mouldings are often continued from the sides of a frame, and planted on the face of the panels to various designs ; under these conditions great care must be taken to fix the mouldings so that the grain of the moulding is in the same direction as the grain of the panels ; this will necessitate a large amount of the moulding being stuck across the grain. If this be not done, those mouldings that are fixed with their grain at right angles to that of the panels will after a short time become detached, owing to their inability to expand or contract with the panel ; they also have the tendency to split the panel.

*Flush Panels.*—Panels are often prepared flush with the framing on one or both sides in order to obtain increased strength. A bead is usually stuck on the edges of the flush side that are in the direction of the grain to save the appearance of the large open joint that occurs when the panel shrinks ; the transverse edges are left square, as the panel does not shrink in the direction of its length. Such panels are termed bead butt. Flush panels of the above description often have the bead carried round the four edges of the panel; this is bad as carried out in ordinary practice, the transverse bead being let into the panel, and having its grain at right angles to same, thus preventing its shrinking freely, the panel being frequently split. If beads are required on the four sides, they should be stuck on the frame. Panels are often made flush both sides, being ornamented with beads or mouldings on both sides, or left plain; the latter usually being done where it is required to cover the door with baize or other material.

*Joints.*—Joints in joinery are classified as follows :—1. Side joints; 2. Angle joints; 3. Cross joints; 4. Joints to prevent warping; 5. Framing joints.

(1.) *Side Joints.*—The side joints that have been used in connection with floors have been explained and illustrated in the chapter on that subject. The remainder are as follows :— The plain or square joint : in this case the edges are shot true, so that if a straight-edge be applied to their common face when the boards are applied one to the other, it shall be in an approximate plane : it is necessary to try the boards in this way, and not attempt to shoot the edges square, as the boards generally cast and twist a certain amount in their rough state. When the joint is prepared the two edges are applied together and are glued ; this accomplished, the edges are placed in position and the top board rubbed backward and forward on the bottom to rub all the superfluous glue out of the same and to cause a certain amount of suction. When the joint is adjusted it is left till it is dry, and then cleaned off and prepared for the purpose for which it is intended.

While the joint is in a wet condition the pieces are often secured by cramps or cleats, to prevent any movement in same till dry. All kinds of glued joints between boards should

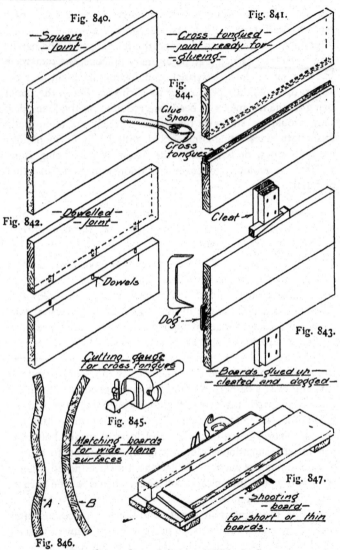

Fig. 840.

_Square joint_

Fig. 841.

_Cross tongued joint ready for glueing_

Fig. 844.

Glue Spoon

Cross tongue

Fig. 842.

_Dowelled joint_

Cleat

Dowels

Dog

_Boards glued up cleated and dogged_

Fig. 843.

_Cutting gauge for cross tongues_

Fig. 845.

_Matching boards for wide plane surfaces_

A　B

Fig. 846.

Fig. 847.

_Shooting board for short or thin boards_

Figs. 840 to 847.　Jointing-up Wide Surfaces.

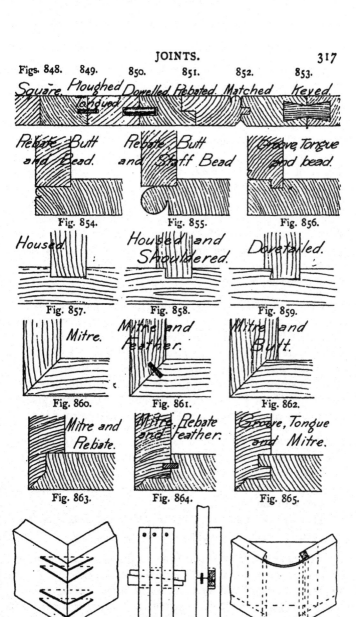

Figs. 848. 849. 850. 851. 852. 853.

Square. Ploughed Dowelled Rebated. Matched Keyed.
Tongued

Rebate, Butt and Bead.

Rebate, Butt and Staff Bead

Groove, Tongue and bead.

Fig. 854.　　Fig. 855.　　Fig. 856.

Housed.　　Housed and Shouldered.　　Dovetailed.

Fig. 857.　　Fig. 858.　　Fig. 859.

Mitre.　　Mitre and Feather.　　Mitre and Butt.

Fig. 860.　　Fig. 861.　　Fig. 862.

Mitre and Rebate.

Mitre, Rebate and Feather.

Groove, Tongue and Mitre.

Fig. 863.　　Fig. 864.　　Fig. 865.

Keyed Mitre　Counter Cramp.　Rounded Corner.

Fig. 866.　　Fig. 867. Fig. 868.　　Fig. 869.

be treated in this way.    Figure 840 shows a plain or square joint.

*The Tongued Joint.* — In this method, as shown in figure 841, the edges are shot as before, a groove is made in the centre of each of same $\frac{1}{8}''$ by $\frac{1}{3}''$ in depth at the most by means of a plough.    Cross tongues 1″ in width are then inserted in one of the grooves, the joint is then tried, glued, and rubbed as before.    The grooves are filled with glue by means of the glue spoon, as shown in figure 844.    Increased strength is obtained in the joint by this method, as a larger glued surface is obtained and the two pieces are to a certain extent interwoven.    It is not good, however, for panels of $\frac{1}{2}''$ thickness or under.

*The Dowelled Joint.* — The edges in this joint, shown in figure 842, are shot in the ordinary manner, the boards are then placed face to face with the shot edges uppermost; they are squared across about every 18″, a gauge mark is scratched about the centre of the edges from the face of each board (thus obtaining the exact centre for each dowel), the holes are then bored about $\frac{3}{4}''$ deep, the dowel inserted, the edges glued, and the boards cramped up.    This makes a very strong joint.

*Keyed Joint.* — This joint, as shown in figure 853, is not much used at the present except in imitation of old work.    It consists of a dovetail shaped key, about 3″ in length, let into the ends and the back of the boards to be joined, and cleaned off level with the back.

There are several side joints in which the pieces are not intended to be glued together, such as the following :—

*The Matched Joint.* — The form of the joint, as shown in figure 872, is a grooved and tongued arrangement, to prevent an open through joint in case of shrinkage, having on the tongued edge a moulding, generally a chamfer or a bead, to emphasize the joint, and so obliterate the ill effects of the open joints that ensue on shrinkage. Where prepared by hand the grooves and tongues are worked by a pair of planes known as matching planes, as shown in figures 870 and 871. Match-boards are employed in the place of framing where large surfaces are to be covered. They should not be more than

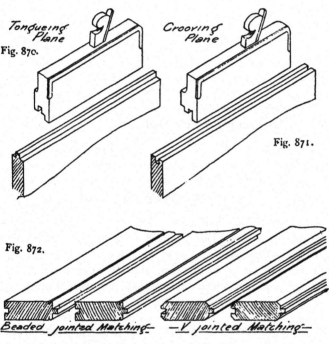

Tongueing Plane

Fig. 870.

Grooving Plane

Fig. 871.

Fig. 872.

Beaded jointed Matching.

—V jointed Matching.—

Figs. 870–872. Preparation of Match Boarding.

4″ in width, and should be nailed on one edge only, in order that no joint shall open too much, and that the whole shall be free to shrink. A similar joint to this is used for the joint of horizontal sliding sashes where it is desired to obtain a dust and draught-proof joint.

*The Rebated Joint.*—This is generally employed for the meeting styles of folding-doors and sashes for mutual support, and to prevent an unobstructed through joint, see figure 799.

(2.) *Angle Joints.*—These joints are divided into three classes—(*a.*) Those in which the joint is parallel to the grain of the pieces joined. (*b.*) Those in which the joint is at right angles to the pieces joined. (*c.*) Those which may be used in both of the above cases.

(*a.*) *Rebate, Butt, and Staff Bead.*—In this, as shown in figure 855, there is a rebate formed, so that the pieces may be easily placed in position, a staff bead being worked at the angle to obtain decorative effect. The two pieces are secured by nailing, or by glueing and blocking.

*Rebate Butt and Bead.*—This is worked similarly to the preceding example, having a small bead at the side instead of a large angle bead; this is shown in figure 854. It is used where one face is required perfectly plain.

*Groove Tongue and Bead.*—This member has a tongue instead of a rebate, as shown in figure 856. These three forms are usually employed where the work is fixed piece by piece, it being difficult to obtain a close joint; the moulding is added to counteract this defect. If the angle is glued, some form of a mitred joint is generally employed.

(*b.*) *Joints across the Grain.*—The principal joints for this purpose are, 1, the common dovetail; 2, the lap dovetail; 3, the secret dovetail, the first being used if the visible joint be of no account, the second where it is desirable to have one face perfectly clear, and the third if both faces are to be seen. They are made as shown in figures 873 to 875.

*The Keyed Mitre.*—The pieces are here mitred, placed into position, and a number of pairs of saw cuts inclined to each other made across the angle. The joint is then glued, put together, and into the cuts are fitted coarse shavings, which are glued and fitted tightly, as shown in figure 866.

(*c.*) *Joint common to both directions of Grain.*—These resolve themselves into some form of the mitre joint; the common mitre, as shown in figure 860, is easily formed but difficult to fix together, and is unreliable unless glued and blocked for its whole length.

*The Mitre and Tongue.*—This is a good and strong joint, but the grooves are difficult to form unless a machine is used.

*The Mitre and Butt.*—This joint, as shown in figure 862, used where the pieces are of unequal thickness.

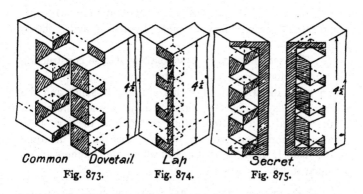

Common  Dovetail.  Lap.  Secret.
Fig. 873.  Fig. 874.  Fig. 875.

*The Mitre and Rebate.*—This joint, as shown in figure 863, is more difficult to prepare, but is a very good form for jointing, especially where nails only are employed.

*Mitre, Rebate, Butt and Tongue.*—This joint, as shown in figure 864, is used where the joint is parallel to the pieces joined; it is used in the best work, for the angles of pilasters, etc. It is glued and blocked, if possible no nails being used.

Y

(3.) *Cross Joints.*—Cross joints are of two forms—(*d.*) Those in which the edge of one piece is fitted into the side of the other; (*e.*) those in which the pieces completely cross.

(*d.*) *The Housed Joint,* as shown in figure 857, is used where the pieces are gauged to a thickness, and it is possible easily to fit the one exactly into the other, the two pieces being secured by nailing.

*The Shoulder and Housed.*—This is used either where the housed member is rough on one side—a tongue is therefore worked as being the most expeditious method of fitting—or where the internal angle is to be seen. In this case the shouldered joint gives the cleaner appearance. This joint is shown in figure 858.

*The Dovetailed Housing.*—This is employed where it would be undesirable to fix the pieces by nailing, and is shown in figure 859.

(*e.*) The second class of cross joints is used in constructing the carcases for drawers, or for nests of shelves, etc.

In many instances the uprights are usually solid, and the horizontal rails are about 3″ wide. The joint is a combination of the halved and housed joint; the idea is to have both members continuous and so increase the tie. The same arrangement is used for positions in which the members are of an equal width.

*Counter Cramp.*—Figures 867 and 868 show a counter cramp which is a convenient arrangement for drawing up the heading joints of wide surfaces such as counter tops.

*Rounded Corner.*—Figure 869 shows a method of con structing a rounded corner showing continuous grain. It consists of a solid block cut to the shape of the curve about which the stuff to be bent, after having been reduced to $\frac{1}{8}$″ in thickness, is bent, glued, wedged, and screwed.

(4.) *Joints to Prevent Warping in Wide Surfaces.*—All boards
have a tendency to warp or cast if left to themselves or with
insufficient fixing, and when several are jointed for such purposes as table-tops, fascias, dados, or sign-boards, the defect is
liable to be greater. To reduce the amount of the casting to
a minimum, the boards should be selected and matched with
the heart side of adjacent boards on opposite faces of the work,
and means must be taken to fix the boards in a flat position,

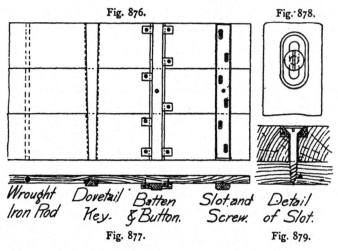

Fig. 876.

Fig. 878.

Wrought Iron Rod    Dovetail Key.    Batten & Button.    Slot and Screw.    Detail of Slot.

Fig. 877.

Fig. 879.

but in a manner that they will be able to shrink and expand
freely. There are four general methods employed for this:—
1, the dovetailed key; 2, the batten and slot screw; 3, the
batten and button; 4, by iron rods.

In the first method, as shown in figure 876, a wood key of
the shape shown in figure, dovetailed on its edges, is fitted to a
cross groove made to receive it. The object of tapering the
key is to render it easier to fit. After the key is driven in, it is
screwed at its wider end.

*The Slot Screw and Batten.*—By this method a batten of

sufficient thickness is prepared with a number of countersunk slots, into which brass slots are fitted ; these are placed on the back of the boarding, one screw is driven in the ordinary way through the centre of the batten to fix the latter to the work, screws are driven into the centre of each slot; the latter fix the batten and boarding together, but allows the work freedom to expand or contract, see figure 876.

*The Batten and Button.*—This method is more effective than the last, but is unsightly.    It consists of a rebated batten screwed to board in centre, and secured by hardwood buttons of the form shown in figure 876.

*The Iron Bar.*—In this case holes are bored through the edges of the boards before they are glued up.    The iron bar is next passed through the holes in the edges, which latter are glued, and the joints are closed by means of cramps.    This method, as shown in figure 876, is very effectual, especially if both sides of the work are exposed to view, or where there is no room for any projections at the back.

(5.) *Framed Joints.*—These joints invariably resolve themselves into some form of the mortice and tenon joint.

The form of the mortice and tenon joint is shown in figure 893.    The typical dimensions for such a joint are as follows :— The thickness of tenon should be $\frac{1}{3}$rd of the thickness of the stuff, and the width should not exceed four times the thickness, and in no case more than $2''$ in width.    The reason for the latter rule is that if the tenon is of a greater width than that stated, it has a tendency to bend under the pressure of the wedges, and in the second place it is necessary to keep the tenon sufficiently narrow to render the shrinkage inappreciable.

*Haunchion.*—Consists of a stump tenon, usually $\frac{1}{2}$in. long, adjoining the tenon proper, fitting into a groove cut to receive it, as in figure 893, to prevent the tenoned piece twisting, also to obviate a through joint, and to keep the surfaces of the rails and styles in the same plane should the joint shrink or become loose. The shortness of the haunchion adds great strength to the root of the tenon, about which, but for the haunchion, it would often snap when exposed to sudden shocks.

The wedges employed should be thick at the narrow end, and have a very slight inclination, in order that when they are driven home they shall compress the tenon with the greatest force at the root, so that should any shrinkage take place in the style, its direction would be towards the shoulder.

*Double Tenon.*—Figure 895 shows a double tenon, which is used if the framing is very thick, or as in this case at that end of a rail where it is intended to fix a mortice-lock into the style; the minimum distance in the thickness of the door between the tenons should be the thickness of the lock. The rule respecting the thickness of double tenons is that the sum of the double thicknesses should equal $\frac{1}{3}$rd the thickness of the stuff. On the other end of the lock rail a double pair of tenons is unnecessary.

*Joint at Lock Rail of Diminished Style.*—This joint, as shown in figure 902, is similar in principle to that of the ordinary lock rail, the difference being that it is moulded at the top, and has a bevelled joint.

*Bottom Rail.*—Where the bottom rail is wide, there are two tenons which are set out, as shown in figure 896, differently from those of the middle rail, owing to the end of style and bottom edge of rail being flush.

*Foxtail Wedging.*—Figure 897 illustrates the method of fixing known as foxtail wedging. The mortice in this case is stopped about $\frac{1}{4}''$ from the back edge of style, and is cut slightly dovetail in shape. The tenon has a saw cut about $\frac{1}{8}''$ from each side, into which a wedge is inserted, and on being cramped up

the wedges are forced home.　There should be a space
allowed of the thickness of the wedge in the mortice at the
end of the tenon, otherwise in the event of a wedge being

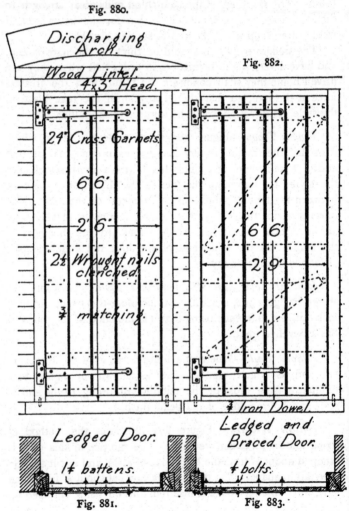

Fig. 880.

Discharging
Arch.

Wood Lintel.
4×3″ Head.

24″ Cross Garnets.

6′6′

2′6′

2½″ Wrought nails
clenched.

¾″ matching.

Ledged Door.
1½″ battens.

Fig. 881.

Fig. 882.

6′6′

2′9′

¾″ Iron Dowel.

Ledged and
Braced Door.
⅝″ bolts.

Fig. 883.

turned over, it would prevent the joint closing, and it would be a matter of great difficulty to withdraw the same. This joint is employed in the best work, where it is undesirable to see the end grain of the tenons penetrating the styles.

The joints at the angles of moulded sashes are similar in principle to the above, but modified, as shown in figures 916 to 918 to meet the different conditions.

*Doors.*—These are classified under the following heads :—ledged, ledged and braced, framed and braced, and framed and panelled doors.

*Ledged Doors.*—These consist of a number of vertical battens, fixed by wrought-iron nails, driven in from the face of the battens, and clenched to horizontal rails or ledges, as shown in figure 880.

This is the commonest description of door, and is used for temporary purposes, outhouses, etc. The battens may be square, chamfered, ploughed and tongued, or grooved and tongued, and are generally fixed to posts by cross garnets or strap hinges with screws or bolts.

*Ledged and Braced.*—These are doors made in a similar manner to the ledged doors, but having in addition braces, which stiffen the doors and keep them from dropping at their free edge; the lower end of braces should always be near the supported edge. The joint with a brace and rail is generally made as in figures 882 and 883, which show the battens grooved, tongued, and V-jointed, and hung by cross garnet hinges.

*Framed and Braced Doors.*—These are a good and strong form of door, and extensively used for external work to withstand rough wear. A framing is made, styles and top rail being of equal thickness, bottom and middle rails with the braces having a thickness equal to the styles, minus the thickness of

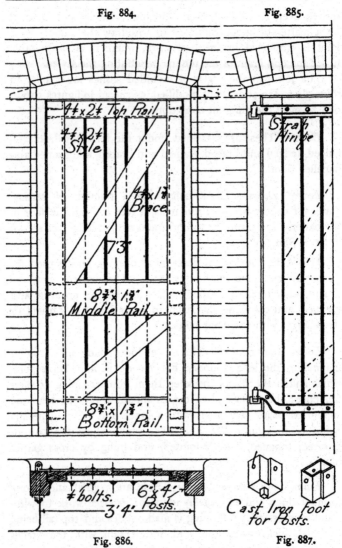

Fig. 884.          Fig. 885.

4⅞×2⅞ Top Rail.

4⅞×2⅞ Style

4⅞×1⅞ Brace

7′3″

8⅞″×1⅞″ Middle Rail.

8⅞″×1⅞″ Bottom Rail.

Strap Hinge

⅜ bolts.     6″×4″ Posts.

3′4″

Cast Iron Foot for Posts.

Fig. 886.                     Fig. 887.

Figs. 884–887.   Framed and Braced Doo

Fig. 888.                               Fig. 889.

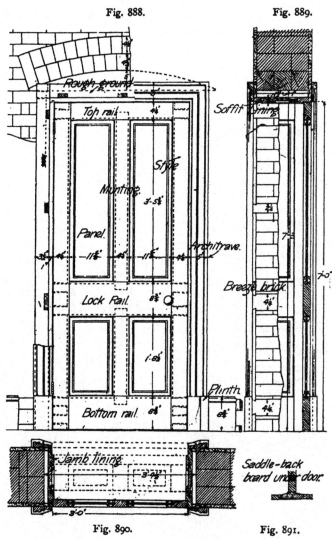

Figs. 888–891.   Four-panel Door and finishing for same.

Fig. 892.

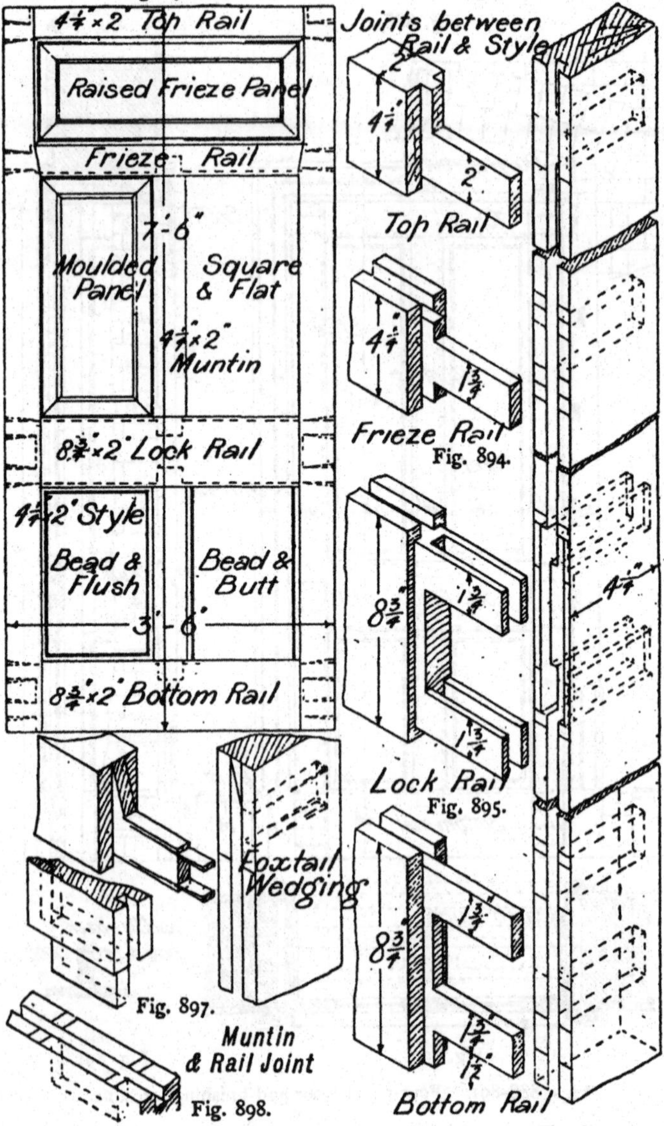

$4\frac{1}{4}\times2$ Top Rail

Raised Frieze Panel

Frieze  Rail

$7'$-$6''$

Moulded     Square
  Panel      & Flat

$4\frac{1}{4}\times2$
Muntin

$8\frac{3}{4}\times2$ Lock Rail

$4\frac{1}{4}\times2$ Style

Bead &     Bead &
 Flush      Butt

$3'$ - $6'$

$8\frac{3}{4}\times2$ Bottom Rail

Foxtail
Wedging

Fig. 897.

Muntin
& Rail Joint

Fig. 898.

Fig. 893.

Joints between
 Rail & Style

Top Rail

Frieze Rail
Fig. 894.

Lock Rail
Fig. 895.

Bottom Rail

Figs. 892-898.  Framing Joints.          Fig. 896.

battens, all members flush on the back, as shown in figure 886. The bottom and middle rails are cut with barefaced tenons, and the braces may be stub-tenoned into rails and styles. The battens are then pushed into their position, and secured to rails and braces by wrought-iron nails hammered in from the face and clenched at the back. The battens extend from the top rail to the ground, more effectively to let any water drain off the door that might fall upon it. Hook-and-eye or strap hinges are suitable for hanging these doors.

*Framed and Panelled Doors.*—These are described by the number of panels they contain, and consist of styles, rails, muntins and panels.

*Four-panel Doors.*—Figure 888 shows a four-panel door 7 feet high, 3 feet wide, and the finishings, with all the necessary dimensions. The height of the top line of middle or lock rail is usually between 3 feet 2 inches and 3 feet 6 inches, so that the handle of the door coming in the centre of the lock rail may be at a convenient height. Doors are usually hung on butts, as shown in figure 888.

*Five-panel Doors.*—Doors 7 feet high and over that height are often divided into more than four panels, the rail between the top and lock rails is called a frieze rail, and the top panel a frieze panel, as shown in figure 892.

*Sash Doors.*—These are doors the upper parts of which are arranged for glass panel or panels, the styles of which are sometimes narrower at their upper part to gain more light, and are called diminished styles, as in figure 899.

*Chamfered Panels.*—These are panels made of narrow battens, chamfered on edge, and together forming a V-joint, as in figure 884. The battens are sometimes cut and arranged diagonally.

*Saddle-back Board.*—In order to allow doors to easily clear carpets without having a wide joint at bottom when closed, a saddle-back board is fixed, as shown in figure 891, which allows a wide joint when the door is open.

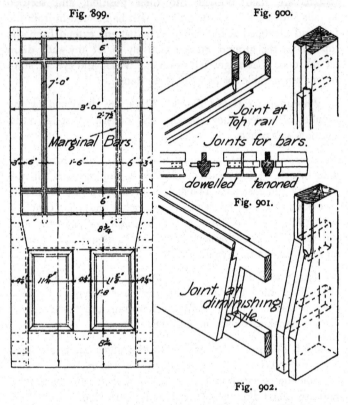

Fig. 899.

Fig. 900.

Joint at Top rail

Joints for bars.

dowelled    tenoned

Fig. 901.

Joint at diminishing style

Fig. 902.

*Door Linings.*—Doors are hung either in solid frames, constructed as shown in figures 880 to 886, or in linings plain or framed. Plain are used in thin partition walls, there not being sufficient width in the linings to frame them. They are prepared from stuff from $1\frac{1}{4}$ inch to 2 inches in thickness, rebated the thickness of the door on both edges to form stops;

or fillets $\frac{1}{2}$ inch in thickness are nailed about the jamb linings to form the stops, as shown in figure 903, which is suitable for a half-brick wall or a stud partition. The joint between the jamb lining and head is shown in figure 904.

Figure 905 shows the method of framing the linings for a 9-inch wall. In this case the panel is made to project to form the door stop.

Figure 906 illustrates the method of constructing the door linings for a 14-inch wall, the width being sufficient in this case to prepare the frames in the ordinary method.

Figure 907 shows the method of preparing the linings where the doors to be hung are very heavy; in this case solid frames are used in conjunction with framed linings to form the jambs and the stops.

There are three general methods of arranging the finishings about the jambs. The framed grounds about the linings are common to all systems; first, the grounds where not visible may be rough and the mouldings forming the architraves mitred about the jamb linings, and fixed to them, as shown in figure 905; secondly, the grounds may be wrought, and form part of the finished architrave, a narrow moulding being used in addition to cover the joint between the wrought ground and the plaster, as shown in figures 903 and 906. Where the architrave is wide it is often built up of two pieces, the door jamb, if heavy, being used as one piece. Into this a second piece is tongued, and is fixed on its outer edge only, allowing freedom to shrink or expand, and generally forming a double-faced architrave, as shown in figures 905 and 907; thirdly, the finishings may consist of pilasters and an entablature or over-door, as shown in the *Advanced Course.*

*Sashes and Frames.*—These may be classified as (*a*) vertical sliding sashes, (*b*) casement, (*c*) lantern lights, and (*d*) skylights. The (*a*) section includes the double-hung, three-light, and circular-headed sashes; (*b*) section in this work includes sashes

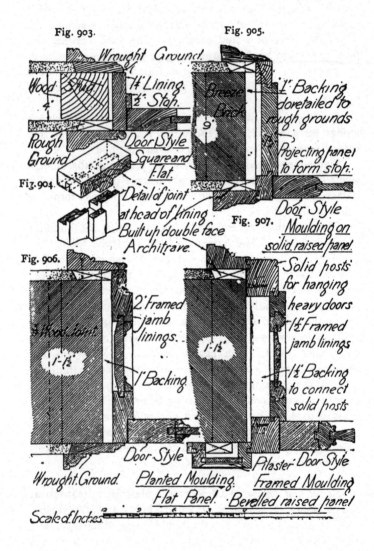

Fig. 903.

Wrought Ground.

Wood Stud

4"

1¼" Lining.
½" Stop.

Rough Ground

Fig. 904.

Door Style
Square and Flat.

Detail of joint
at head of lining
Built up double face
Architrave.

Fig. 905.

Breeze Brick
9

1" Backing
dovetailed to
rough grounds

1½" Projecting panel
to form stop.

Door Style
Moulding on
solid raised panel.

Fig. 907.

Solid posts
for hanging
heavy doors

1½" Framed
jamb linings

1½" Backing
to connect
solid posts

Fig. 906.

⅜" Wood Joint

1"-1½"

2" Framed
jamb
linings.

1" Backing

1"-1½"

Door Style
Planted Moulding.
Flat Panel.

Wrought Ground.

Pilaster  Door Style
Framed Moulding
Bevelled raised panel

Scale of Inches

hung on centres, hospital lights, casement, and casement with sliding upper sash; (*c*) and (*d*) includes all forms of lantern lights and skylights, for which refer to the author's *Advanced Course.*

*Vertical Sliding Sashes.*—These may be single or double-hung, the latter being the method adopted permitting more efficient ventilation. In this system there are always two sashes; in the single-hung the top sash usually is fixed, in the double-hung both arranged to slide. The sashes slide in a vertical plane in grooves formed in the casings. They are suspended by flax cords passing over axle pulleys and counter-balanced by iron or lead weights. This renders it easy to raise heavy glazed sashes. The advantages of this form of sash is that all movement takes place in its own plan space and does not interfere with the draping of windows; also the arrangement lends itself to be constructed wind and watertight.

A frame consists of pulley styles, head and sill. The pulley styles are tongued to the head, the joint being secured by nailing through the head; they are housed to the sill and secured by wedging or nailing. The pulley styles are grooved in the centre to receive parting beads to form the grooves for the sashes to work in, as shown in figures 908 to 913. A portion about eighteen inches long is detached from one side of each style; this is known as the pocket piece, its purpose being to form an access to the weights. The sill is best prepared from oak or teak to resist dampness. It is prepared from stuff about three inches in thickness, sunk, weathered and throated on its top surface to throw off water, and grooved on its underside to receive an iron water bar. The sides of the frames are prepared from $\frac{7}{8}$-inch linings, fitted, grooved, tongued, and nailed to the pulley styles and head, the oak sill cut and sunk to receive them. The outside linings project $\frac{1}{8}$ inch beyond the pulley styles and head, the projecting portion forming one side of the groove in which the top sash

Fig. 908.                    Fig. 909.   Fig. 910.

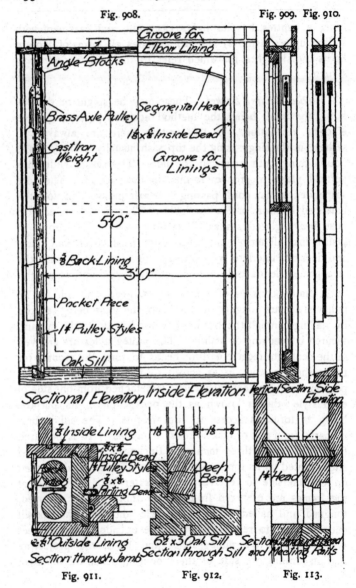

Groove for Elbow Lining

Angle Blocks

Brass Axle Pulley

Segmental Head

$1\frac{6}{8} \times \frac{6}{8}$ Inside Bead

Cast Iron Weight

Groove for Linings

5'0"

$\frac{3}{8}$ Back Lining

3'0"

Pocket Piece

1½ Pulley Styles

Oak Sill

Sectional Elevation    Inside Elevation.   Vertical Section. Side Elevation.

$\frac{7}{8}$ Inside Lining

$\frac{5}{8} \times \frac{5}{8}$ Inside Bead

$1\frac{6}{8}$   $1\frac{6}{8}$   $\frac{3}{8}$   $1\frac{6}{8}$   $\frac{7}{8}$

1½ Pulley Styles

$\frac{3}{8} \times \frac{3}{4}$ Parting Bead

Deep Bead

1½ Head

$\frac{6}{8} \times \frac{1}{3}$ Outside Lining

Section through Jamb

6½ × 3 Oak Sill

Section through Sill and Meeting Rails

Section through Head

Fig. 911.         Fig. 912.         Fig. 113.

works. The inside lining is made flush with the pulley style, the side of the groove for the lower sash being made by a movable bead fitted about the frame. The inside linings should be grooved for the reception of elbow linings. The

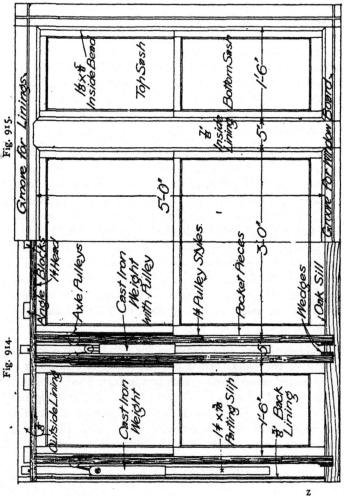

Fig. 915.

Fig. 914.

Groove for Linings.

Groove for Window Board.

18 × ⅝″ Inside Bead

Top Sash

Bottom Sash

1′-6″

⅞″ Inside Lining

5″

5′-0″

3′-0″

Angle Blocks

¼″ Head

Axle Pulleys

Cast Iron Weight with Pulley

¼″ Pulley Styles

Pocket Pieces

Wedges

Oak Sill

Outside Lining

3″

Cast Iron Weight

1¼ × ⅞ Parting Slip

1′-6″

⅜″ Back Lining

z

dotted lines on figure 912 show the usual method of fixing
bead to the sill, the firm lines a deep bead allowing the sash
to be raised for the admission of air through the meeting
rails. The joints of the sashes are shown in figures 916 to
918. The meeting rails are made to fit tightly with a splayed
joint to prevent rattling when the sashes are closed, as shown
in figure 913.

The weights in the casings are separated by a thin fillet
known as a parting slip, suspended from the head, and the
casing is enclosed at the back by a thin lining known as a back
lining, one edge of which is let into a groove in the inside
lining, the other edge being nailed to the back edge of the
outside lining.

*Three Light Windows.*—They are constructed in a similar
manner to ordinary double-hung sash frames, with the exception that they have two mullions, the centre light usually being
made larger than the two side lights, which are of equal width.
The two mullions are made hollow to contain the weights,
only light weights being needed for the three pairs of sashes,
the arrangement being as follows :—The weights in the mullions
have a pulley at their upper end, about which the cord that
suspends it is passed ; this cord is taken through the axle
pulleys and secured to the sash on each side of the mullion.
Thus the one weight in each mullion does for two sashes, as
shown in figures 914 and 915.

*Circular-headed Sash Frame.*—The circular head is formed
in one of two ways : first, a piece of veneer is bent about a
cylinder and staves about $1\frac{1}{4}$ inch by 1 inch are fitted and glued
to the back of the veneer. They are fitted at the springing and
screwed to the pulley styles as shown in figure 919 ; secondly,
the head is built up of a series of vertical laminæ glued and
nailed or screwed to each other, the soffit being wrought. They
are fitted and fixed to the pulley styles as in the previous case.
The parting bead is carried round the soffit, a block is fitted

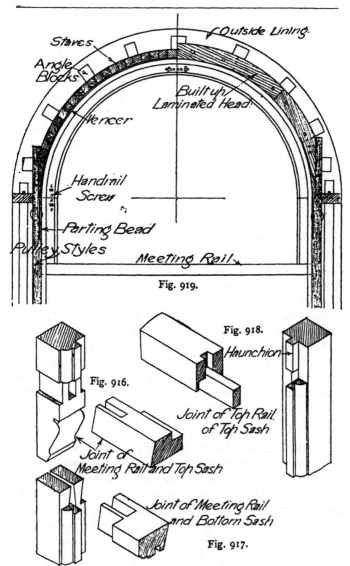

Staves

Angle Blocks

Outside Lining.

Built up Laminated Head.

Hencer

Handrail Screw

Parting Bead

Pulley Styles

Meeting Rail

Fig. 919.

Fig. 918.

Haunchion

Joint of Top Rail of Top Sash

Fig. 916.

Joint of Meeting Rail and Top Sash

Joint of Meeting Rail and Bottom Sash

Fig. 917.

z 2

inside the casing at the springing level to receive the parting slip. The remaining construction is similar to the preceding frames. The joints in the circular sash-head are dowelled and secured with handrail screws, as shown in figure 919.

*Sashes hung on Centres.*—For factories, stables, lantern lights, and other positions where economy is an object, or in positions that would be inconvenient to be hung with weights on cords, sashes hung on centres in solid frames are employed, and are especially suitable to obtain light and air in high positions that are not easy of access, as these sashes are hung on iron centres or pivots, higher up than their centres of gravity to allow for the weight of the cord attached to it, so that the window may be closed by its own weight from below by simply taking the cord off the cleat on which it is usually wound. In public buildings or in buildings that admit of the expense, metal lever fanlight openers are employed as being more certain and efficient in the action. These sashes require carefully fitting and the beads cut as in figure 923, otherwise the wind and rain will find a passage through the space between the beads fixed on the sashes and the frames. To admit sash being taken out, grooves are sunk in the beads, as shown in figure 923.

Sometimes the outside edge of the style is plough grooved, so that when the sash is opened horizontally it may be taken out of the frame. This avoids the grooving of the beads that are fixed on the sashes, and, if the groove on the edge is no objection, greatly simplifies the work.

*Hospital Lights.*—These consist of fanlights; they are used for hospitals and factories, and are suitable for buildings, the rooms of which are desired to be ventilated without draught, where the appearance is a secondary consideration. They consist of a number of sashes fitted into a solid rebated frame, the lowest of which is hung with butts fixed to its bottom rail and to the sill of the frame. The remainder of the sashes are fitted with a rebated joint, one above the others, as shown in

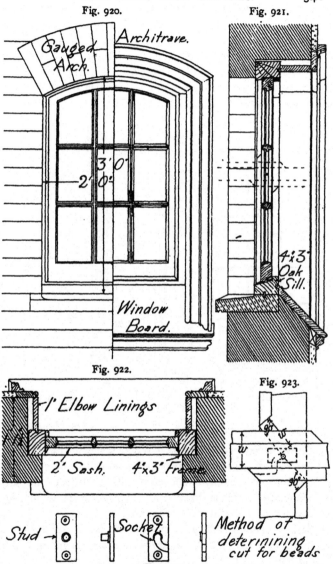

Fig. 920.

Fig. 921.

Fig. 922.

Fig. 923.

Fig. 924.

Figs. 920–924.    Sash rotating upon Centres.

figure 927, and are hung by pivots fixed to the two lowest corners of the sashes and the centres fixed in jamb posts. A lining is fixed to each post, at right angles to the frame (the window linings in thick walls are used for this purpose), having beads fixed in an inclined position, on which the sashes rest when open, thus preventing a down draught.

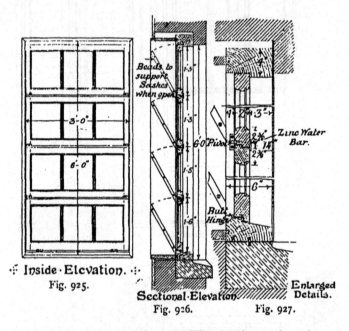

·:· Inside · Elevation. ·:·
Fig. 925.

Sectional·Elevation.
Fig. 926.

Enlarged Details.
Fig. 927.

*Casement Sashes and Frames.*—Where sashes are hung by one of their vertical edges, they are called casement sashes. It is a method most commonly employed on the Continent; they are convenient to be used as doors. The sashes and frames may be constructed to open outwards or inwards.

Sashes opening outwards present no great difficulty in obtaining watertight joints, but have this objection, that if a sash should happen to be left open, without being secured, the wind may act upon it and smash the glass, which in high situations would be dangerous. The joint between the bottom rails of sashes and sill in sashes opening inwards are very difficult to get wind and watertight; but for this objection they would undoubtedly be more frequently used, as they possess the advantage of being much easier to clean than any other method of fixing sashes, and the danger consequent thereon is reduced to a minimum.

*Sashes opening outwards.*—The members of frames for these sashes are rebated from the outside. With the exception of this and the joint at the sill, they are similar to the members of sashes opening inwards.

*Sashes opening inwards.*—Figures 928 to 935 give an illustration of a pair of casement sashes, opening inwards, in solid frames, showing a metal water bar on bottom rails; it is very effectual in withstanding a driving rain. At the seaside or in places exposed to the weather the use of a metal water bar becomes a necessity; in sheltered situations a wood water bar may be used. The outer meeting style has a projecting moulding worked on it, or in inferior work this may be planted on, to protect the vertical joints between the meeting styles. The hanging styles have a tongue worked on to fit into grooves sunk in the posts, and they are secured on the inside by bolts. Sometimes an espagnolette bolt is employed, which combines handle, top and bottom bolts, and the sash may be loosened or closed by one turn of the handle.

*Hook Joint.*—Figure 931 shows a draught and weather-resisting arrangement worked upon the meeting styles, and is termed the hook-joint.

*Fanlights.*—Where the frames for casements are large a transome is framed to the posts, the lower portion of the frame being fitted with a pair of casement sashes hung as

doors, whilst the upper opening is fitted with a sash hung
by its top or bottom rail according as to its opening, either
outwards or inwards : these are termed fanlights.

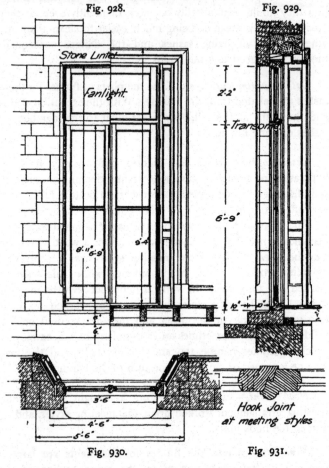

Fig. 928.        Fig. 929.

Fig. 930.        Fig. 931.

*Casements with Sliding Upper Sash.*—These are fully
described and illustrated in the *Advanced Course.*

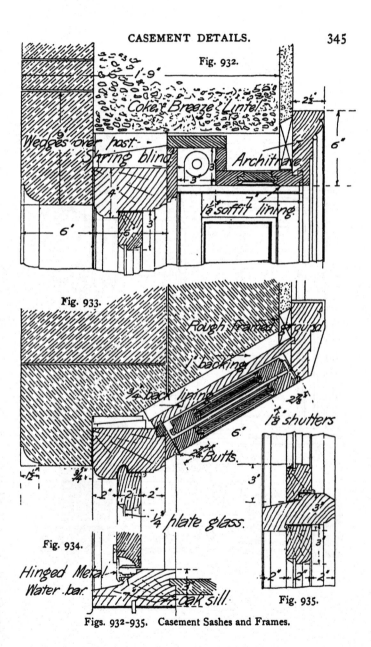

Fig. 932.

1'-9"

2½"

Coke Breeze Lintel.

6"

Wedges over post.

Spring blind.

Architrave.

1⅛ soffit lining.

7"

3"

3"

6"

Fig. 933.

Rough framed ground.

1" backing.

¾" back lining.

2⅜"

6"

1⅛ shutters.

2⅜ Butts.

3"

¼ plate glass.

Fig. 934.

Hinged Metal Water bar.

Oak sill.

Fig. 935.

Figs. 932-935. Casement Sashes and Frames.

Fig. 936.

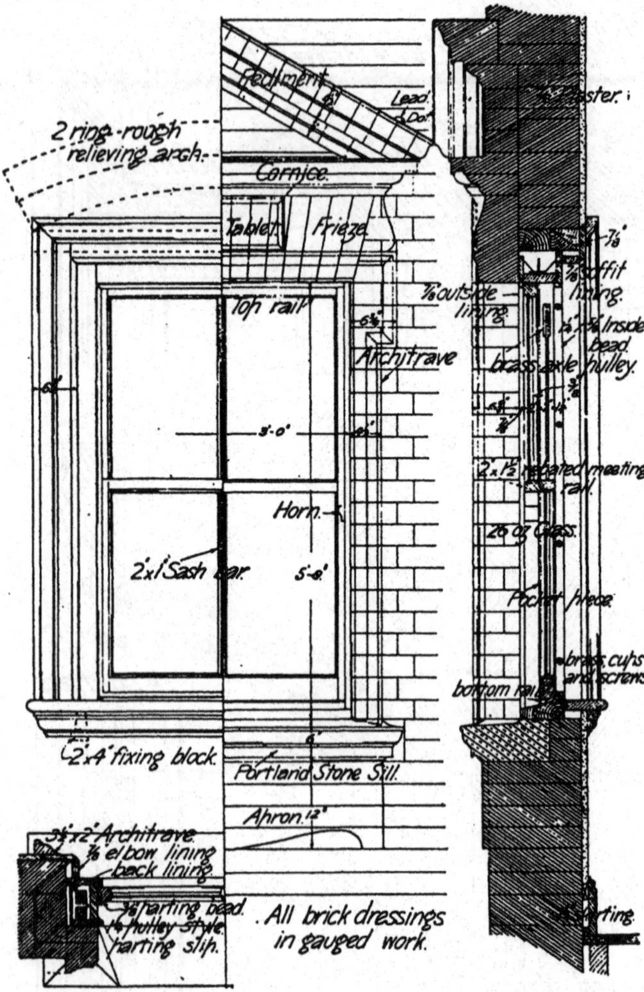

2 ring rough relieving arch.

Pediment

Lead.

Pilaster.

Cornice.

Tablet. Frieze.

Top rail.

Architrave

6"

3'-0"

Horn.

2"x1" Sash bar.

5'-8'

6"

¾ outside lining.

¾ Soffit lining.

½ x ¾ Inside bead.

brass axle pulley.

2 x 1½ beaded meeting rail.

26 oz Glass.

Pocket piece.

brass cups and screws.

bottom rail.

2"x4" fixing block.

Portland Stone Sill.

Apron. 1½"

3½x2"Architrave.

¾ elbow lining.

back lining.

⅝ parting bead.

⅜ pulley stile.

parting slip.

. All brick dressings in gauged work.

Fig. 938.

Fig. 937.

Figs. 936–938.   Double-hung Sashes and Cased Frame.

*Window Finishings.*—These usually consist of a window board, elbow linings, rough or wrought grounds and architraves.

These may be classified as, first, plain; secondly, framed linings; thirdly, with shutters, boxing or lifting.

Figures 936 to 938 illustrate the finishings with plain linings and consist of window board, elbow linings, rough or wrought grounds and architrave moulding. These are used for thin walls.

Secondly, if the linings exceed seven inches in width they are framed and panelled; to facilitate the dispersion of the light internally these linings may be splayed.

Thirdly, to prevent ingress through the windows, the latter are frequently fitted with wood shutters, which are either boxing, lifting or revolving.

*Boxing Shutters.*—These consist of a number of narrow pannelled wood frames the full height of the window, as shown in figures 928 to 935. They are rebated together on their vertical edges, and connected with each other by hinges, the two side shutters being hung to the vertical members of the sash frame; this allows them to be folded back one against the other into a small space arranged to receive them at the side of the opening termed the box, the shutters when closed having the appearance of ordinary framed linings; when open the recess is exposed, as shown in figure 933.

*Lifting Shutters.*—These are fully illustrated in my Advanced Course.

*Soffit.*—The soffit is usually framed to match the framed linings. Where there are folding shutters it is necessary to arrange a recess in the soffit, termed the blind box, sufficiently large to contain the blind, as shown in figure 932.

*Fixing of Sash and external Door Frames.*—To comply

with the Model Bye Laws all external frames should be
fitted in reveals. The inside face of the reveal, in other
than rubbed brickwork, is usually irregular. The frames
are bedded on plaster screeds against these reveal faces.
The underside of the wood sill and the upper side of the
stone sill are thickly coated with white-lead or thick paint,
and an iron tongue, termed a water bar, is inserted in a
groove made in the stone sills to receive it. The frame is
placed on the stone sill, carefully levelled, pressed against
the plaster screeds, which have been plumbed; the frame is
then nailed or screwed to wood joints, plugs or other fixings.
The elbow linings are then fixed, thus effectually preventing
any movement in the frame. Solid door frames that have
no wood sill are first scribed to the stone sill, an iron dowel
is let into the bottom of the post, and is inserted in a mortice
prepared to receive it in the stone sill. The solid frames
are wedged downwards from the lintel and fixed to the
wood joints as before. In places such as stables, subject
to considerable dampness, cast-iron shoes about ⅜-inch thick
and 6 inches deep are made to enclose the feet of posts,
as shown in figure 887. They are sunk flush into the feet
of the posts, they are cast with a joggle which is let into
the stone sill, and run in with Portland cement grout as
before.

*Tools.*—The ordinary tools of the carpenter and joiner
comprise setting out, cutting, boring and miscellaneous tools.
Figures 939 to 949 show the setting-out tools, the square,
bevel, compass, gauges, rule, level, plumb rule, chalk line, striking
knife and straight edge; figures 950 to 977 show those used
for cutting purposes, the axe, adze, chisels, bench planes, rebate
planes, fillister, plough, shoulder, bull nose, and bead planes;
spokeshave, table saw, pad saw, hand saw, tenon saw and bow
saw. Figures 978 to 989 show the boring tools, consisting of
brad-awl, gimlets, auger, centre bit, nose bit, spoon bit, quill bit,

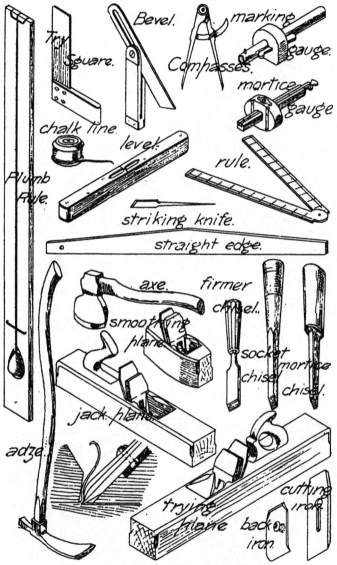

Bevel.

Tri square.

marking gauge.

Compasses.

mortice gauge

chalk line.

level.

rule.

Plumb Rule.

striking knife.

straight edge.

axe.

firmer chisel.

smoothing plane.

socket chisel.

mortice chisel.

jack plane.

adze.

trying plane.

cutting iron.

back iron.

Figs. 939–960. Carpenters' and Joiners' Tools.

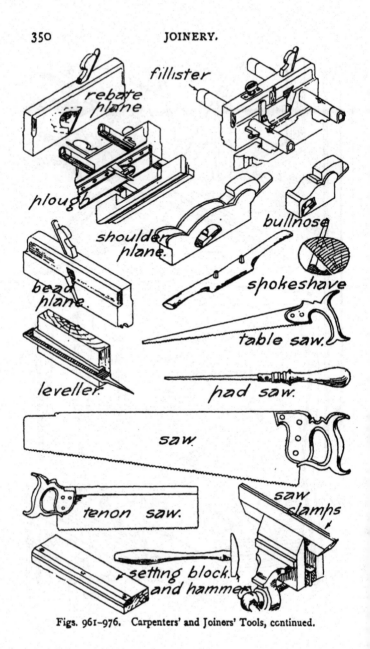

Figs. 961–976.  Carpenters' and Joiners' Tools, continued.

screw-driver bit, metal countersink, rose countersink, wood countersink, and the brace. Figures 990 to 1,004 show the miscellaneous, and include the hammer, mallet, screw-driver, pincers, oil stone, leveller, clamps, setting block and hammers for the sharpening of saws, bench, handscrews, cramps, hold-fast, mitre cut and shoot, mitre-cutting machine.

For a detailed explanation of each tool the reader is referred to the Author's " Carpentry Workshop Practice." *

*Hinges.*—Figures 1005 to 1114 show the types of hinges in common use.

Figure 1005 shows the butt hinge used for hanging doors that are sufficiently thick to have the hinge let in to the edge of the door.

Figure 1006 is the rising butt hinge; the wearing surfaces of the two leaves are formed as helical curved surfaces, the pur-pose being to cause the door to rise as it is opened, and thereby clear any small obstruction, such as carpets, etc., also to allow the door to close automatically.

Figure 1007 shows the pin hinge, used for the hanging of heavy doors. The centre pin may be withdrawn; this renders it convenient to fit the two halves of the hinge without lifting the heavy door at each time it is offered up during fitting.

Figure 1008 illustrates the back flap hinge, employed where the doors, shutters, etc., to be hung are thin, and consequently no room on the edge for a butt hinge.

Figure 1009 shows the cross garnet, largely used in the hanging of thin match-board doors. They are screwed on the faces of the doors.

Figure 1010 shows a strap hinge, used in heavy external work, such as stable doors, gates, etc. They are bolted on to the face of the work. The two parts are loose, thus permitting the work to be lifted off.

Figure 1011 shows a counterflap formed in three parts and

* Cassell & Co., Ltd. ; price 1*s*. 6*d*.

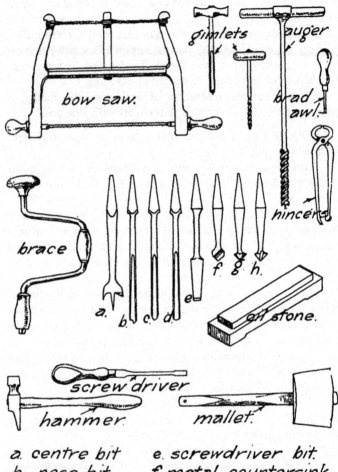

a. centre bit
b. nose bit.
c. spoon bit
d. quill bit.

e. screwdriver bit.
f. metal countersink
g. rose countersink.
h. wood countersink.

Figs. 977–994. Boring and Miscellaneous Tools.

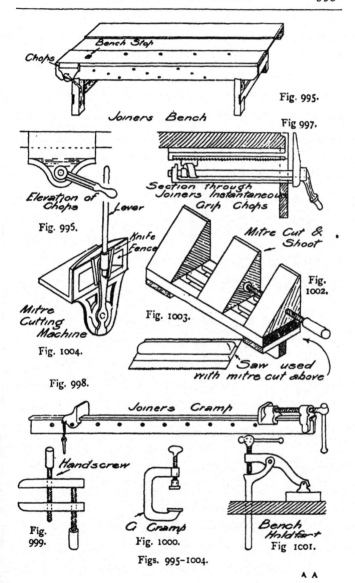

Bench Stop

Chops

Fig. 995.

Joiners Bench

Fig 997.

Elevation of
Chops

Section through
Joiners Instantaneous
Grip Chops

Fig. 995.

Lever

Knife
Fence

Mitre Cut &
Shoot

Fig.
1002.

Fig. 1003.

Mitre
Cutting
Machine

Fig. 1004.

Saw used
with mitre cut above

Fig. 998.

Joiners Cramp

Handscrew

G Cramp
Fig. 1000.

Bench
Holdfast
Fig 1001.

Fig.
999.

Figs. 995–1004.

A A

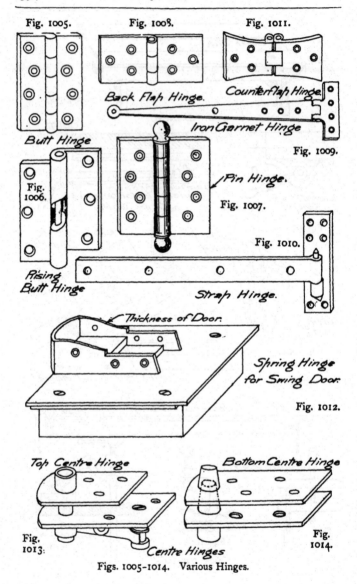

Fig. 1005.

Fig. 1008.

Fig. 1011.

*Back Flap Hinge.*

*Counterflap Hinge.*

*Butt Hinge*

*Iron Garnet Hinge*

Fig. 1009.

Fig. 1006.

*Pin Hinge.*

Fig. 1007.

Fig. 1010.

*Rising Butt Hinge*

*Strap Hinge.*

*Thickness of Door.*

*Spring Hinge for Swing Door.*

Fig. 1012.

*Top Centre Hinge*

*Bottom Centre Hinge*

Fig. 1013.

*Centre Hinges*

Fig. 1014.

Figs. 1005–1014.  Various Hinges.

having two centres. This arrangement allows the two leaves to be folded back to back.

Figure 1012 shows a spring hinge for a swing door. It consists of a metal box, containing a strong steel spring; this actuates a metal socket into which the heel of the door is fitted; the spring always tends to keep the socket in a central position and the door closed.

Figure 1013 shows the top centre hinge used in connection with the previous example. It consists of two parts, an adjustable pivot which is let into the frame and a socket let into the head of the door. The centre is arranged to be adjusted by means of a screw worked from the face at the opposite end of the plate from the pivot. The pivot and screws are attached to the opposite ends of a rocker fixed to a centre on the back of the plate, as shown in the figure 1013. To hang a swing door, the bottom spring hinge and the top centre are fitted to the door. The door is then pushed into the bottom socket and revolved on the bottom centre to a position at right angles to that occupied by the door when closed; the top socket is then carefully centered under the pivot, the screw attached to the rocker is then turned, causing the centre to protrude into the socket, thus fixing the door.

Figure 1014 shows a bottom centre hinge used for swing doors where it is unnecessary for the door to close automatically. It consists of a metal pivot and socket.

*Locks and Fastenings.*—Figure 1015 shows a tower bolt used for fixing to the back faces of external doors.

Figure 1016 illustrates a flush bolt. These are let into the doors either upon a face or an edge of a door, and flush with the surface. They are used where the projecting tower bolt would be objectionable. Figure 1017 illustrates the espagnolette bolt, used for securing casement windows. They consist of two long bolts, one of which secures the top and the other the bottom of the sash. Both bolts act simultaneously by

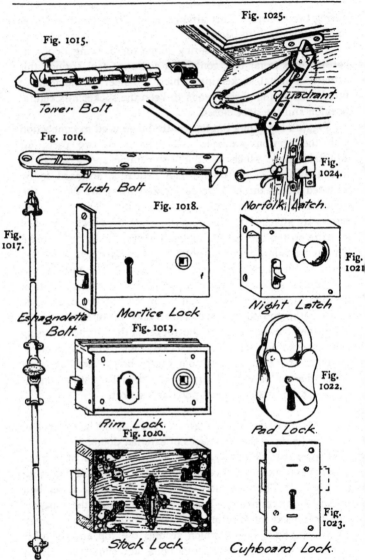

Fig. 1025.

Fig. 1015.

Tower Bolt

Quadrant.

Fig. 1016.

Flush Bolt

Fig. 1024.

Norfolk Latch.

Fig. 1017.

Espagnolette Bolt.

Fig. 1018.

Mortice Lock

Fig. 1021.

Night Latch

Fig. 1019.

Rim Lock.

Fig. 1022.

Pad Lock.

Fig. 1020.

Stock Lock

Fig. 1023.

Cupboard Lock.

Figs. 1015-1025.   Locks and Fastenings.

turning the handle or lever in the centre. Figure 1018 shows a mortice lock. These are employed for doors 2 inches and above in thickness. They are let into the edges of doors. Figure 1019 shows a rim lock. These are used for thin doors, and are screwed on to the face and the edge of the door.

Figure 1020 illustrates a stock lock, which consists of a rim lock encased in hard wood. They are used for stables, church work, etc., and are frequently bound with iron of a more or less ornamental character. They are usually dead-shot·locks; that is, to be opened by a key only. Figure 1021 illustrates a night latch, used for external doors. They are a form of rim lock, opened on the inside by a drawback knob and from the outside by means of a key. Figure 1022 shows a padlock and hasp used for securing doors in temporary and common work. Figure 1024 shows a Norfolk latch used in inferior doors; but ornamental types of same are frequently used in ecclesiastical work. Figure 1025 illustrates a quadrant and wheels employed for the opening of skylights and sashes in positions not easily accessible.

# CHAPTER XII.

# PLUMBING.

_Coverings._—External plumbing work consists of covering roofs entirely with sheet lead, or those parts on slated or tiled roofs which cannot be made watertight with the latter materials. For flats, gutters, flashings, etc., lead far surpasses any other material, and combines lasting and waterproof properties. In towns, lead coverings are more economical than zinc. Zinc lasts well in the pure air of the country, but when exposed to the deleterious effects of sea air, or the smoke-laden atmosphere of large towns, it is readily destroyed.

Sheet lead, for external work, is obtainable in two forms, cast and milled.

_Cast Lead._—The lead is melted and run into sand-covered moulds to form sheets of the required size, and may be obtained up to 16 feet in length and 7 feet in width. Cast lead is very durable under great changes of temperature, and, being very soft, is suitable for positions where subjected to traffic, but it is subject to flaws and sand holes.

_Milled Lead._—Slugs or thick cakes of lead are cast and then passed between rollers to reduce them to the desired thickness. Sheets may be obtained in lengths of 35 feet and widths up to 9 feet. Milled lead is more uniform in thickness than cast lead, and is freer from the sand holes and flaws, but is not nearly so durable for roof coverings, nor is its resistance to wear equal to that of good cast lead.

_Laying Sheet Lead._—The boarding to receive sheet lead should be perfectly smooth. All the joints should be traversed by a plane to take away any projecting arrises which would, sooner or later, show through the lead, especially after wet weather. There should be no sharp angles, and all projections should be rounded off. The grain of the boarding on flats and gutters should be laid in the direction of the current, in order that any corrugation formed by the casting of the boards should not retain pools of water.

The soles of gutters are recommended to be formed by narrow boards 4½ inches wide, nailed with the heart side upwards, so that the edges will press tightly against the bearers if the boards cast; the boards should be well nailed, about 1¼ inch from each edge. The width of joints, caused by shrinkage, is obviously much less in narrow than in wide battens.

In fixing lead, the sheets should be free to contract or expand, or pieces used be of such small dimensions that the contraction or expansion will be inappreciable; this is practically satisfied at ordinary temperatures when the lengths of the pieces do not exceed 7 feet. The joints made between the edges of lead must be arranged so that no water can pass through, or the covering be blown up by the wind. If the above conditions are not taken into consideration the force of expansion and contraction will cause the sheet lead to slide down if fixed in inclined positions, or buckle and rise in the centre if it be laid on a flat.

*Expansion of Metals.*—The following table gives the expansion of the metals used in construction, due to heat, between the temperatures of 32° and 212° Fahr. To find the expansion due to an alteration of temperature of one degree, divide the tabulated number by 180:—

| | | | | | | |
|---|---|---|---|---|---|---|
| Cast Iron | ... | ... | ... | ... | 1 part in | 889 |
| Wrought Iron | ... | ... | ... | ... | 1 ,, | 819 |
| Steel, Untempered | ... | ... | ... | 1 ,, | 927 |
| Steel, Tempered | ... | ... | ... | 1 ,, | 807 |
| Lead | ... | ... | ... | ... | ... | 1 ,, | 349 |
| Zinc | ... | ... | ... | ... | ... | 1 ,, | 340 |

*Specification of Lead Coverings.*—The following are the weights of lead recommended for the various parts of external coverings :—

| | | |
|---|---|---|
| Roofs, Flats, and Main Gutters ... | 7 lb. to 8 lb. | lead |
| Hips, Ridges and Small Gutters ... | 6 lb. to 7 lb. | ,, |
| Flashings ... ... ... ... | 5 lb. | ,, |
| Cisterns, bottom ... ... ... | 7 lb. to 8 lb. | ,, |
| ,, sides ... ... ... | 6 lb. to 7 lb. | ,, |
| Sinks, bottom ... ... ... | 7 lb. to 8 lb. | ,, |
| ,, sides ... ... ... ... | 8 lb. to 10 lb. | ,, |
| Soil Pipes ... ... ... .. | 8 lb. to 10 lb. | ,, |

Lead is usually described and specified by its weight in pounds per superficial foot. The following table gives the thickness of milled lead in common use. The thickness, it may be noticed, is nearly 17 thousandths of an inch for each pound in weight per superficial foot :—

| Weight in lbs. per foot super. | Thickness in inches. | Weight in lbs. per foot super. | Thickness in inches. |
|---|---|---|---|
| 1 | 0·017 | 7 | 0·118 |
| 2 | 0·034 | 8 | 0·135 |
| | 0·051 | 9 | 0·152 |
| 4 | 0·068 | 10 | 0·169 |
| 5 | 0·085 | 11 | 0·186 |
| 6 | 0·101 | 12 | 0·203 |

The lengths of pieces of lead generally used in practice for gutters or flats should not be more than 9 feet, and the fall or inclination of the gutters or flats should not be less than 1 inch in each 9 feet of length. Cover and step flashings should not exceed 6 feet, and ridge pieces not more than 7 feet in length. Dormer cheeks, if very large, should be put on in two or more pieces. Large dormer tops should have a roll fixed upon them, but for small dormers the lead can expand over the edges.

*Nailing.*—Copper nails should be used, but nailing should not be resorted to unless absolutely necessary.

*Bossing.*—That is, working the lead to the required form with boxwood tools—is preferable to soldering such angles as the returned ends of gutters, drips, or cesspools, or any position where the perpendicular part does not exceed 6 inches in height.

*Lead Tacks.*—These are narrow strips of lead from 2 to 4 inches in width, used for fastening the free edges of flashings, ridge coverings, apron pieces, etc. They are termed *tingles* in the north of England.

*Laps* or *Passings.*—The distances pieces of lead lap over the adjoining pieces in aprons for gutters, stepped flashings, ridge coverings and other situations where it would be unwise to have the lead in one continuous length. The usual length of passings is 4 inches for upright and 6 inches for horizontal work.

*Raglets.*—The grooves or chases, usually 1½ inch deep, cut into stone walls to receive the upper edges of the lead flashings.

*Fixings for Flashings.*—Where fixed to brick walls a small slightly inclined fillet of cement should be made on the lower side of the horizontal joint, and the lead flashing secured in it by lead wedges varying from 3 to 9 inches apart. The open joints between the wedges should be pointed in cement in brick walls, or with mastic, or be run with lead if in stone walls.

*Cover Flashings.*—The name given to the lead coverings fixed over the turned-up parts of lead gutters or flats : they are usually about 6 inches wide, the object being to keep water from passing between the turned-up lead and the wall, and to allow the covered sheet freedom to contract or expand.

*Apron Pieces.*—The name applied to the lower horizontal flashing of a chimney shaft, skylight, dormer, or wall penetrating a roof, as shown in figures 1047 to 1050.

*Tilting Fillets.*—These are pieces of wood, triangular in section, and are used where inclined surfaces abut against walls in order to tilt the slates, and so convey the water away from the walls. Wood fillets are also fixed under the eaves courses of slates, so that they may lie close, and thus prevent the wind getting under them, as shown in figures 1046 to 1053.

*Cement Fillets.*—To save the expense of lead flashings, fillets composed of equal portions of Portland cement and sand are run along the junction of lean-to roofs with walls, but the cement sometimes shrinks or breaks away, resulting in an open joint, thus failing to answer the purpose for which it is intended. Zinc soakers, together with cement fillets, are more effectual and are extensively used in cheap buildings, but are not nearly so durable as lead soakers and flashings.

*Joints.*—The joints most extensively used for lead coverings for external work may be classified as follows :—

*Joints across the Flow or Current.*—Lap and drip joints.

*Joints parallel with the Flow.*—Rolls, hollow rolls, and seams or welts.

*Lap Joints.*—These are horizontal joints on inclined surfaces of pitched roofs covered with lead. The boarding should be placed at right angles to the inclination. The sheets of lead should be placed between the rolls, the upper ends being turned over the edge of the boarding and nailed, so that the sheets are secured along the entire length of the top edge. This effectually resists the tendency to crawl down the slope of the roof, and is far better than nailing to the face of the boarding. The lower edge of the upper sheet laps over the top edge of the lower one. If the covering is on a vertical surface, as

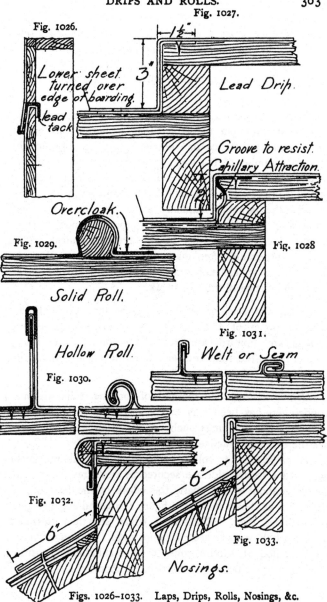

Fig. 1027.

Fig. 1026.

$1\frac{1}{2}''$

$3''$

Lead Drip.

Lower sheet turned over edge of boarding.

lead tack

Groove to resist Capillary Attraction.

Overcloak.

Fig. 1029.

Fig. 1028

Solid Roll.

Fig. 1031.

Hollow Roll.

Welt or Seam

Fig. 1030.

Fig. 1032.

$6''$

$6''$

Fig. 1033.

Nosings.

Figs. 1026–1033.    Laps, Drips, Rolls, Nosings, &c.

shown in figure 1026, a 4-inch lap would be sufficient; but if
the surface has an inclination of not less than 45°, a 6-inch lap
is necessary. If the inclination is less than 45° the lap would
require to be increased considerably, or a drip would have to
be formed.

*Drips.*—Large flat roofs and very long gutters are constructed of a number of plane surfaces slightly inclined and
raised a short distance one above the other, forming when
finished a number of low steps called drips, which should not
be at a greater distance than 9 feet apart. Drips should be
made for preference 3 inches in depth, as shown in figure 1027,
to resist the power of capillary attraction; but when made
2 inches or less, they should have a groove formed, as shown
in figure 1028, to resist that force. For economical reasons,
drips in gutters are often made less than 2 inches deep, but
this results in water being drawn between the laps of the lead
and in the rotting of the woodwork.

*Rolls.*—On flats, or at the ridge or junction of the two
opposite slopes of a gutter or roof, wooden rolls $1\frac{3}{4}$ inch
diameter and upwards are fixed at the joint. Sheet lead is
dressed round the roll, and well into the angles, to obtain a
firm grip, as in figure 1029.

*Hollow Rolls.*—The method of forming rolls without a
wooden core has been very extensively used on steep-pitched
roofs on large buildings, such as cathedrals and abbeys, and
was the common practice in the mediæval ages.

To make a hollow roll, copper or lead tacks about 6 inches
long and 2 inches wide are secured by two screws each to the
boarding, about 2 feet apart, the edges of the lead bay are
turned up, as shown in section figure 1030, and are then ready
to be folded over. The folded edges are then dressed to
enclose a hollow, as in section figure 1030. Although this is a
good method of forming rolls, it is not suitable for positions
where any traffic is likely to come upon them.

*Nosings.*—At the boundaries of flats adjoining sloping roofs, or at the intersection of two differently inclined surfaces, as at the curb of a mansard roof, the lead covering may be terminated as a nosing, the object being to secure the ends of the lead laid on the flat, and yet allow them to shrink or expand freely. Figure 1032 shows a flashing of 6-lb. lead, with the lower half laid on the slates, and the upper part dressed against the vertical boundary of the flat. A wood nosing is nailed over the top edge of the lead, thus firmly securing the flashing, and the ends of the lead bays are then dressed round the nosing. The lower edge of the flashing is secured from being torn up by the wind by means of lead tacks, which are better than soldered dots, and allow the flashing to expand and contract.

Figure 1033 shows a flashing secured by copper nails, instead of being covered by a wood nosing, and flat welted nosing turned, for preventing the wind getting under the ends of the lead bays.

Figure 1034 is an elevation, and 1035 a section showing the method of forming nosing for the curb of a mansard roof.

*Welts or Seams.*—The joints for sheet lead when running with the current on vertical surfaces may be the flat welts or seams. These are made by fixing lead or copper tacks about 2 feet apart at the junction of the lead sheets; the edges of the bays are bent up and turned over together, and then dressed flat, as shown in figure 1031. For flat pitched, or horizontal surfaces, seams are not so good as rolls.

*Ridge Coverings.*—Six-lb. lead is the thickness usually adopted for ridge coverings; the lengths of the pieces should not exceed 7 feet.

The lead should be dressed over a wood ridge roll, which should not be less than 2 inches in diameter, and may be fixed in one of two ways: First, by double-pointed nails, in which case the lead tacks that support the lower edge of the ridge coverings pass beneath the roll, and are secured by it, as

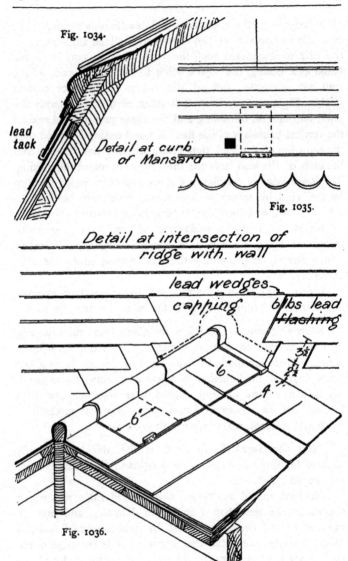

Fig. 1034.

lead
tack

Detail at curb
of Mansard

Fig. 1035.

Detail at intersection of
ridge with wall

lead wedges

capping                    bibs lead
                           flashing

$3\frac{1}{2}$
$1\frac{1}{2}$

6"                    4"

6"

Fig. 1036.

Figs. 1034–1036.   Nosings and Ridge Coverings.

shown in figure 1037; secondly, by nailing to the ridge piece, as shown in figure 1036, the lead tacks being secured to the ridge piece before the ridge roll is fixed. The lead wings should be dressed close to the bottom of the roll, and extend 6 inches down the slope on either side of the ridge, the free edges being secured by lead tacks.

Lap joints are formed at the junctions of the ridge pieces. This has been objected to on the grounds that water is liable to be drawn between the sheets by capillarity or driven in by the wind, or drain in, if the ridge is not perfectly level. These objections may be obviated by forming a water groove in the lap, the under piece of lead being dressed into the groove. Any water getting between the laps on the sloping sides would fall on the slates and drain off.

*Difference between Valleys and Hips.*—If the upper surfaces of the slopes of two adjacent roofs include an angle less than 180°, or, in other words, form a hollow, it is termed a valley; but if the angle be greater than 180°, thus forming a ridge or projection, it is called a hip.

*Hips.*—These may be made watertight in two ways. First, by dressing a piece of lead, of the required width and not longer than 7 feet, over a roll, and letting the sides lie 6 inches on the slates similar to a ridge. To prevent the lead sliding down the hip, the lowest piece of roll is fixed, the first piece of lead is laid on with its upper end extending 6 inches beyond the upper end of the roll, as shown in figure 1038, the lead is then bossed down. The next length of roll is then fixed in position, and the lead put on with its lowest extremity resting on the upper end of the first piece of lead. The bottom end of the second piece of lead thus laps over the first piece 6 inches. Lead tacks should be fixed about 3 feet to 3 feet 6 inches apart, and the lower piece of the lead to be fixed should be clipped over the upper at each lap joint. Secondly, the hip may be formed by soakers, which may be fixed over a

Fig. 1037.

Fig. 1039.

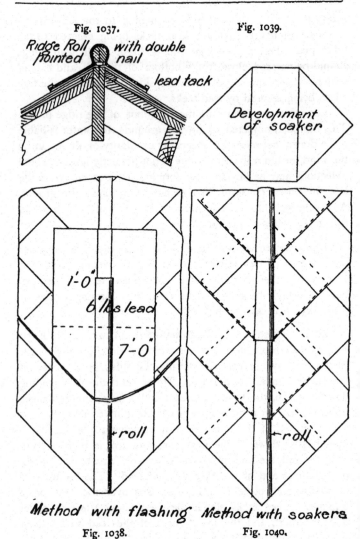

Ridge Roll with double pointed nail

lead tack

Development of soaker

1'-0"

6 lbs lead

7'-0"

roll

roll

Method with flashing　Method with soakers

Fig. 1038.

Fig. 1040.

Figs 1037-1040. Hip Coverings.

roll, as shown in figure 1040, where they are visible, or the roll may be dispensed with, in which case there is a saving of lead. The roll is the best, as it forms a wind guard to the slate edges on the hip.

Figure 1039 shows a plan of a hip soaker.

*Valley Coverings.*—In valleys, the coverings may be arranged to form gutters, small fillets being fixed on the slopes of the roofs to tilt the slates, as shown in section figure 1041.

*Stepped Flashings.*—The joints between sloping roof surfaces and end walls built of brickwork or stone rubble are best protected by means of pieces of sheet lead, called stepped flashings, 6 inches lying on the roof and 6 inches against the wall. The upper edges are turned 1 inch into the raglets or joints of the brickwork prepared to receive them, and are fixed as described in the paragraph on flashings. Figure 1047 refers to this kind of flashing. To set out stepped flashings, first roll out the piece of lead, which should be 13 inches wide, then fold it lengthways in the centre, like the letter ∟. Draw a line, 2½ inches distant from the angle, on the stand-up side. This line is usually called the "water-line." The piece of lead should then be laid in position on the roof, and the horizontal joints in the brickwork transferred to the lead between the top edge and the water-line. Next draw lines from the points where the horizontal lines cut the water-line to the point where the horizontal line immediately above cuts the outer edge of the piece of lead. Draw other lines 1 inch distant, parallel with and above the horizontal lines. The small triangular pieces of lead above these last lines to be cut out with a knife. The pieces between the parallel lines to be folded and wedged into the raked-out joints of the brickwork.

*Raking Flashing.*—The name given when the turned-up edge of a flashing is secured to a chase or raglet, cut parallel to the slope of a roof. This method is adopted for stone walls adjoining sloping roofs.

B B

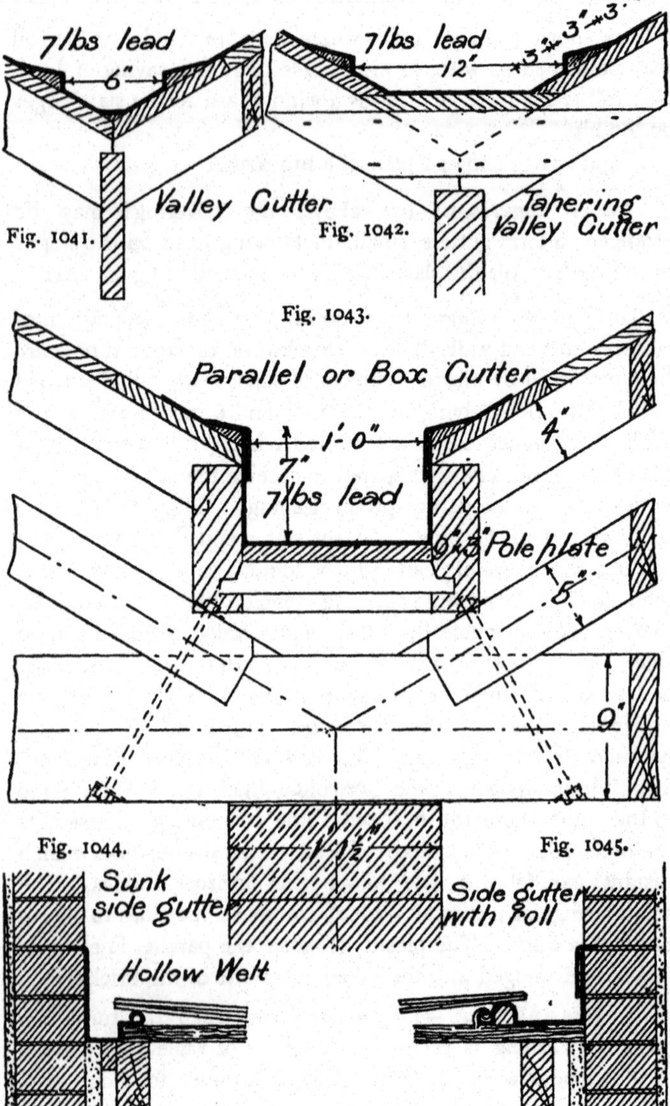

7lbs lead

6"

Valley Gutter

Fig. 1041.

7lbs lead

12"

3" 3" 3"

Tapering
Valley Gutter

Fig. 1042.

Fig. 1043.

Parallel or Box Gutter

1'-0"

7"
7lbs lead

4"

6×5"Pole plate

8"

9"

Fig. 1044.

Sunk
side gutter

Hollow Welt

1'1½"

Fig. 1045.

Side Gutter
with roll

Figs. 1041–1045.  Gutters.

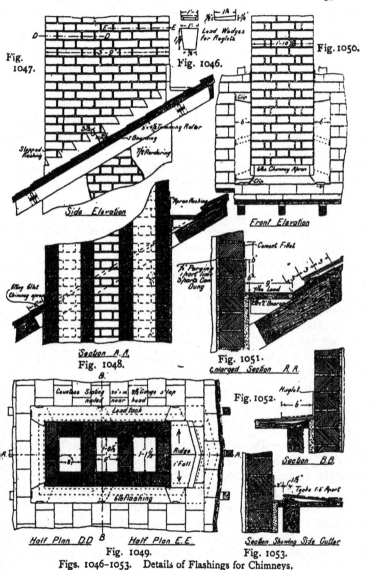

Fig. 1047.

Fig. 1046. Lead Wedges for Raglets

Fig. 1050.

Side Elevation

Front Elevation

Section A.A.
Fig. 1048.

Fig. 1051.

Enlarged Section A.A.
Fig. 1052.

Section B.B.

Half Plan D.D    Half Plan E.E.
Fig. 1049.

Section Showing Side Gutter
Fig. 1053.

Figs. 1046–1053.  Details of Flashings for Chimneys.

*Flashings for Chimney Stacks.*—Figures 1046 to 1053 show the mode of preventing any leakage through the joints on the four sides of a chimney stack, at the upper end by means of a gutter, at the sides by fixing stepped flashings or secret gutters, and at the lower end by an apron-piece fixed with lead wedges and tacks, as shown in figures 1046 to 1050.

*Secret Gutters.*—When flashings are fixed beneath instead of lying on the slates, secret gutters are constructed, as shown by figure 1053.

These secret gutters derive their name from their being hidden from view by the slates. In some situations the slates are carried over the secret gutter so as to nearly touch the wall. This protects the lead from the sun, but the arrangement is bad in any position where leaves of trees or any rubbish might drift into and choke the gutter. To obviate this, the gutter is constructed as shown in figure 1053, where the tilting piece forms the depth of the trough, and the edge of the lead under the slates is bent to form a small hollow welt to guide away any water that might pass over the fillet.

Figure 1044 shows the boarding cut short on the rafters to give depth to the gutter, and a hollow welt is turned on the edge of the lead for the purpose given above. This is better than the method shown in figure 1053.

*Side Gutters.*—Water is often conveyed from the upper gutters of mansard roofs by side gutters, as shown in figure 1045.

*Soakers.*—The intersections of sloping roof surfaces with end gables or penetrating walls are sometimes made weather-tight by pieces of lead termed soakers, fixed parallel to the walls, and resting on the slopes of roofs about 4 inches. They are turned up against the walls from 2 to 3 inches, and a cover stepped or raking flashing is fixed over the turned-up edges. The soakers should be the length of the slates minus the margin, and plus an inch for clipping over the heads of the slates. Figure 1054 shows a soaker cut for Countess slates; and figure 1054 is an elevation of soakers, fixed with stepped cover flashings over them to prevent the water leaking between

Fig. 1054.

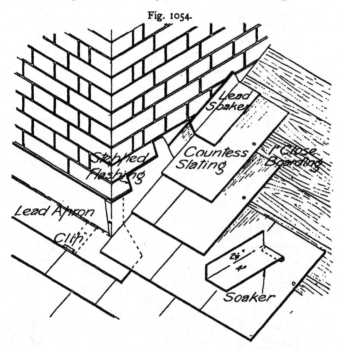

Lead Soaker

Stepped Flashing

Countess Slating

1" Close Boarding

Lead Apron

Clip

Soaker

the soakers and the wall. This is better than ordinary stepped flashings, as the wind cannot blow the rain between the roof and wall, and neither can the wind get under the edges as when ordinary stepped flashings are used.

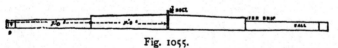

Fig. 1055.

In some cases each soaker turns up 6 inches against the wall, and steps are cut and the edges tucked into a raglet, as

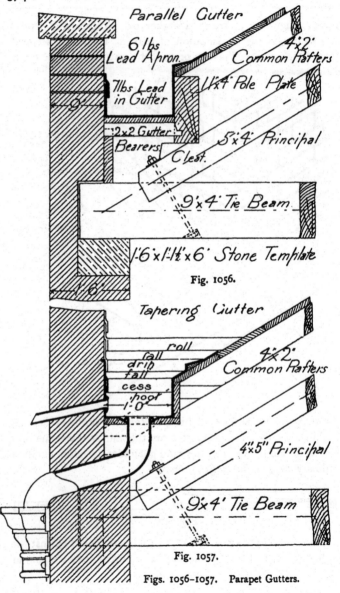

Parallel Gutter

6 lbs Lead Apron.

4"x2" Common Rafters

7 lbs Lead in Gutter

1½"x4" Pole Plate

6"

2"x2" Gutter Bearers

5"x4" Principal

C. lest.

9"x4" Tie Beam

1'6"x1'-1½"x6' Stone Template

Fig. 1056.

1'6"

Tapering Gutter

roll
fall
drip
fall
cess
shoot
1'-0"

4"x2" Common Rafters

4"x5" Principal

9"x4" Tie Beam

Fig. 1057.

Figs. 1056–1057.   Parapet Gutters.

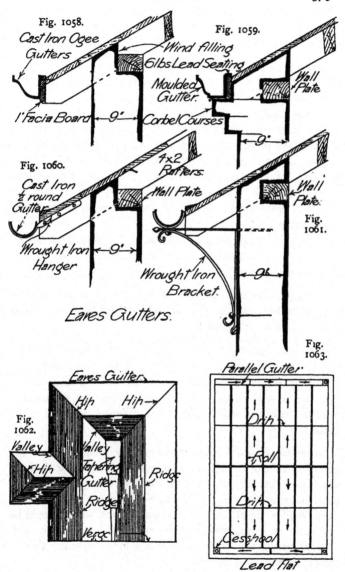

Fig. 1058.

Cast Iron Ogee Gutters

1" Facia Board

9"

Fig. 1059.

Wind filling

6lbs Lead Seating

Moulded Gutter.

Corbel Courses

Wall Plate

9"

Fig. 1060.

Cast Iron ½ round Gutter

Wrought Iron Hanger

4×2 Rafters.

Wall Plate

9"

Wrought Iron Bracket.

Wall Plate.

Fig. 1061.

9½"

Eaves Gutters.

Fig. 1062.

Eaves Gutter.

Hip

Hip

Valley

Valley

Hip

Tapering Gutter

Ridge

Ridge

Verge

Fig. 1063.

Parallel Gutter.

Drip

Roll

Drip

Cesspool

Lead Flat

Figs. 1058-1063.

in ordinary stepped flashing. This is a good plan, but takes a longer time to execute.

*Gutters.*—Rain water that falls upon roofs naturally runs to the lowest part, and provision must be made to carry the water away. This is done by means of gutters.

Gutters may be either eaves or parapet gutters.

Eaves gutters consist of small iron troughs, as illustrated in figures 1058 to 1061. In a storm or great downpour of rain they do not, however, accomplish the purpose for which they are intended, and it is not advisable to fix them on that side of a roof which overhangs the public highway. They are usually of cast iron, but should be fixed under the direction of the plumber.

If a roof slopes towards the public highway, it is recommended that a parapet gutter be constructed to guard against slates, tiles, or snow falling into the street.

*Parapet Gutters.*—Gutters on the outer side of which a parapet wall is constructed are named parapet gutters, and may be parallel or tapering.

*Parapet Tapering Gutters.*—Parapet gutters are of irregular width when their plans are as shown in figures 1055 and 1062. This suits the construction of roofs where the common rafters rest on the wall plates. Figure 1055 shows how to find the required widths of sheet lead at the different positions in the gutters.

*Parapet Parallel Gutters.*—Roofs constructed with the common rafters resting on a pole-plate fixed some distance away from the parapet, as shown in figure 1056, are termed parallel or box gutters. The pole-plate must be sufficiently deep to allow for all necessary falls and drips, which should be the same as when ordinary tapering shaped gutters are used. The timbering for this construction is more expensive than in the ordinary shaped gutters.

*Tapering Internal Gutters.*—Gutters constructed between the slopes of two adjacent roofs are called internal gutters, and may be tapering when the common rafters of adjoining slopes rest on the same plate. Figure 1042 shows a section giving all necessary details for such gutter, and figure 1062 shows a plan.

With the exception of the construction of the timber-work, the same section shows a flat sole valley in the slope on an internal angle formed by two roofs intersecting. In this case no drips are required.

*Parallel Valley* or *Box Gutters.*—When a roof is constructed so that the common rafters of opposite slopes are fixed to separate pole-plates, parallel gutters may be made. If the depth of the gutter is more than 12 inches at the deepest end, it is usual to fix a lead apron, as shown in figure 1043, instead of turning the gutters under slates, as in figure 1056.

Trough gutters are constructed in M-shaped roofs where it is necessary to convey the water from the internal gutter to one of the side gutters by timber troughs lined with lead, as shown in figure 688. They are sometimes miscalled box gutters.

*Snow Boards.*—These should be provided in all gutters to preserve an uninterrupted channel for draining away the water as the snow melts after a heavy snowstorm. The snow thaws on the underside first, through being in contact with the warm roof; and if the gutters and outlets be choked with frozen snow, that which is thawed will run through the joints in the roofing material. The snow boards, as shown in figure 1064, consist of a number of strips of wood 2 inches by 1 inch, placed in the gutter with their length parallel to the current of the gutter and spaced about ¾ inch apart; they are supported by and fixed to cross-pieces 4½ inches by 2 inches. From the underside of these bearers a piece at least 2 inches wide is taken out from the centre, and extends to within 3 inches from each end to form an arch for the melted snow to flow through. These are also advantageous on lead flats over which foot traffic passes.

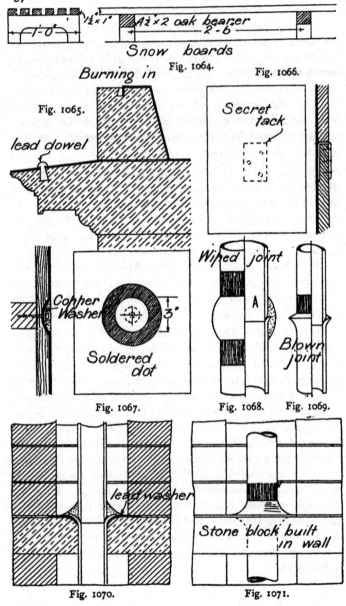

½"×1"
1'-0"
4½"×2 oak bearer
2'-6"

Snow boards

Fig. 1064.

Fig. 1066.

Burning in

Fig. 1065.

lead dowel

Secret tack

Copper Washer

Soldered dot

3"

Wiped joint

A

Blown joint

Fig. 1067.

Fig. 1068.

Fig. 1069.

lead washer

Stone block built in wall

Fig. 1070.

Fig. 1071.

*Snow Guards.*—These consist of low iron railings fixed at the lower edges of all sloping roofs, for preventing large masses of snow slipping down the roof surface and falling over the eaves; they should be at least one foot high, and the rails should be large enough to thoroughly break the masses. In severe climates on pitched roofs these should never be omitted.

*Burning In.*—Lead cover flashings for parapet gutters are sometimes secured to the stone blocking course by being turned into a groove or raglet, cut dovetail shape, on the top or front surface of the stone; molten lead is then poured into the groove, which thoroughly secures the flashing to the stonework, as shown in figure 1065. As the molten lead shrinks on cooling it is necessary to " set it up," that is to expand the surface with caulking tools or blunt chisels with a rounded end, to fill the groove, and thus ensure the lead being thoroughly secured.

*Soldered Dots.*—Where large vertical surfaces have to be covered with lead, as on dormer cheeks, they are sometimes secured in the following manner:—Round hollowings, 3 inches in diameter, are dished out of the boards, the lead is dressed into these hollowings and screwed to the boards and studs (the latter are best for screwing to if they come in the required positions). A tinned copper washer should be placed under the head of the screw to distribute the fixing over a greater surface of lead, so that when shrinkage or expansion takes place, the lead shall not be torn by the screw. The hollow is encompassed by a ring of soil (which is a composition of size, lamp black and chalk) to prevent the solder from adhering to the flush surface of the lead covering. The hollow is then filled up level with solder, and is then known as a soldered dot. Figure 1067 is a section showing the construction. Lead fixed in this manner is liable to crack near the soldered dots.

*Secret Tacks.*—A better mode of fixing dormer cheeks is shown in figure 1066, where a lead tack is soldered to the back side of the lead covering. The loose end of the tack is passed through a slot cut in the boarding, and secured by copper nails on the inside. The tack, or secret tack as it is called, fixes the covering securely, is not so unsightly as is the former method, and allows more freedom for the lead cheek to expand.

*Lead Dowel* or *Dot.*—Where exposed parts of stone cornices, strings, etc., are covered with sheet lead, the latter should be protected from high winds. The lead should be laid before the wall or blocking course is built, as shown in figure 1065, or turned up and built in the mortar joint, if any. Or the lead may be passed through the wall, and turned up on the inner side. If laid after the wall is built, the lead may be dressed against the outer face of the wall and tucked and wedged in a mortar joint that has been raked out to receive it. To prevent the free edge of the flashing being torn up by the wind, circular holes exactly opposite each other are made in the lead and surface of the stone; the edges of the lead round the hole should be turned up slightly. An iron mould with a small hole through the top is held over the hole in the lead and stone, and molten lead poured in to fill up the holes prepared to receive it, and also the cup-shape hollow of the mould. This fixing is called a lead dowel or dot.

*Wiped Joint.*—The joints of all lead pipes for conveying water under pressure should be wiped, as shown in figure 1068. The process of making the joint is as follows:—The end of one piece of pipe is opened, as shown in figure 1068, by means of a tan pin (a conical-shaped piece of box-wood). The end of the other piece of pipe is rasped to a sharp edge on the internal face. The pipes are then fitted together, care being taken to keep the internal surfaces true. The two ends are covered with soil for about 5 or 6 inches from the extremities, and then scraped bright with a tool called a shave hook for a distance of about 1½ inch. The pipes are then placed together and secured in position; molten solder is then poured on from a ladle and wiped round the joint with a piece of greased cloth. The solder adheres to the part that has been shaved, and forms a sharp line about the pipe at its junction with the soil. The covered end marked A should be tapered in the direction of the current. Plumbers' solder is composed of two parts, by weight, of lead to one of tin.

*Taft Joint.*—This is an inferior kind of joint, used for connecting lead pipes together, similar to that shown in figure 1069.

*Blown Joint.*—The ends are prepared and fitted, as shown in figure 1069, as for the taft joint and secured in position; a little flux, usually resin, is applied to the joint to aid in the fusion of the solder. Stick solder is used, and is melted by means of a blow-pipe flame, the solder only filling up the part of the pipe that has been opened out. This is the usual joint made in composition pipes for gas, these being much lighter in section than lead, and the maximum pressure on such joints is very small.

*Block Joint.*—Where large pipes are fixed in vertical chases of walls, the following soldered joint is sometimes constructed, and may be made as follows :—The top end of the lower length of pipe is passed through a circular hole, cut in a wood or stone block built in the wall, and the end of the pipe turned over a lead collar, which is placed upon the block. The upper length of the pipe, with the ends prepared, is then arranged in its proper position over the lower one, and the joint soldered, as shown in figures 1070 and 1071. These joints frequently break round the arris formed by turning the end of the lower pipe over the collar.

# CHAPTER XIII

# SLATING.

—

SLATES are used to cover roofs. They are considerably lighter than tiles and are not so absorbent. A good slate absorbs $\frac{1}{200}$ of its dry weight, and an ordinary pressed tile 9 per cent. of its dry weight; consequently the scantlings for a slated roof need not be so large as those for a tiled roof. They may be obtained in the market, of various dimensions as given:—

### Dimensions of Slates.

| | Inches. |
|---|---|
| Smalls ... ... ... ... ... | 12 × 6 |
| Doubles ... ... ... ... ... | 13 × 7 |
| Ladies (large) ... ... ... ... | 16 × 8 |
| Countesses ... ... ... ... ... | 20 × 10 |
| Duchesses ... ... ... ... ... | 24 × 12 |
| Princesses .. .. .. ... ... | 24 × 14 |
| Empresses ... .. ... ... ... | 26 × 16 |
| Imperials ... ... ... ... ... | 30 × 24 |
| Rags ... ... ... ... ... ... | 36 × 24 |
| Queens ... ... ... ... ... | 36 × 24 |

But the sizes in common practice are as follows:—Duchess slates, 24 inches by 12 inches, for moderately flat roofs of $\frac{1}{5}$ pitch; Countess slates, 20 inches by 10 inches, for roofs of $\frac{1}{4}$ pitch; and Ladies, 16 inches by 8 inches, for roofs of $\frac{1}{3}$ pitch—the practice being the steeper the pitch the smaller the slate, and the more exposed the building the greater the pitch of the roof.

The following terms are used in slaters' work :—

Head   ... The upper or top edge of slate.

Back   ... The upper or exposed surface of slate when laid.

Bed   ... The lower or under surface of a slate.

Tail   ... The lower or bottom edge of slate.

Margin   ... The part of each course of slates exposed to view.

Gauge   ... The distance apart nails have to be inserted on battens or boards; this is required to determine the position of the battens to receive the nails by which the slates are secured.

Bond   ... Where a joint of two adjacent slates is immediately in the centre of the slate, the tail end of which rests upon them, the slates are said to bond.

Lap   .... The distance the tail of one slate overlaps the head of second course below, when slates are nailed near the centre, or the distance the tail of slate overlaps the nail-hole of second course below when slates are nailed near the head.

     The lap in practice ranges from $2\frac{1}{2}$ inches to 4 inches ; in this book it will be taken as 3 inches.

Holing   ... The piercing of slates to receive nails.

Sorting   ... Where the roof is to be covered by slates of different lengths, they are regulated to proper dimensions so that the largest slates may be nailed near the eaves, and the smallest at the ridge.

Eaves   ... The lower part of slating hanging over a wall.

Verge   ... The finished edge of slating overhanging a gable wall or bargeboard.

There are two methods of laying slates :—

     1st.—By nailing near the head.
     2nd.—By nailing near the middle.

The groundwork to receive slates may be prepared in one of the following methods :—

    1. *With Wood Battens* nailed horizontally across rafters fixed to the required gauge.

    2. *With Close Boarding*—that is, the rafters are boarded over, and the slates nailed direct to the boards; this is a better method than battening.

    3. *With Close Boarding and Asphalted Felt.*—The felt is a non-conductor and prevents radiation, and thus preserves a more equable

temperature in the interior of buildings; it is also waterproof, and thus forms an extra precaution against damp.

4. *Close Boarding Felt and Battens.*—Where the slates are laid direct on the felt, the latter is liable to decay from want of ventilation; to prevent this, battens are laid horizontally on the felt to which the slates are fixed, thus allowing a circulation of air over the surface of the felt.

5. *Close Boarding, Felt, Vertical and Horizontal Battens.*—The last method is open to the objection that should any water find its way between the slates, as in the case of a broken slate, the water would lodge upon the horizontal battens and cause them to rot. To avoid this, battens are first laid on the felt to the slope of the roof, and fixed one over each rafter; horizontal battens are nailed to these again, and then the slates, so that should any water get beneath the slates it can run away. This leaves a larger air space, which is better for ventilating the battens and felt, also for preserving the temperature inside the building. This method should be adopted on all monumental buildings, and on roofs not easily accessible.

*Gauge for Nailing Slates near the Head.*—In this method of fixing slates, the nail holes are placed one inch from the head. The calculations to ascertain the gauge are as follows:—
One inch plus the lap is deducted from the total length of the slate, the first being due to the material above the nail-hole not being included in the lap, and the remainder is divided by two, which will give the gauge, and may be stated thus:—

$$\text{Gauge} = \frac{\text{length of slate} - 1'' - \text{lap}}{2}, \text{ which}$$

In Duchess Slates would be $\frac{24 - 1 - 3}{2} = \frac{20}{2} = 10$ inches.

,, Countess ,, ,, $\frac{20 - 1 - 3}{2} = \frac{16}{2} = 8$ ,,

,, Ladies ,, ,, $\frac{16 - 1 - 3}{2} = \frac{12}{2} = 6$ ,,

Figure 1072 shows slates fixed by this method.

*Gauge for Fixing Slates near the Middle.*—The gauge is determined as follows:—The lap is deducted from the length,

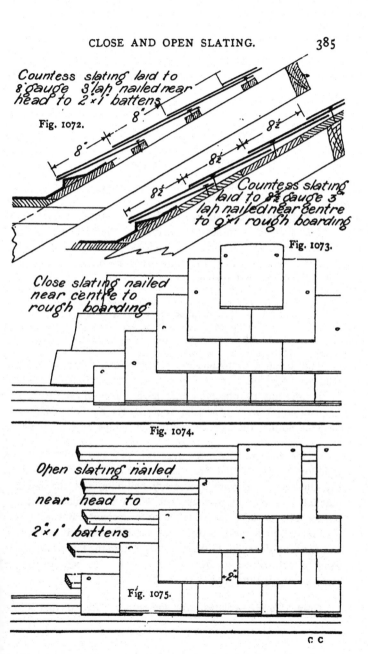

Countess slating laid to 8 gauge 3" lap nailed near head to 2×1 battens

Fig. 1072.

8"   8"   8"

8½   8½   8½

Countess slating laid to 8½ gauge 3" lap nailed near centre to 9"×1 rough boarding

Fig. 1073.

Close slating nailed near centre to rough boarding

Fig. 1074.

Open slating nailed near head to 2"×1 battens

2"

Fig. 1075.

and the remainder is divided by 2, and may be stated thus :—

$$\frac{\text{Length of slate} - \text{lap}}{2}$$

In Duchesses would be $\frac{24-3}{2} = 10\frac{1}{2}''$.

,, Countesses ,, $\frac{20-3}{2} = 8\frac{1}{2}''$.

,, Ladies ,, $\frac{16-3}{2} = 6\frac{1}{2}''$.

Figure 1073 shows slates fixed by this method.

The slater determines the position of the nail-hole in the following manner: the lap is added to the gauge, and $\frac{1}{2}$ inch for clearance is allowed—measurements being taken from the tail. It may be stated thus :—

Gauge + lap + $\frac{1}{2}''$ = distance of nail-hole from tail—
In Duchesses would be $10\frac{1}{2} + 3 + \frac{1}{2} = 14''$.
,, Countesses ,, $8\frac{1}{2} + 3 + \frac{1}{2} = 12''$.
,, Ladies ,, $6\frac{1}{2} + 3 + \frac{1}{2} = 10''$.

The distance of nail-holes from the long edges in either method is $1\frac{1}{4}$ inch.

It can thus be observed that fewer slates will be required to cover equal areas of similar roofs by the method of nailing near the middle, than by the method of nailing near the head.

*Advantages and Disadvantages* of the two methods are as follows :—Slates that are nailed near the head always have two thicknesses of slate over every nail-hole, and the lap is practically 1 inch more than that which is calculated if the portion above nail-holes is taken into account. But as this method requires more slates, it is therefore more expensive; and the long distance the nails are fixed from the tail allows the wind to act with a greater leverage, which, in bad weather, sometimes strips the roof.

*Open Slating.*—Slates are sometimes laid so that the

adjoining slates of the same course are a distance apart, about 2 inches, as in figure 1075. A roof covered in this manner would not require nearly as many slates as the former methods; it is largely used for sheds and temporary buildings.

*Doubling Eaves Course.*—Slates are laid commencing at the eaves; the length of the slates should be as follows :—Gauge + lap + 1 inch in slates nailed near the head. In Countess slates this would be 8 + 3 + 1 = 12 inches, and nail-hole would be 11 inches from tail, as shown in figure 1072. Where slates are nailed near the centre, gauge + lap is the length; this would be in Countess slates $8\frac{1}{2}$ + 3 = $11\frac{1}{2}$ inches, and the nail-hole would be at the centre of tilting piece, but if the tilting piece is covered with lead, the nail-holes would be made near the head, as shown in figure 1073.

*Cutting of Slates.*—In the cutting of slates, they are laid on an iron straightedge or cutting-dog, and the edge is trimmed with a zax. The face in contact with the iron straightedge is true and regular, but the upper surface is jagged and rough. The doubling eaves course is laid with its regular edge up; the course above and all succeeding courses are laid with their regular edge down, so as to obtain a close joint to guide away the rain and wind.

*Lead and Copper Tacks.*—These are used for repairing defective slated roofs, the method being as follows :—The broken slates are removed, the nails being cut or drawn by a tool called a ripper, which has a thin blade about 2 feet long and 1 inch wide, being about 2 inches broad at the end, with a notch each side to receive nail; the tack is then hooked over the head of the slate below, the new slate is inserted, and the lead clip turned up over the tail. Two tacks should be used to each slate.

These lead or copper hooks are usually about $\frac{1}{2}''$ wide, the distance between the hooked ends being equal to the length of the slate minus a gauge.

*Tilting Fillets.*—To enable the tails of slates to fit closely against the slates below, and to form a close joint to keep out wind and wet, a tilting fillet or springing piece, as it is sometimes called, must be nailed under the tail of the first slate or doubling course. This will give all the slates a tilt, and cause them to bend on their tails. The slates will consequently be slightly apart, the less the distance apart the better, so that, during repairs, they may be less likely to be broken. The tilting pieces are about $\frac{1}{2}$ to $\frac{3}{4}$ inch thick, and tapered, as in section figure 1072.

When roofs are battened, instead of a separate tilting piece, the first batten may be $1\frac{1}{4}$ to $1\frac{1}{2}$ inch thick, and tapered.

Tilting fillets are also used against chimney breasts, to guide water away from walls, as in section figure 1048, or at ridge, to compensate for the tilt of slate lost by its length being shorter than others, as in section figure 1078, or in any position where slates are desired to be slightly raised.

*Tails of Slates.*—Slates should be laid with their tails horizontal, but wherever taper-shaped gutters occur in roofs the side of the gutter nearest the ridge is not parallel with it, as shown in plan, figure 1055. So the tails of doubling courses form an exception to this rule.

*Slate Hips and Ridges* are now much used in the place of lead. The slate ridges are holed and bedded in hair-mortar, the holes filled with white lead and secured with copper screws, and the adjoining pieces dowelled together with small slate dowels, as shown in figure 1078.

*Stone Slab Roof Covering.*—Stone slabs, generally from $\frac{1}{2}''$ to $1''$ in thickness, and of dimensions up to $2' \ 0''$ square are used for the coverings of roofs, being very durable and giving a pleasing effect.

They are quarried in various sizes, after which they are sorted in lengths, as all slabs laid in the same course should be

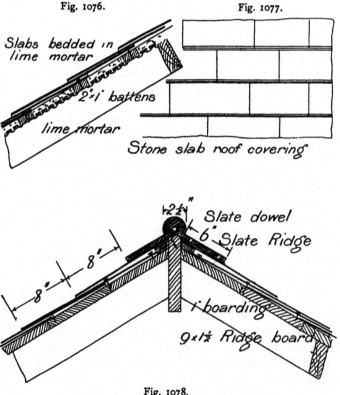

Fig. 1076.

Fig. 1077.

Slabs bedded in lime mortar

2×1″ battens

lime mortar

Stone slab roof covering

Slate dowel

6″ Slate Ridge

boarding

9×1½ Ridge board

Fig. 1078.

of the same length and where possible of the same width. The largest slabs are placed in the courses nearest the eaves, and the courses are usually set out to diminish as they approach the ridge.

Stone slabs may be obtained suitable for roofing purposes from Collyweston, near Stamford, Naunton in Gloucestershire, and Stonefield, Oxfordshire.

Taking the Collyweston slabs as typical, they are laid as follows :—

The battens are fixed longitudinally to the common rafters, the necessary diminishing gauge having first been determined by the size of the material to be used. The spaces between the battens are then lathed and filled in solid with ordinary lime mortar brought up flush with the battens. The slabs are then fixed in position by nailing near the head with one galvanized iron nail, having previously been bedded on stone lime mortar and pointed with the same material. The stone lime is prepared from an upper layer of stone from the same quarry as the slabs.

The slabs are cut and mitred at hips and valleys, and finished with lead in a manner similar to ordinary slating.

The slabs on being tested for absorption are found to collect little more than an average roofing tile. Stone slabs usually absorb about $\frac{1}{13}$ of their weight.

This method of roof covering has been used at the Indian Institute, Oxford, and on many of the modern buildings of Oxford and at the Henley New Town Hall.

Figures 1076 and 1077 show a part section and part elevation of slabs laid by this method.

*Nails.*—Slating nails have flat circular heads, have a sharp point on shank, and are made from $1\frac{1}{4}''$ to $2''$ long. The five following materials are in general use for the manufacture of nails, viz., iron, zinc, copper, composition and lead.

(1). *Iron* nails may be either cast or malleable. Cast nails resist oxidation better than the malleable, but being brittle are inferior to the above, and are only used for cheap work. Malleable nails are first cast and then made malleable ; they are often galvanized or painted to resist oxidation, but they are better when dipped while hot into boiled linseed oil, which method is frequently and successfully adopted. Iron nails are now not much used.

(2). *Zinc* nails are relatively to iron nails soft and easily broken, but with care they may be used without any appreciable waste, and are very durable as they do not corrode. These are extensively used.

(3). *Copper.*—These may be obtained as wrought or cast, are very soft, relatively expensive, they are non-corrosive, and are not much used.

(4). *Composition.*—These are a mixture of zinc, copper and tin. The alloy is much harder than either the copper or the zinc; does not oxidise to any extent, and is better adapted for driving. They should be employed on all important work.

(5). *Lead.*—These, as shown in figure 594, are similar to the ordinary slating nails, and about 4 inches in length. They are used for securing slates direct to iron battens. The nail is passed through the hole, and is bent about the small tee or angle iron batten. They are sometimes used for boiler houses and similar work, where a great measure of fire resistance is of more consequence than appearance.

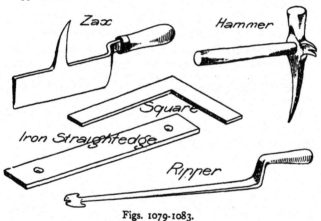

Figs. 1079-1083.

*Specification of Slates.*—The whole of the slates to be of the best quality, properly squared, free from spots, and of such quarries and colour as shall be approved by the architect. Each slate, when nails are used, to be holed near the centre of its length (or near the head) and secured with two strong nails for each slate, the third slate to overlap the first not less than 3 inches, and to bed flat upon the previous course. Upright slating to have 4-inch laps and be fixed to walls with four nails each.

In all descriptions of roof slating the eaves to be a double course of slates, the under course being cut to the required length.

The nails for Countess and Duchess slates to be not less than 1½ inch long, and for larger slates to be not less than 2 inches long.

If iron nails are used they are to be dipped while hot into boiled linseed oil.

*Slaters' Tools* —The principal tools distinctively used by the slater are the zax, iron straightedge, hammer, and ripper, as shown in figures 1079 to 1083.

(1). *The zax* is used for trimming the edges and holing the slates.

(2). *The iron straightedge* is used for trimming slates, and is fixed to the edge of the bench or block by means of screws or sharp iron points projecting from the underside. Another tool similar to the dog used in rough carpentry work is sometimes used on works in progress in place of the iron straightedge as it is easier to fix, being driven into any convenient plank.

(3). *The hammer* with hammer face at one end, and sharp point at other for holing, with a projection from side with slot for drawing nails.

(4). *The ripper* is a thin steel blade about 2 feet in length, provided with a handle; the other extremity is flattened out to a semicircular shape and has three slots as shown in figure 1083; it is used for removing broken slates, the nails being cut by the slotted semi-circular end.

In addition to the above the slate mason is provided with squares, saws, mallets, and chisels similar to other masons.

# CHAPTER XIV.

# BUILDING QUANTITIES AND MEMORANDA.

*Necessary knowledge.*—To be able successfully to take off quantities from. drawings, and prepare estimates preparatory to pricing, three things are necessary—first, a thorough knowledge of arithmetic and mensuration; secondly, a sound knowledge of the details of building construction in all trades; thirdly, practical experience in the application of the two foregoing qualifications, to take off the measurements and state them in the manner usually adopted and understood.

The detailed dimensions are first of all "*taken off*" the drawing, and written down in columns, in the order of the length, breadth and depth, with an abbreviated description of the work to which the dimensions refer. When this has been done for all trades, the next operation is "*squaring the dimensions*," to obtain either the cubic or superficial measurements resulting from each set of dimensions. These results are collected together in the "*abstract*" for each trade; the total "*collections*" in each abstract are then placed in their proper order in a " bill of quantities;" the bill is ruled so that a price and a total may be placed against each item; the sum of all the totals will give the total proposed cost of all the work mentioned and described in the bill of quantities. This is termed the "*estimate.*"

The student will not be able to proceed far without knowing the usual methods adopted for measuring the various kinds of builders' work, both labour and material, and the proper descriptions to be placed in the bill; the following are some general notes relating to this matter, under the proper trade headings.

## EXCAVATORS' WORK.

*General digging*, at per yard cube. Beyond 6 feet deep, keep separate and divide into stages of 6 feet each; wheeling over 20 yards keep separate, and divide into runs of 20 yards each; basketing out, keep separate, if no concrete take digging 6 inches wider than base of footings; describe if dig and cart away; dig, wheel, and spread, or if dig and part fill in and ram. In trenches the space occupied by the concrete and wall is dig and cart or dig, wheel, and spread. The remaining space is dig, fill in, and ram.

*Concrete* at per yard cube generally; if under 12 inches deep, at per yard superficial. State proportions of materials to be used, if in lime or cement, &c.

*Wells*, at per yard cube, stating diameter, depth, distance to be removed and description of soil.

*Surface digging*, at per yard superficial, if 12 inches deep or under.

*Spread and level*, at per yard superficial, with description of average depth.

*Hard brick rubbish*, at per yard superficial, if 12 inches deep or under; above, at per yard cube.

*Clay puddle* at per yard super, describing thickness.

*Shoring and planking* to sides of excavation at per foot run of length of surface supported, stating the depth.

*Strutting and planking* to trenches at per foot run of length of trench, stating the width and depth, this lineal dimension to include both sides of the trench. Strutting and planking should be taken to all trenches over 4 feet in depth.

*Drains* at per foot run, to include digging, laying, jointing, concreting, filling, and ramming. Take a running dimension over all pipes and connections. When the latter are taken they are charged as extra only upon the cost of this item. State size of pipe, with description, how jointed and enclosed, and the average depth of the trench.

Lengths of drain upon which there is any additional labour should be kept separate and so described, such as "deep digging," "tunnelling," "strutting to trench," &c.

*Bends, junctions, traps and gulleys* are numbered, and having been measured in the length of the drain are described as "extra only" upon the cost of the drain.

*Manholes* are numbered and described.

*Piles* are numbered. State size and description of timber and length to be driven.

Number the rings, shoes, pointing, and cutting off heads when driven ; state weight of ironwork.

*Memoranda.*—The following is the ratio of the increase in bulk of the undermentioned earths when excavated :—

| | | | | | |
|---|---|---|---|---|---|
| Earth and clay ... | ..: | ... | ... | about | $\frac{1}{4}$ |
| Sand and gravel | ... | ... | ... | „ | $\frac{1}{12}$ |
| Chalk ... | ... | ... | ... | ... | „ $\frac{1}{3}$ to $\frac{1}{2}$ |
| Rock ... | ... | ... | ... | ... | „ $\frac{1}{2}$ to $\frac{4}{5}$ |

In the mixing of concrete, the final bulk is about $\frac{2}{3}$ of the volume of dry materials before mixing ; for instance, a cubic yard of concrete *in situ* requires about $1\frac{1}{2}$ cubic yards of cement, gravel and sand.

### BRICKLAYERS' WORK.

Brickwork, where used in large masses as for engineering works, is sometimes measured per foot cube. The usual London method is as below :—

*The following is measured superficially :*

*Brickwork in mortar*, at per rod of 272 superficial feet, $1\frac{1}{2}$ brick thick, all thicknesses to be reduced to this standard.

All half-brick walls are measured similarly but priced separately.

*Brickwork in cement*, measured as above, but usually included in the brickwork in mortar, and described again separately as "*Extra only in cement.*"

Keep separate all work above 50 feet from the ground, also all work built in small pieces.

*Facings* are measured at per foot superficial, as "*extra only*" upon the cost of the general brickwork; allow 3 inches below the ground and 3 inches at junction of roof with gable wall, add the length by 2¼ inches for each internal angle.

*Fair face and point*, at per yard superficial, where the same kind of bricks are used in face and body of wall.

*Fair and rough cutting* all over 6 inches wide at per foot superficial.

*Damp course* at per foot superficial.

*Arches in walls* in lime mortar are measured in with the ordinary brickwork. Rough arches in cement, axed or gauged arches, are described and charged as "*extra only*" upon common brickwork, all extra necessary labour and material above those required in the common brickwork are charged for in all arches, such as fair or rough cutting, turning pieces, &c.

*Trimmer arches and vaulting* at per foot super, describing and stating thickness.

*The following work is all measured by the superficial yard*:—Brick-paving, brick-nogging, rough rendering, cement floated face, cement or tar paving, asphalte-paving, and tiling on brick.

*The following is measured per foot run* :—

Fair and rough cutting, if under 6 inches wide.

Birdsmouth, squint quoin, cutting and pinning.

Chase for pipes, cutting, toothing, splays, chamfers.

Cement fillet, lime and hair filleting, joints of groined arches, rake and point flashings.

*All the following to be numbered, with a proper description* :—

Mitres and stopped ends to cornices, strings, &c., bedding and pointing door and window frames; flues, parge, and core, stating average length; ends of sills cut and pinned, holes cut for pipes, coppers, and setting; all stoves and setting; chimney pieces and fixing; brick seatings to w.c. traps, trimmer arches levelled up for hearths, rendering chimney backs in cement, air bricks and fixing.

*Memoranda.*—To reduce cubic feet of brickwork to rods, 1½ brick thick, deduct ⅛th, and divide by 272.

Facings require about 7 bricks per foot superficial, a rod of brickwork with a ⅜″ joint requires 4,356 bricks and 3$\frac{1}{80}$ cubic yards of mortar. Allowing for waste and consolidation, this will mean 4,500 bricks, 1½ yard of chalk lime, and 3 yards of sand, or 4,500 bricks, 1 yard of stone lime, and 3½ yards of sand, or 36 bushels of cement and 36 of sand to bricks as before, or 30 bushels of blue lias lime to 60 bushels of sand. A ton of Portland cement is 11 sacks. There are 21 09 striked bushels to 1 yard, 30 bushels of lump lime to 1 ton of quicklime, 26·6 bushels of ground lime to 1 ton of quicklime.

In the mixing of lime and cement mortars of the usual proportions the final volume may be taken as about ⅔rds of the volume of the dry materials before mixing, as may be deduced from the quantities for 3$\frac{1}{80}$ cubic yards of mortar required for a rod of brickwork.

A striked bushel = 1·28 cubic feet, therefore a cubic yard = 21·09 striked bushels.

### MASONS' WORK.

Take all stone at per foot cube including hoisting and setting. If the hoisting is over 40 feet in height, take an extra to, and divide into heights of 20 feet, as 40′ to 60′, 60′ to 80′, &c.

*All labours to be taken separately and to be measured at per foot super.* The final labours include all preparatory labours :—

*Half sawing or half-plain.*—This labour is taken to all faces upon which no other labour is taken, such as the back of ashlar work, the back of a cornice, &c.

*Half-plain work.*—This labour is taken to beds and joints ; each of the two worked surfaces to each joint is paid for as half plain work.

*Plain work.*—This labour is taken to all exposed plain surfaces upon which no further labour is placed.

*Plain work to beds and joints.*—Sometimes taken in London, the area of the joint made up by the two worked surfaces in contact being measured as the area of one of the surfaces in lieu of being taken as half-plain work to the two surfaces in contact.

*Sunk beds and joints.*—Taken in all cases where the face of the bed or joint is sunk below the general face.

*Sunk face.*—In all cases where the face of the work is sunk back or recessed, and the surface is equal to plain work.

*Moulded work.*—All mouldings where the girth exceeds 6 inches.

*Circular face; circular sunk face; circular moulded face.*— When the work is circular on plan as well as circular in elevation, the work is described as *circular circular.*

Pavings, landings, treads and risers, hearths, shelves, large templates, and cover-stones over girders measured superficially with description.

*The following labours are measured at per foot run :—*

*Moulded work,* under 6 inches girth, rebate and chamfer, joggle joint, groove, throating.

*The following at per foot run will include both labour and material :—*

Curbs, steps, window-sills, and thresholds, not exceeding 18″ × 6″ in section.

*The following items are numbered :—*

Stopped ends and mitres to mouldings, splays, chamfers, &c., with proper description; mortices, dowels, lead plugs, circular perforations, holes for pipes, cramps for copings, sinks, gully stones, chimney pieces, small templates, corbels.

## STONE WALLING.

*Rubble work* is measured at per foot cube to walls 2 feet or above in thickness. Walls less than 2 feet in thickness are kept separately and their thicknesses stated.

The extra labours on face work should be taken separately.

## SLATING.

*Slating.*—This is measured by the square of 100 feet superficial; allow the length × 6″ for all straight cuttings and × 12″ for circular cuttings, the length × 6″ to each side of hips and valleys; allow the length × half the length of the slates used for the double course at eaves. Allow 6″ for verges.

Circular and conical work to be kept separate.

Slate in thicknesses is measured at per foot superficial, the labours upon some being measured separately.

Slate cisterns, lavatory tops, chimney pieces, &c., are usually numbered with a description.

The slater purchases slates, obtaining 1,200 for every 1,000 invoiced.

*Slaters' memoranda.*—

Ladies.—16″ × 8″ × 6½″ gauge nailed near head, 278 to cover a square of 100 feet super.

Countesses.—20″ × 10″ × 8½″ gauge nailed near head, 170 to cover a square of 100 feet super.

Duchesses.—24″ × 12″ × 10½″ gauge nailed near head, 115 to cover a square of 100 feet super.

Each slate requires two nails, 1⅜″, 1½″, 1¾″ long respectively, for the above dimensions. To the above quantities it is usual to allow $\frac{1}{10}$th for cutting and waste.

### TILING.

*Plain Tiling* is measured at per square of 100 feet superficial; allow the length × 6″ for all straight cuttings, × 12″ for all circular cuttings. Allow the length × 6″ for eaves also to each side of hips and valleys. Allow 6″ by the length for working to a straight line, bedding and pointing to verges.

Measure at per foot run for ridges, hips, or valley tiles, bedding eaves in cement, filleting in cement.

Number apex tiles, hip hooks.

Tiles fixed vertically should be kept separate and the gauge
described. Circular tiling to be kept separate and described.
Where the pitches of intersecting slopes vary, attention must
be called to the different gauges necessary if special valley or
hip tiles be used.

*Pan Tiling.*—Allow the length × 12″ for cuttings to each
side of hips and valleys, and measure the hip tiling and bedding
in mortar by the foot run, in other respects measured similarly
to plain tiling. The tiler purchases tiles, obtaining the exact
number per thousand, no allowance being made as is done for
slates.

*Tilers' memoranda.*—Hurst's Handbook :—

A plain tile measures 10½″ × 6½″ × ½″, and weighs about
2½ lbs.

A pan tile measures 13½″ × 9½″ × ½″, and weighs about 5¼ lbs.

Tiling requires—

Per square of 100′ super 4″ gauge 600 plain tiles.

|  |  |  |  |  |  |  |
|---|---|---|---|---|---|---|
| ″ | ″ | ″ | 3½″ | ″ | 700 | ″ ″ |
| ″ | ″ | ″ | 3″ | ″ | 800 | ″ ″ |
| ″ | ″ | ″ | 10″ | ″ | 180 | pan tiles. |
| ″ | ″ | ″ | 11″ | ″ | 164 | ″ ″ |
| ″ | ″ | ″ | 12″ | ″ | 150 | ″ ″ |

A square of plain tiling weighs on the average 15 cwts.

    ″   ″ pan     ″     ″     ″   8 ″

A plain tile lath is 1″ wide and ¼″ thick.

A pan   ″   1½″   ″   1″   ″

100 plain tile laths 5′ long equal 1 bundle.

12 pan   ″   10′   ″   ″   1   ″

1 bundle of laths, 1½ cwt. }
  of nails, 1 peck of tile } to 1 square of plain tiling.
  pins }

1 bundle of laths, 1¼ cwt. }
  of nails, 3 hods of mortar } to 1 square of pan tiling.

## CARPENTERS' WORK.

Keep separate the different kinds of wood, placing the least expensive first, such as fir, pitch pine, oak, teak, &c.

*The following are taken at per foot cube:—*

Fir rough and fixed in plates, lintels, and wood bricks.

Fir rough and fixed in ground joists and sleepers.

Fir rough and fixed in floors.

Fir rough framed and fixed in floors.

Fir rough and fixed or rough framed and fixed in girders.

Fir framed in roofs and in roof trusses.

Fir framed in stud partitions, and fir framed in trussed partitions.

Oak in templates. Oak in curbs.

All work is considered as framed where it is necessary to use chisels in forming the joints. Allow 6" for laps. The number of laps for fir plates equals the (length in feet ÷ 12) − 1, for oak plates the (length in feet ÷ 12) − 1. Allow for each scarf three times the depth of the timbers lengthened. The number of scarfs for fir plates equals the (length in feet ÷ 15) − 1.

### CARPENTERS—MEMORANDA.

*The following work is measured at per square of* 100 *feet superficial:—*

Boarding to roofs and flats, and including firring to the latter.

Battening for slates or tiles. Wall battening, stating distance apart and including plugging.

Weather boarding, sound boarding, centering to vaults, &c.

*The following work is measured at per foot superficial:—*

Gutter boards and bearers.

Eaves boards, valley boards.

Centering to trimmers and apertures.

Bracketing for cornices, cradling for entablatures, girting the faces of the two latter items.

*Measure the following work at per foot run:—*

Mouldings, rebates, grooves, featheredge springing pieces, herring-bone strutting, tilting fillet, turning pieces to arches, 4½" soffit, deal rolls for flats.

The length of all scantlings in lineal measured work or reduced to superficial or cubic measures, when measured in detail, to be taken at the extreme points, including tenons and scarfs. The length of a scarfed joint, unless shown specifically, to be taken as three times the depth of the pieces joined.

*Number the following and describe* :—Cleats, shaping ends of rafters, rebated drips in gutters, cesspools, extra to angle brackets, extra to fixing ironwork to trusses. Short rolls under 2 ft. 6 in. long.

*Buying of Timber.*—Pine and spruce timber is sold by the standard hundred, the load, or by the square of 100 feet super. There are several standard hundreds in use, as follows :—

London    ...  120 pieces 12 feet long  9 in. by 3  in.
Petersburg ...  120   ,,    6   ,,    ,,  11 ,,  ,, 3  ,,
Christiania ...  120   ,;  11  ,,    ,,   9 ,,  ,, 1¼ ,,

The Petersburg standard is the one most generally followed, and equals 165 feet cube; a load of timber is 50 cubic feet (hewn), so that there will be 3 3/10 loads in a standard. It will simplify calculation to commit to memory one or two facts relating to these measurements :—165 feet cube is 165 feet run of 12 in. × 12 in.

In dealing with a scantling of 12 in. × 4 in., its section is one-third of the 12 in. × 12 in., so that to make up a standard in that scantling it would require 165 × 3 = 495 feet run.

The price per foot run can be obtained in this way if the value of a standard is known, the cost of cutting being added.

Classification of timber according to size :—

|  |  |  |  | in. |  | in. |  | in. |  | in. |
|---|---|---|---|---|---|---|---|---|---|---|
| Balk | ... | ... | ... | 12 | by | 12 | to | 18 | by | 18 |
| Whole Timber | ... | ... |  | 9 | ,, | 9 | ,, | 15 | ,, | 15 |
| Half | ... | ... | ... | 9 | ,, | 4½ | ,, | 18 | ,, | 9 |
| Scantling | ... | ... | ... | 6 | ,, | 4 | ,, | 12 | ,, | 12 |
| Quartering | ... | ... | ... | 2 | ,, | 2 | ,, | 6 | ,, | 6 |
| Planks | ... | ... | ... | 11 | to | 18 | by | 3 | to | 6 |
| Deals | ... | ... | ... |  |  | 9 | ,, | 2 | ,, | 4½ |
| Battens | ... | ... | ... | 4½ | ,, | 7 | ,, | ¾ | ,, | 3 |
| Strips and Laths | ... | ... | 4 | ,, | 4½ | ,, | ½ | ,, | 1½ |

Pieces larger than planks, generally called timber; but sawn all round, called scantling; when equal dimensions, called die square.

### JOINERY.

Joinery, including material, is measured by the foot superficial or foot run and special labours are numbered.

*The following items are measured by the foot superficial:—*

Skylights, Sashes and Cased Frames with detailed description, any extra labours to be measured as they occur.

Window linings and boards, backs, elbows and soffits, jamb linings, spandril linings, outside shutters, boxing shutters, lifting shutters, doors and gates, framings, w.c. fittings, bath fittings, cisterns and sinks, sashes in solid frames. Grounds over 3 inches wide.

Staircase work—treads and risers, wall strings, apron linings, outer strings.

Floors are measured by the square of 100 feet superficial. Allow a dimension of 3″ for all raking cuttings, keep separate all small pieces, and describe as including bearers.

For all superficial dimensions, such as doors, windows, &c., the tenons are to be included in the price of the superficial quantity when put together and fixed complete.

Joinery may be described of the finished thicknesses, when no variation is permissible. If the finished dimensions are not definitely stated, then it is usual to describe the stuff of the thickness out of which it is produced. An allowance of one-sixteenth to one-eighth of an inch should be made for every wrought surface, the maximum allowance varying directly as the thickness increases.

In work measured superficially, the separation in the measurement and value of all work consisting of partly straight and partly curved, such as segmental sash frames and sashes, is to be made precisely at the springing or commencement of the curve, taking the extreme height by the extreme width, each division net and the prices to include all materials and labour.

The same method is to be observed in dividing curved parts from such as are wreathed, groined from straight parts of centering, cradling, &c.

*The following items are measured at per foot run :—*

Grounds under 3 inches wide, and describe labours on same.

Skirtings, including backings, with description; also scribings of floors and skirtings.

Handrails, balusters (including tenons), deal cornices, members of solid door and window frames, and architraves, noses to landings, window nosings, capping to gates, mouldings: extra labours as they occur.

In work measured lineally, such as solid door frames, when the head is circular, keep separate and describe.

*The following items are numbered :—*Tongued and mitred angles, housings, returned ends, &c.

On account of the number of items in joiners' work, the work usually numbered is placed in connection with the leading item to which it refers. For example, all the extra labours numbered on skirtings would follow after the item for skirting, instead of being placed at the end of the bill together.

The tendency of modern practice in the measurement of joinery is to take by the foot run with description, and take every labour, such as mitres, fitted ends and housings, as numbers.

### IRONMONGERY

Number all articles with full description.

State if fixing is included.

Where screws are used charge as an extra.

Separate all brass work and describe; any provisions should state if prime cost or to include fixing.

*Hinges.*—Measure the length along the knuckle for butts and back flaps; or from knuckle to point for cross-garnet hinges. All patent hinges according to description.

*Bolts.*—Measure length of the rod of bolt and describe.

*Locks.*—Measure length and describe; take sets of furniture separately. Cupboard locks are measured by their height, not length; take finger plates separately.

### SMITH AND FOUNDER.

*The following items are measured and reduced to weight :—*

All heavy cast, wrought iron and steel constructional work; measure all labours on the iron separately, such as drilling, planing, fitting, &c.

In riveted work an allowance must be made for weight of rivet heads, usually 5 per cent. added on in ordinary riveted girders.

The length of round steel having the same weight as a rivet of equal diameter is equal to the thickness of the plates joined plus $2\frac{1}{2}$ times the diameter of the rivet.

Describe any necessary labour in the hoisting of constructional work.

Chimney and bearing bars measure at per foot run, and bring to cwts.; straps and bolts measure at per foot run, and bring to cwts.; also saddle bars, guard bars.

Flitch plates at per foot superficial, and bring to weight in cwts.

Hand rails and balusters are measured at per foot run, and brought to weight; describe if *"framed."* Circular and wreathed work keep separate.

*The following items are measured per foot run :—*

Eaves gutters measured over all, and charge as *extra to* for all outlets, angles, stopped ends, mitred ends, &c.

Rain-water pipes measured over all, and *extra to* shoes, bends, heads, &c., separately.

Railings wrought and cast. Gates taken separately.

*The following items are numbered :—*

All outlets, angles, stopped ends, mitred ends, hangers and brackets of eaves gutters ; all shoes, bends, heads, &c., of rain-water pipes ; gates wrought and cast ; grates, ranges, sets of copper fittings, soot doors, ventilators, manhole covers, and all patterns for cast-iron work.

### PLUMBERS' WORK.

Sheet lead is measured by the foot superficial, and brought to weight.

The following allowances must be taken into account :—

For $1\frac{1}{2}''$ drips—6" in length.

For 2" drips—8" in length.   The same for rolls.

Flashings, add 6" for every angle, and about $\frac{1}{2}''$ for every foot allowance for tacks.   Allow for each passing 4".

Ridges allow as for flashings, and about 18" wide ; valleys and hips are taken rather wider.

Allow a width of 12" for stepped flashings, with same allowances as for straight flashings.

Soakers may be taken the length of slate minus the length of the margin of the slate, by a width of 6 inches.

*The following are taken at per foot run :—*

Lead or oak wedging to flashings.

Copper · nailing, welted edge, burning in, serrated edges, and labour to secret gutter.

Lead pipes at per lineal yard, stating weight.   State if including all soldered joints, tacks, wall hooks, &c.

Number all branch joints, soldered ends.

Bath and w.c. fittings are described, care being taken to specify what labours are to be included, and what are taken separately in other parts of the bill.

### GAS FITTING, BELLHANGING, AND ELECTRIC LIGHTING.

*Gas Fitting.*—Measure by the foot run over all, describing material and diameter ; opening of ground, stating depth and filling in, numbering and taking extra only for all connections, bends, elbows, tee pieces, &c.

Number all perforations and making good, brackets, pendants, and all fittings, for which allow a provisional amount, or select from manufacturer's list and state catalogue number.

*Hot-Water Pipe and Fittings.*—Measured similarly to gas fitting.

*Bellhanger.*—Number all bells, pulls, and cranks. Allow a provisional amount, or quote manufacturer's list, bell boards painted complete.

Copper wire by weight, describing gauge.

Zinc tubing per foot run.

*Electric Bells.*—Measure by the lineal yard wiring, tubing, and casing.

Number the pushes, bells, fittings, indicators, batteries, &c., with description of same.

Allow provisional amount for perforations and making good.

*Electric Lighting.*—Measure by the lineal yard wiring, tubing, and casing, with description of same.

Number the meter, switches, distributing boards, cut-outs, resistances, lamps and all fittings.

Where electricity is not supplied by a company, describe and allow provisional amount for the erection of a motor, dynamo, and accumulators.

Allow a provisional amount for perforations and making good.

## PLASTERERS' WORK.

Internal and external work are measured in a similar manner.

*The following items are measured at per yard superficial :—*

Render
Render and set
Render, float, and set  } to brick walls.
Rough cast

Render in Portland cement to walls and floors; describe if jointed.

Lath, plaster, and set ⎫
Lath, plaster, float, and set ⎬ to partitions and ceilings.
Colouring and whitening. ⎭

*The following items are measured at per foot superficial :—*

Hearths, cornices, and mouldings over 12″ girt, and generally small work over 12″ girt, such as pilasters, &c., in which the labour is more costly than ordinary work.

*The following items are measured at per foot run :—*

All work in narrow margins, cornices and mouldings under 12″ girt, angles, beads, enrichments, quirks, &c.

*The following are numbered :—*

Mitres, splays, stopped ends, extra labours, pateræ. modillions, caps to columns, trusses, centre ornaments, &c.

All circular work to be kept separate.

The dimensions for ceilings with solid cornices are taken between the walls; to those with bracketed cornices, the measurements of ceilings are taken between the cornices. The lengths of cornices run *in situ* or bracketed, are taken as equal to the sum of the lengths of the walls minus two projections of the cornice; but if run on ground or rag and stick work, then take round of wall plus two projections for each of the external angles.

*Memoranda.*—One cubic yard of coarse stuff will require ½ yard of chalk lime, 1 yard of sand, and 9 lbs. of hair.

This should cover in rendered work :—

| 36 superficial yards | 1″ thick |
|---|---|
| 42 ,, ,, | $\frac{7}{8}$″ ,, |
| 48 ,, ,, | $\frac{3}{4}$″ ,, |
| 60 ,, ,, | $\frac{5}{8}$″ ,, |
| 72 ,, ,, | $\frac{1}{2}$″ ,, |
| 96 ,, ,, | $\frac{3}{8}$″ ,, |

but in the contraction and waste there will be a loss of from 12 to

15 per cent., therefore the quantities in the above table must be reduced by that amount to obtain the actual surface covered.

The pricking up coat for lath work will require about 10 per cent. more than that necessary for rendering.

### GLAZIERS' WORK.

Glass is measured by the foot superficial, taking the extreme dimensions each way, and describing thickness as 15 oz., 21 oz., 26 oz., &c. in sheet glass, and the thicknesses in fractions of an inch in plate glass, and state whether patent plate, rough-cast plate, rolled plate, polished plate, or coloured glass, and describe method of fixing as sprigged back and front puttied, beaded and screwed, and if bedded on wash-leather.

In shaped or circular panes, the circumscribing rectangle is measured, and an extra is charged for cutting to shape ; a sketch of the shape should be given. Bent work is measured at per foot super, and charge for all necessary patterns. Lead lights are measured at per foot superficial, with a description and sketch of ornamental work.

### PAINTERS' WORK.

*Painters' work is measured by the yard superficial*, and described as two, three, or four oils, according to the number of coats, and the area covered by the brush is the exact dimension to be paid for.

The finishing coats are described as of Common, Superior, or Delicate tints. Common colours include lamp black, red lead, Venetian red, English or Turkey umber, and all other common ochres as grays, buffs, stones, &c. Superior colours include bright yellows, reds, indigo, prussian blue, mineral green. Delicate tints include verditer, pea greens, rich reds, pinks and bright blues.

In measuring from executed work it is usual to allow for edges in height only, and returns in the width ; in measuring from drawings, add to the surface dimensions for edges to plain work, $\frac{1}{6}$th; to framed and moulded work, $\frac{1}{4}$th, to allow for edges.

Flatting should be described as "*extra to.*"

Staining, graining, sizing and varnishing are measured by the superficial yard.

*The following items are measured at per foot run :*—Skirtings, handrails, gutters, rain-water pipes, iron bars, edges of shelves, strings, cornices, square balusters and newels.

Cutting-in should be described, and the item in which the labour occurs priced accordingly.

*The following items are usually numbered :*—Sash and door frames, each side measured as one, and if with mullions or transomes each division to be considered as a frame. They are taken as 10′ and under 25′; 25′ and under 50′. Sash squares, each side measured as one, by the dozen. They are taken as 1′; 1′ and under $2\frac{1}{2}$′; $2\frac{1}{2}$′ and under 5′; and above 5′. If sash has only one pane it is termed a sheet. Chimney-pieces, door scrapers, ornamental balusters, and newels to staircases, iron air gratings.

*French polishing* on plain surfaces per foot superficial to surfaces covered by the polish.

Mouldings.—Girt and measure by the foot run; as in the example of a moulded spandril, framing panels would be measured over all before framing up, per foot super; styles, muntins, and rails are measured by the foot super; mouldings girt and measured by the foot run.

### PAPERHANGERS' WORK.

The paper is charged at per piece, including hanging, and may be sized and varnished; if throughout, this may be included in description, if only in part measure separately.

To find the number of pieces required (allowing $\frac{1}{7}$ for waste), divide the number of superficial feet in the space to be papered by 54; the result gives the number of pieces of paper required.

The superficial quantity may often be taken from the plasterers' bill.

A piece of paper is supposed to measure 36 feet long, 1¾ feet wide; this equals 63 superficial feet, but it is often found to measure 35 feet long, 1⅝ feet wide. This equals 58⅓ superficial feet.

### GENERALLY.

The tendency of modern quantity surveying, in order to approach accuracy in estimating, is, 1st, to cube all the material and to pay in addition for the separate labours; 2nd, where this method is too tedious, the work is described and paid for by the superficies; again, 3rd, where the superficies is relatively small and the material has no fixed ratio to the cost of labour, the material and labour are fully described and paid for by the lineal dimension; again, 4th, where the labour is an unknown quantity or has a fancy value, then the detail is numbered; but it must be borne in mind that for similar work the accuracy obtained will be in the order respectively of the methods stated, though the time taken to obtain the answers will be in the reverse order, that is, the 1st method will prove the most accurate, but it will occupy the longest time.

*Approximate Estimating.* — Approximate estimates are obtained by finding the cubic dimensions of a building, measuring from the underside of the brick footings to halfway up the slope of the roof, adding any important projections and pricing at the cost this work may be judged to be worth. The success of this method will depend upon the experience and good judgment of the estimator.

Churches, chapels, &c. are often estimated per seat of the seating accommodation; hospitals, per bed; schools, per head of the number of scholars accommodated; and dwelling houses, per foot super of the floor area.

The cubing is the method usually adopted.

# EXERCISES.

## DRAWING.

| No. | Question. | No. | Question. | No. | Question. | No. | Question. |
|---|---|---|---|---|---|---|---|
| 1 | (46) | 2 | (62) | 3 | (71) | 4 | (83) |

## BRICKWORK.

| No. | Question. | No. | Question. | No. | Question. | No. | Question. |
|---|---|---|---|---|---|---|---|
| 1 | (1) | 7 | (16) | 13 | (38) | 19 | (74) |
| 2 | (5) | 8 | (17) | 14 | (56) | 20 | (75) |
| 3 | (9) | 9 | (30) | 15 | (58) | 21 | (88) |
| 4 | (12) | 10 | (36) | 16 | (64) | 22 | (93) |
| 5 | (13) | 11 | (41) | 17 | (69) | | |
| 6 | (15) | 12 | (54) | 18 | (72) | | |

## MASONRY.

| No. | Question. | No. | Question. | No. | Question. | No. | Question. |
|---|---|---|---|---|---|---|---|
| 1 | (12) | 4 | (26) | 7 | (52) | 9 | (67) |
| 2 | (17) | 5 | (37) | 8 | (58) | 10 | (91) |
| 3 | (18) | 6 | (44) | | | | |

## GIRDERS.

| No. | Question. | No. | Question. | No. | Question. |
|---|---|---|---|---|---|
| 1 | (7) | 2 | (79) | 3 | (94) |

## JOINTS IN CARPENTRY.

| No. | Question. |
|---|---|
| 1 | (6) |
| 2 | (20) |

## FLOORS.

| No. | Question. | No. | Question. | No. | Question. |
|---|---|---|---|---|---|
| 1 | (2) | 3 | (28) | 5 | (54) |
| 2 | (8) | 4 | (39) | | |

## WOOD ROOFS.

| No. | Question. | No. | Question. | No. | Question. |
|---|---|---|---|---|---|
| 1 | (10) | 3 | (21) | 5 | (70) |
| 2 | (11) | 4 | (59) | 6 | (78) |

## IRON ROOFS.

| No. | Question. | No. | Question. | No. | Question. | No. | Question. |
|---|---|---|---|---|---|---|---|
| 1 | (14) | | | | | | |
| 2 | (22) | | | | | | |

## JOINERY.

| No. | Question. | No. | Question. | No. | Question. | No. | Question. |
|---|---|---|---|---|---|---|---|
| 1 | (3) | 5 | (27) | 9 | (47) | 13 | (81) |
| 2 | (6) | 6 | (35) | 10 | (51) | 14 | (82) |
| 3 | (23) | 7 | (36) | 11 | (61) | | |
| 4 | (25) | 8 | (44) | 12 | (66) | | |

## PLUMBING.

| No. | Question. | No. | Question. | No. | Question. | No. | Question. |
|---|---|---|---|---|---|---|---|
| 1 | (4) | 4 | (48) | 6 | (65) | 8 | (90) |
| 2 | (19) | 5 | (49) | 7 | (76) | 9 | (92) |
| 3 | (40) | | | | | | |

## SLATING AND TILING.

| No. | Question. | No. | Question. | No. | Question. | No. | Question. |
|---|---|---|---|---|---|---|---|
| 1 | (10) | 4 | (34) | 6 | (50) | 8 | (80) |
| 2 | (24) | 5 | (42) | 7 | (60) | 9 | (87) |
| 3 | (29) | | | | | | |

## MISCELLANEOUS.

| No. | Question. | No. | Question. | No. | Question. | No. | Question. |
|---|---|---|---|---|---|---|---|
| 1 | (31) | 4 | (44) | 7 | (53) | 10 | (63) |
| 2 | (32) | 5 | (45) | 8 | (57) | 11 | (68) |
| 3 | (43) | 6 | (49) | 9 | (58) | 12 | (86) |

## MATERIALS.

| No. | Question. | No. | Question. | No. | Question. | No. | Question. |
|---|---|---|---|---|---|---|---|
| 1 | (30) | 4 | (72) | 7 | (84) | 9 | (89 |
| 2 | (33) | 5 | (73) | 8 | (85) | 10 | (92) |
| 3 | (55) | 6 | (77) | | | | |

# APPENDIX

—◦◦—

## BOARD OF EDUCATION SYLLABUS.

—

(*Published September 3rd, 1906.*)

### SUBJECT III.

## BUILDING CONSTRUCTION AND DRAWING,

The instruction given should be so arranged that by the time the Student finishes his course of study, he should have acquired a knowledge of building materials, plant and construction sufficient for the work upon which he is likely to be engaged. That he may be able to make free use of this knowledge in practice, he must also be a good draughtsman ; good drawing is an essential part of the course, but it must always be borne in mind that drawing is a *means* and not an end in itself; drawings of work to be carried out should be such as to give full information and exact guidance to workmen who may have to use them. In the higher stages of the subject students should acquire proficiency in making finished drawings as well as what may be called descriptive and explanatory drawings.

A larger number of questions will be set in the examination papers than the Candidate will be allowed to attempt, so that he may have some range of selection of questions which bear upon branches to which he has given special attention.

Compulsory questions may be set at the examinations.

It should be seen that candidates are fairly provided with pens, ink, pencils, and drawing instruments (including tee and set squares, drawing boards, Indian ink, &c.) when they present themselves for the examination. The use in examination of the ordinary box-wood, ivory, or paper scales and protractors, and slide rules is permitted. Tables will be supplied to candidates in Stage III. and Honours to assist them in calculation.

### STAGE I. (ELEMENTARY).

In Stage I., the drawing exercises should not extend beyond descriptive and explanatory drawing, but they should aim at cultivating a fair degree of skill in pencil drawings. All lines should be neat and clear. Students who are quick in executing their pencil drawings should practise making ink tracings with clear lettering and figuring.

All students should practise freehand drawing of details, so that they may be able readily to make a neat dimensioned sketch from which a drawing to scale might afterwards be prepared or which may itself be sufficient for purposes of explanation. The use of squared paper may be introduced with advantage in exercises of this kind.

The course should include elementary instruction, with reference to the various materials used in building. Each group of materials should be taken up in the class as introductory to a series of drawing exercises, illustrating their use in buildings so far as suitable for discussion in a first year's course. There would fall to be considered in this way: the nature and properties of sand, lime, and cement; the composition of mortar or concrete and its application in floors, walls, &c.; the properties of bricks, stones, tiles, and slates; the various kinds of timber in ordinary use; the constituents of cast-iron, wrought-iron, and steel, and the essential or characteristic differences of their properties.

Instruction should be given as to foundations in ordinary soils, footings for walls of moderate height; the construction of simple scaffolding; the various bonds of brickwork in plain walling, flues, arches, and fire-places; varieties of simple masonry such as rubble and ashlar walling and the plain masons' work on sills, reveals, &c.; plain carpentry in floor joists, stud partitions, ordinary roofs of span not exceeding that for a King-Post truss; firrings of flats; simple joiners' work in floor laying, skirtings, deal-cased frames and double-hung sashes, and solid frames for simple casements, panelled doors and jamb linings, door frames and ledged and braced doors; ordinary plastering on walls, partitions, and ceilings, and the composition of the various coats; slating, including the dressing, cutting, and nailing of the slates; plain tiling and pan tiling and the various methods of hanging the tiles, and the treatment of valleys, hips, ridges, and eaves; roof plumbing, including the laying of flats with rolls, drips, &c., lead gutters, and flashings; simple glazing. Students should also be taught how to draw the sections of rolled joists, channels, angles and tees.

In all these subjects practical examples of the materials used and the various operations of dealing with them should be brought before the student, either in the class room or elsewhere; in as many cases as possible, he ought actually to see and handle full-size examples of everything in which he is being instructed theoretically. He should also familiarize himself with the nature and use of all the tools used in elementary building operations. Students should lose no opportunity to inspect any building opera-

tions going on in their locality.    Every student ought to examine
in detail the structure of the houses in which he lives and works
and attends classes.

## STAGE II. (ADVANCED).

Before proceeding to Stage II., Students should have a good
knowledge of the subjects included in the Syllabus for a Preliminary
Course for Trade Students, as well as of those subjects included in
the above Syllabus for Stage I.

The Course of Instruction in this Stage should cover a more
advanced knowledge of all the subjects enumerated for Stage I.,
together with simple exercises in calculating quantities of materials,
not such calculations as a Quantity Surveyor would make, but such
as would fall to be made by a Foreman of Works who has to order
sufficient materials for the amount of work which he knows has to
be done.

The class lessons and drawing practice should include the
following subjects : Excavation in various kinds of soils, including strutting and planking, concrete foundations for walls and
piers, the use of damp courses and the materials employed for
them; gauged brickwork; hollow walls and the various methods
of bonding them together; junctions of walls of various thicknesses and at different angles ; chimney breasts and flues; irregular
bonds ; fireproof construction in floors and roofs ; the best
known building stones, their quarrying, bedding, cutting and
dressing ; characteristics of timber, its conversion and seasoning.
Attention should be given to the increasing use of machinery in
treating timber for carpenters' and joiners' work ; advanced
carpentry and joinery ; ordinary forms of staircase construction
with close strings and bent strings ; two and three-light windows
with cased frames, and hung sashes, and also with solid frames,
mullions and transoms and casements, outside doors with bolection
mouldings, sash doors and the finishings of door and window
openings ; finishing in eaves, hips, ridges, &c.; the nature,
qualities and weights of various kinds of slates ; elementary
drainage ; the laying and jointing of glazed stoneware pipes ;
advanced constructional plumbers' work, including cold water
supply to cisterns, and the position of the same in a house, baths,
sinks, water-closets and their connections, waste pipes, soil pipes,
ventilation pipes, &c. ; scaffolding for large buildings, shoring,
strutting, needling, and under-pinning ; centring for arches up to
15 feet span ; the general principles of loaded beams ; bending
moments due to concentrated and distributed loads ; the use of
the triangle and polygon of forces in order to practically determine the resultant force in direction and magnitude, and to resolve
such a resultant into its component forces ; the determination of
the stresses in simple braced structures ; elementary exercises in
the calculation of strength of materials.

## STAGE III. (OR HONOURS, PART I.).

The Course of Instruction should include the consideration of buildings of all kinds and sizes. In the examination the Candidate will be expected to show that he has a fair knowledge of the principles of Physical Science as illustrated in relation to building construction. He should be able to design simple roof trusses and beams, and to draw their stress diagrams; he should know the elements of the theory of arches, how to provide for the stresses in various parts of a building, and the methods of inspecting and testing cement, timber, iron and steel, and the use of formulæ.

In the various sections of the course exercises in calculating quantities of materials should be continued as in the preceding Stage.

The class lessons and drawing practice should include the consideration of:—

Foundations—natural and artificial, upon land and under water, damp sites and their treatment. Brickwork, including all kinds of bonding, setting out bond in frontages, etc.

Terra cotta and artificial stone; their manufacture and uses.

Principles of sanitation; drains, traps, gulleys, disconnecting chambers, sewers, their ventilation and drain connections, iron drains. Drain testing and ventilation.

Masonry. Character of various stones used in building and localities where found, how to test for quality and bed, fitness of various stones for different atmospheres, weight generally, and approximate strength; stone stairs, composite walls, arches.

More detailed knowledge of scaffolding, including gantries, elaborate centring, framing for concrete walls and modern methods of hoisting materials, roofing up to 60 feet span. Timber: its seasoning, diseases, cause of decay, and means of preserving it. Roof timbering, open, hammer beam, and composite trusses. Modern iron trusses, including trussed purlins; all roof finishings, including slating, tiling, plumbing, etc., sky lights and lanterns. Wood stairs of all kinds, including handrailing.

Cast iron, wrought iron and steel, properties, uses, strength, weight and preservation. Iron and steel columns, stanchions and girders, including riveting, bolting, etc. The calculation of bending moments and shearing stresses.

Ventilation and heating; hot water supply; provisions for gas and electric supply, in so far as these may affect the structure of the building; water supply; lightning conductors; preservation of timber; various kinds of glass and glazing; plastering in all its branches.

Attention should be specially directed to the increasing use of skeleton construction in steel, and to ferro-concrete construction.

E E

## HONOURS.

No candidate will be credited with a success in Honours who has not obtained a previous success in Stage III. or in Honours of the same subject under previous regulations, and who does not qualify in the Board's examination in Architecture. The qualification in Architecture need not be obtained in the same year as that in the Honours examination in Building Construction and Drawing.

The Examiners will have in mind in setting the questions the actual practice of architects in designing buildings, and in their guidance of assistants and clerks of works, to ensure that orders will be properly carried out, the dealings with contractors, and also the actual erections of buildings and carrying out of building operations. Candidates will be asked to make sketch designs and to give instructions to draughtsmen for careful scale drawings and specifications. The questions may deal with any part of the subject and with any kind of building, and may require a knowledge of any materials or construction in use in good practice.

Those candidates whose answering of the paper is sufficiently satisfactory will be summoned to South Kensington or some other centre for a practical examination. This further examination will last for two or more days; the time will not exceed seven hours each day. Candidates will be asked to design a building suitable for a definite purpose, and they will be called upon to give such plans, elevations, and sections, and such details and notes for a specification, as shall be required by the Examiner. An estimate of cost may also be demanded.

Intimation concerning the general nature of the building to be designed will be sent in advance to candidates, together with the notice to attend for this second part of the examination.

For this practical examination candidates must provide T-squares, set squares, drawing instruments, ink, and colours. Drawing paper and drawing boards will be supplied by the Board of Education.

No candidate can be classed in Honours who is not successful in the practical examination.

# EXAMINATION PAPERS

OF THE

# BOARD OF EDUCATION,

SOUTH KENSINGTON, LONDON.

———

## SUBJECT III.—BUILDING CONSTRUCTION AND DRAWING.

———

### STAGE I.
### GENERAL INSTRUCTIONS.

IF THE [RULES ARE NOT ATTENDED TO, YOUR PAPER WILL BE CANCELLED.

*Immediately before the Examination commences, the following*

## REGULATIONS *are* TO BE READ TO THE CANDIDATES.

Before commencing your work, you are required to fill up the numbered slip which is attached to the blank examination paper

You may not have with you any books, notes, or paper other than that supplied to you for use at this examination.

You are not allowed to write, draw, or calculate on your paper of questions.

You must not, under any circumstances whatever, speak to or communicate with another candidate. Those superintending the examination are not at liberty to give any explanation bearing upon the paper.

You must remain seated until your papers have been collected, and then quietly leave the examination room. No candidate will be allowed to leave before the expiration of one hour from the commencement of the examination, and none can be re-admitted after having once left the room.

All papers, not previously given up, will be collected at 10 o'clock.

If any of you break any of these regulations, or use any unfair means, you will be expelled, and your paper cancelled.

---

Before commencing your work, you must carefully read the following instructions :—

The questions may be answered in any order, but each answer must be clearly marked with the number of the question to which it refers.

Drawings must be made on the single sheet of drawing paper supplied, beginning on the side marked with your distinguishing number, which must face you at the right-hand top corner. *Sketches* may be made by hand on the squared paper attached to the drawing paper. Where the instruction in the question is "sketch," the sketch may either be in pencil or the pencil sketch may be gone over with the ordinary writing pen and ink to make the lines clearer. Additional foolscap will, if necessary, be supplied to you by the Superintendents.

The *tracing* is to be drawn on the piece of tracing paper attached to the drawing paper.

Answers in writing must be as short and clearly stated as possible, and the references to drawings and sketches must be made absolutely clear by letters or numbers.

The value attached to each question is shown in brackets after the question. But a full and correct answer to an easy question will in all cases secure a larger number of marks than an incomplete or inexact answer to a more difficult one.

Questions marked (*) have accompanying diagrams.

*The examination in this subject lasts for four hours.*

*The numbers in brackets relate to the general numbering of all the questions printed in this volume, and are those referred to in the List of Exercises (pp. 412, 413).*

## MAY, 1900.

### ELEMENTARY STAGE.

#### INSTRUCTIONS.

You are permitted to answer only *seven* questions.

[*Note.—The questions starred have an accompanying illustration.*]

(1.)  1.  Briefly define ordinary English brick bond.

Draw, to a scale of $\frac{3}{4}''$ to a foot, a plan showing the arrangement of the bricks in a $2\frac{1}{2}$ brick wall built in English bond, the bricks in the course below being indicated by dotted lines.                                       (11.)

(2.)  *2.  Section of part of a common floor, showing $9'' \times 3''$ joists, and $1\frac{1}{2}''$ boarding with a heading joint.

Draw, to a scale of $\frac{1}{8}$, making any alteration you think necessary, and adding pugging and a lath-and-plaster ceiling.

State the object of pugging, and give the name of the heading joint.                                       (11.)

(3.)  *3.  Interior elevation of a $7' \times 3'$ door.

Draw to a scale of $\frac{3}{4}''$ to a foot, making any alteration you think necessary,

Write against the door its name, and the names of all its different members.  The joints need not be shown.   (11.)

(4.)  4.  Draw one-half full size a section of an ordinary $2''$ lead roll, and of a hollow roll, as used for lead flats.   (11.)

(5.)  *5.  Plans of two successive courses of brickwork at the junction of a party wall with the main wall of a building, the latter being built in single Flemish bond.

Draw, to a scale of $\frac{3}{4}''$ to a foot, showing the joints of the brickwork.                                       (12.)

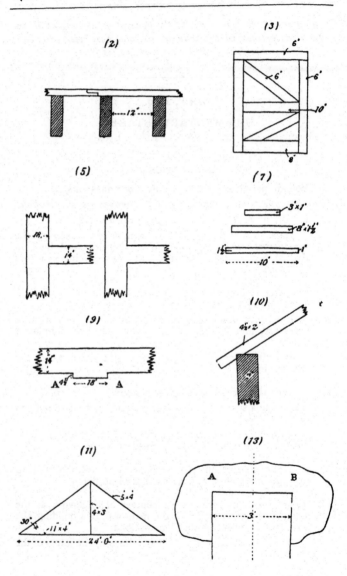

(6.)  6.  Give sketches explanatory of the following terms :—
                Flitch girder.
                Tusk tenon.
                Squint quoin.
                Transom.                          (12.)

(7.)  *7.  Sections showing the different members of a cast-iron
cantilever.

   Draw the section of the cantilever, ⅓ full size.    (12.)

(8.)  8.  Show clearly by sketches the difference between a double
floor with rebated and filleted battens, and a framed floor
with ploughed and tongued boards.                 (14.)

(9.)  *9.  Horizontal section through part of a brick boundary
wall built in English bond.

   Draw to a scale of 1½ inches to a foot, showing the
joints of the bricks in English bond in two courses, the
lower one by dotted lines ; also an elevation of the face *A*,
showing the four top courses of the wall, finished with a
stone saddle-back coping 6″ deep, 3″ wider than the wall,
and weathered down 3″.
   Add a cross-section of the coping over the 14″ brickwork,
showing it properly treated so as to throw rain clear of the
face of the wall.                                 (14.)

(10.)  *10.  Section of part of a small span roof.

   Draw to a scale of 1″ to a foot, adding Countess slates
with 4″ lap, centre nailed on 2″ × ¾″ battens, a 3½″ half-
round eaves gutter, and a 1″ beaded fascia board. The
rafter to project 9″ from face of wall.  Pitch of roof 30°.

   Make any other addition or alteration you think necessary.
Name the different parts.                          (14.)

(11.)  *11.  Line diagram of an ordinary timber truss for a 24-ft.
span.

   Draw at least half the truss to a scale of ¼₀, making any
addition you think necessary.

   Give the name of the truss, and enlarged drawings, to a
scale of 1½″ to a foot, of the joints at the head and foot of
the principal rafter.                             (14.)

(12.)  12.  Draw to a scale of 1½″ to a foot, the plan and central
cross-section of an 11″ × 6″ stone window sill for a 3′ 6″
opening.  The sill to be weathered, throated, and grooved,
and to project 2½″ from a 14″ brick wall.

At right end of plan, show the course of brickwork, in English bond, immediately above the sill, with a 4½″ reveal; and at left end the brick work upon which the sill rests.

(14.)

(13.)  *13.  External elevation of a window head in a 14″ brick wall finished externally with a 14″ gauged camber arch, and internally with a 9″ × 3″ wood lintel and common arch.

Draw to scale of 1½″ to a foot, showing at *A* the details of the external elevation, and at *B* the internal elevation, the brickwork being in single Flemish bond.        (15.)

(14.)  14.  Draw to a scale of 1/24, a little more than half of an iron king-rod roof over a 28-ft. span.  Rise ¼ span.  Show :—

Principal rafters and struts, of tee iron.

King and tie rods, of round iron.

Common rafters and their supports, of wood.

The scantlings and joints can be assumed.        (15.)

## ELEMENTARY STAGE.

### INSTRUCTIONS.

You are permitted to answer only *seven* questions.

(15.) 1. Why is a brick made so that its breadth is less than half its length? Can you explain any advantage in "perforating" bricks? What are gauged arches? (12.)

(16.) 2. Draw, to the scale of $\frac{1}{16}$, the plans of two courses of a brick pillar in English bond; the pillar to be square in section, four bricks in the side. To what height could you safely build a pillar of this section, assuming the safe load on a course of bricks to be 8 tons per square foot and taking the weight of a cubic foot of brickwork at 120 lbs.—neglecting wind pressure? (12.)

(17.) 3. What are:—a skew arch; a skewback; a pillar; an abutment; a column; weathering of a window sill; weathering of stone; voussoirs; chamfer; plinth? (12.)

(18.) *4. A rubble wall with half-round concrete coping; draw it to the scale of $\frac{1}{16}$. Showing the stones and mortar. (12.)

(19.) 5. Show three soakers with cover flashing in position against a brick wall—slates 24″ long. Show clearly to the scale of $\frac{1}{4}$ all details, with such drawings and sketches, accompanied by explanations and dimensions, as you think sufficient. (12.)

(20.) 6. Draw or sketch, to the scale of $\frac{1}{4}$ :—

(1.) A tenon and housed joint, mortised piece 4″ × 4″, tenoned piece 3″ × 4″.

(2.) A tusk-tenoned joint, 9″ × 3″ timbers.

(3.) A dovetailed notch, timbers $4\frac{1}{2}$″ × 3″. (12.)

(21.) 7. Draw or sketch, to the scale of $\frac{1}{8}$, the top and bottom of a king-post showing attachments of principal rafters and of the struts and tie-beam, and showing ironwork:—rafters 8″ × 4″, shank of king post 6″ × 4″, struts 4″ × 4′ tie-beam 12″ × 4″. (12.)

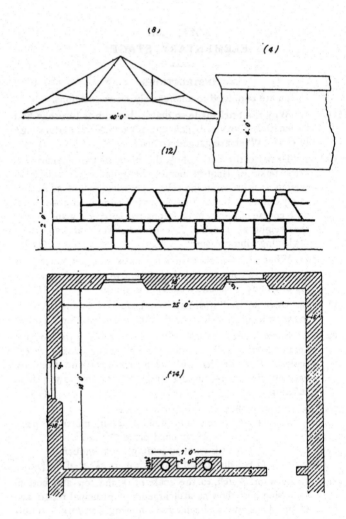

(22.) *8. A skeleton drawing of an iron roof truss. The rafters are formed of ∟ irons. Draw, to the scale of ½, the joint at the apex. Repeat the diagram on your paper, and putting reference numbers to the different members, sketch cross-sections of them. (14.)

(23.) 9. Draw or sketch, to the scale of ¼, a cross-section showing bottom rail of sash, wood sill of window frame, stone sill, and window back; showing elbow. The window back is 2 feet high and the wall of the recess is 10 inches thick. (14.)

(24.) 10. Draw or sketch, to the scale of ⅛, cross-section of eave extending 4 feet up the roof showing slates resting over a cut stone eave course: the wall is 18″ thick, rafters 5″ deep, without roof trusses,—ceiling joists at wall plates; slates 20″ long. Explain fully. (14.)

(25.) 11. Draw, to the scale of 1/12, the inside elevation of a ledged and braced door, and door frame; the door is 7′ × 3′. Show hinges, latch, stock lock; sketch cross-section showing jambs (wall 1½ bricks thick). (14.)

(26.) *12. Two courses of masonry: how would you describe it? Sketch these two courses on your paper, and put reference numbers on the stones of the top course showing the order in which you think a mason would set them, giving reasons. (14.)

(27.) 13. Draw, to the scale of ⅛, elevation of a double casement or French window 6′ × 3′ 6″: draw cross-section of bottom bar of window and sill of frame; sketch hinges and fastening bolt. What is this kind of bolt called? (15.)

(28.) *14. A floor, of fir timber, joists lathed and plastered below to form ceiling of lower room. Draw plan, to the scale of 1/48, showing by single lines complete joisting, figure scantlings. Draw, to the scale of ⅛, cross-section through hearth reaching above camber-bar of fireplace. Show details of bearing of joists and of trimmer. (15.)

## *MAY*, 1902.

## ELEMENTARY STAGE.

### INSTRUCTIONS.

You are permitted to answer only *seven* questions.

(29.) *1. Sketch this tool upon your paper, showing the pick better placed, and explain why you alter it. For what is this tool used? (12.)

(30.) 2. What are:—Reveals, Jambs, Collar braces (Collar beams), Battens, Studs, Deals, Planks, Perpends, Screeds, Kingcloser? (12.)

(31.) *3. The figure shows in skeleton a bracket used by builders for pointing brickwork or for outside plastering: sketch the bracket so that an exact drawing could be made from your sketch (marking dimensions). Show clearly how its parts are connected (the parts being of red deal); show how it is supported when hung against a wall. (12.)

(32.) *4. What is the name of the bracket of the previous question? If we assume that it carries a uniformly distributed load of 8 cwt. on *AB*, so that we may imagine a downward force of 4 cwt. at *A*, what kind of stress is in *AC*, and what is its amount? (12.)

(33.) 5. Given sand, lime, and hair, as delivered at the building, describe in detail how you would prepare "*coarse stuff*" for plastering—you are not supposed to have a mortar mill. (12.)

(34.) 6. Sketch neatly to the scale of about ⅛ a slate of the dimensions 24″ × 12″ dressed and holed: the lap is 4″. What are the dimensions of the "weather" (or margin) of this slate? What is the distance from the hole to the tail? (12.)

(35.) *7. What is the name of the lock shown? Describe how you would fix it to a door. Explain and illustrate by sketches the mechanism of a common single tumbler lock. What are the "*wards*"? (12.)

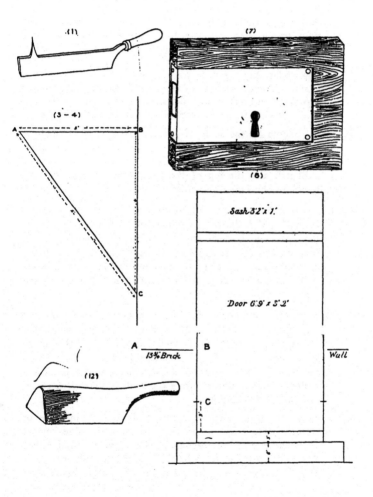

(1)

(7)

(3 – 4)

A       5'       B

4'

C

(8)

Sash 3'2" x 1".

Door 6'9" x 3'2"

A   13% Brick   B      Wall

(12)

C

(36.)   *8.   The sketch shows a door frame for an outside door:
it is set upon door-blocks and the brick walls are being built
to it.

   Draw carefully to the scale of $\frac{1}{12}$ an elevation of the
door frame, step, sill, and brickwork (showing the joints
by double lines to the left of the door for, say, three bricks
from the door): show temporary bracing: describe how
you would stay the door frame temporarily.   Draw, to the
scale of $\frac{1}{4}$, cross-section of frame at $AB$, showing the plan of
top course of brickwork, and a short piece of vertical section
at $C$, showing the connexion with the door block.     (14.)

(37.)   9.   Sketch to the scale of $\frac{1}{12}$ a sample of snecked rubble
masonry face (say about 4' by 4').   Show the mortar joints
with double lines.

   Sketch also the top of the sample, as a plan, showing
how you bond the wall across.                       (14.)

(38.)   10.   A brick wall, 21' high (measured from the soil on which
it rests) 13$\frac{3}{4}$" thick, carries a load of 1 ton per foot run on
its top.   Say what is the approximate weight of a cubic
foot of brickwork.   Draw, to the scale of $\frac{1}{24}$, a cross-
section of the footings on the assumption that the soil is not
to be stressed to a greater amount than 1 ton per square
foot.                                               (14.)

(39.)   11.   Describe exactly the laying of batten width tongued
and grooved common Baltic flooring; how would you
manage when the floor has been finished so far that there
is no longer room for the cramps between the wall and
finished floor?   Sketch the usual flooring nail.   What is it
called?   Where do you drive the nails?   Sketch a cross-
section of a heading joint.                         (14.)

(40.)   *12.   For what purpose is this tool used?   You have to
cover a right circular cone with 6 lbs. lead.   The cone is
3' in diameter at the base, and it is 2' high.   Assuming that
the joints are butted, what is the weight of the lead?
Such a cover being made, if a straight cut is made from
the apex to the circumference of the base, the cover may
be made to lie flat; draw to the scale of $\frac{1}{12}$ its outline
when thus flattened.                                (14.)

(41.)   13. What is bond in brickwork? When you say that a
certain wall is built in Flemish bond, to what do you refer?
You have to build a 1½ brick wall, showing Flemish bond
in *one face*—the appearance of the other face is of no consequence as it is to be plastered. *No bats are allowed.*
Sketch the plan of a course, say 5 bricks long, in full lines
and show the joints of the course below in dotted lines.

(15.)

(42.)   14. Describe carefully the work of laying a kitchen floor
with 6″ × 6″ tiles, ½″ thick, in two colours.   The floor is
14′ × 13′ : how many tiles ought to be *ordered* for the
work?                                                    (15.)

*May*, 1903.

ELEMENTARY STAGE.

You are permitted to answer only *seven* questions.

(43.)  *1.  For what work is this tool used? Describe how the workman holds it when he uses it for ordinary work.

(12.)

(44.)  2.  What are:—meeting-rail, muntin, tread, riser, going, lacing course, quirked ogee, coarse stuff, gauged stuff, droved work, rusticated? (Full marks will be given for *nine* correct definitions.)        (12.)

(45.)  *3.  In a certain town electric wires are borne on brackets, like that shown: given the weight at *A* as being 40 lbs., assuming that this pressure may be transferred to the point *B* where the bracket bears against the slate, as shown, and assuming that the directions of the reaction and of the third force are as shown; what is the amount of the reaction in lbs.?        (15.)

(46.)  *4.  Make a tracing, in ink, of this drawing and of the writing and figures. (The Indian ink should be sufficiently thick to give opaque lines, suitable for photographic printing; the lines should be well defined, uniform in breadth, having firm unbroken edges; and they should neither stop short of nor go beyond the proper points.)        (15.)

(47.)  5.  Sketch (approximately) to the scale of $\frac{1}{2}$, a good strong thumb latch for a door; also the keeper, or catch, to be fastened to the door frame. Describe what you consider to be important points in a good latch, such as you sketch.

(12.)

(48.)  6.  (*a*.) What is a bib-cock? (*b*.) Describe how a "wiped-joint" is made.        (12.)

(49.)  7.  A rectangular rain-water tank weighs 300 lbs., it is 6′ × 4′ and it is 2′ 6″ deep (internal dimensions), it is supported with its bottom horizontal. Owing to a stoppage of the overflow pipe, it is filled with water. (*a*) What is the

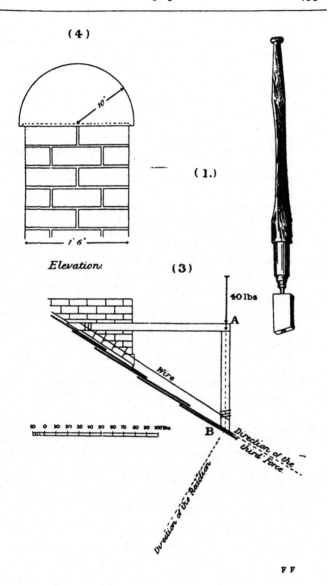

( 4 )

10'

1' 6"

Elevation.

( 1.)

( 3 )

40 lbs

A

Wire

Direction of the third Force

B

Direction of the Resultant

10  0  10  20  30  40  50  60  70  80  90  100 lbs

pressure of the water, per square foot, on the bottom of the tank? (*b.*) What is the weight of the tank and water? (*c*) How many gallons of water does the tank contain?

(14.)

(50.) 8. The slates on a roof are 24″ × 12″, an average slate weighs 7 lbs., the slates are laid with a lap of 4 inches; what is the weight of slates which cover an average square of roof? (14.)

(51.) 9. Draw, to the scale of $\frac{1}{12}$, the elevation of a casement window (not a *pivoted* sash) 4′ × 2′ 6″ (sash size); a single sash in 4 panes; show reveals and stone sill; sketch, approximately to the scale of $\frac{1}{4}$, the essential details of the hinges and fastener. (14.)

(52.) 10. Sketch, to the scale of $\frac{1}{12}$, the face (elevation) of a completed course of rubble masonry (the course being, say, 15″ deep); sketch on top of this completed course portions of a second course in process of building; explain why the stones shown—of the second course—are placed where you sketch them. How does the mason keep his work truly in line? How does he use his plumb rule in rock-faced work? (14.)

(53.) 11. A plasterer is laying on the first coat of coarse stuff on lathed work, and you see him driving the trowel at right angles to the direction of the laths. This is wrong; can you explain why it is wrong? (12.)

(54.) 12. Sketch to the scale of $\frac{1}{8}$ (approximately) a fireplace—grate, oven, and boiler—suitable for the kitchen or living-room of an artizan's cottage; explain the setting and flues. (14.)

(55.) 13. Describe how you would prepare (from materials in the usual commercial conditions) a pot of light-coloured paint for fourth coating, inside work; give a name to the shade of colour you produce. (12.)

(56.) 14. Draw plans of two successive courses of a half-brick built chimney stack of three flues (scale $\frac{1}{12}$); show mortar joints as double lines, show pargetting. (12.)

*MAY*, 1904.

ELEMENTARY STAGE.

———

INSTRUCTIONS.

You are permitted to answer only *seven* questions.

(57.)  *1.  What workman uses this tool?  Can you give a reason
for the grooving of the hammer face *a*?  What purpose is
served by the slot *b*?  Why is the head of this hammer
attached to the handle by cheeks *d*, not by a hole in the
head?                                                    (12.)

(58.)  2.  What are :—Faucit, stopped bead, bolection moulding,
banker, tuck pointing, rusticated, priming, joggle, going
(of stairs), lintel, sectional elevation, architrave?  Nine
correct definitions will get full marks.                (12.)

(59.)  *3.  The skeleton drawing shows two rafters tied at the
feet by a ceiling joist : copy it upon your drawing paper.
Assuming that the joint at *A* is held by a smooth pin
which is at right angles to the paper, find the stresses
produced by the applied force, making use of the triangle
of forces, and write the amount of the resulting stress in
each rafter in cwts. and tenths, along the line of the
drawing corresponding to the particular rafter.         (15.)

(60.)  4.  Answer either (*a*) or (*b*), not both :—

(*a*.) A workman is dressing roofing-slates.  Describe
how he divides a slate into two parts.      (12.)

(*b*.) Sketch on your squared paper a straight line extending over 12 spaces : this line represents the edge
of a roofing-slate which is 24″ long ; the lap being
4″, mark on the line the position of the hole and
dimension its distances from the head and from
the tail of the slate (the slate is not to be " nailed
at the head ").                            (12.)

**F F 2**

(61.)  5.  Answer either (*a*) or (*b*), not both :—

>  (*a*.) Assume the distance between two lines of your
>  squared paper to represent 1″. Sketch upon it
>  a tower bolt fastening attached to a door and
>  door-post : it should be such a bolt as may be
>  secured by a padlock.                          (12.)

>  (*b*.) To the same scale as for *a*, sketch on the squared
>  paper a hook-and-eye hinge for a stable-door which
>  opens outwards and lies flat against the wall when
>  fully turned to it : the door-post is 4″ × 3″, the
>  reveal is 6″ deep.                              (12.)

(62.)  *6.  Make a tracing, in ink, of this ornament ; trace also
the writing and dimensions. (The Indian ink should be
sufficiently thick to give opaque lines suitable for photographic printing, the lines should be well defined, each
line uniform in breadth, having firm unbroken edges ; and
they should neither stop short of nor be carried beyond
the proper points.)                                 (14.)

(63.)  7.  Common window glass is specified as of 15 oz., 21 oz.,
etc. What do these weights refer to? What is this kind
of glass called ?                                   (12.)

(64.)  8.  What is a rod of brickwork? Sketch on your squared
paper (assuming the distance between two lines to represent 3″) a top angle of a window opening, showing, say,
3 courses of bricks down the reveal and about the same
distance along the soffit, and extending 3 or 4 bricks'
length horizontally from the reveal and the same distance
upwards from the soffit. What do you call the bond
which you show in the general facing ?              (15.)

(65.)  9.  Answer either (*a*) or (*b*), not both :—

>  (*a*.) Assuming the gauge of your squared paper to
>  represent 1″, sketch a side view of the ball valve
>  for a water cistern. Show by sketches (enlarged
>  if you think necessary) section of valve from which
>  may be seen how the rising and falling of the ball
>  shuts and opens the valve.                       (12.)

*(1)*

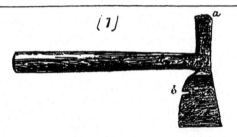

*(3)*

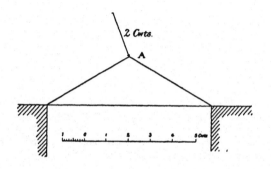

*(6)*

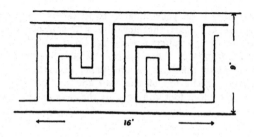

*Greek Fret.*

(*b.*) Assuming the gauge of your squared paper to represent ½″, sketch a section showing clearly a union connexion of the brass ferrule to a 1″ lead pipe.                                                    (12.)

(66.) 10. Draw to the scale of ½ a cross-section of the pulley style and casing of a superior sash frame, 2″ sashes: show the joints clearly; mark on the parts the end grain of the wood; show the weights and parting slip, back lining, etc., complete. How do you fix the guard beads (window slips)? What advantage is claimed for forming a shallow rabbit (rebate) on the pulley styles to form a seat for the guard beads?                                              (14.)

(67.) 11. Describe how you would : (*a*) divide a block of Bath-stone into two useful pieces; (*b*) divide a block of granite into two useful pieces.

Sketch on your squared paper the tools used in each case.                                                              (14.)

(68.) 12. Describe fully the operations of opening a flagged yard; sinking a trench; laying 9″ glazed stoneware spigot and socket pipes jointed with Portland cement mortar; refilling the trench and reinstating the flagged yard (the ground is easy ground, no timbering is required).      (14.)

(69.) 13. Under what circumstances would you put inverted arches in walls just over the foundations: and in what circumstances would you use arches not inverted in the foundations of a building?                                   (14.)

(70.) 14. Sketch on your squared paper (assuming the lines to represent 1′ apart) the elevation of a pair of collar-braced (collar beam) rafters : walls 14′ apart. Show wall-plates and portions of the walls in section. Sketch to a larger scale (take the lines on the paper as being to 3″ gauge) portion of the top of one wall and wall-plate in section : showing rafter foot and showing the eavesgutter in section.
                                                              (14.)

*MAY,* 1905.

ELEMENTARY STAGE.

---

INSTRUCTIONS.

You must not attempt more than *seven* questions in all, and of these No. 1 must be one ; that is to say, you are allowed to take not more than six questions in addition to No. 1.

(71.)  *1.  Make a neat tracing in ink of the drawing given, with the writing and figures : the lines should be firm and solid and should finish exactly at the proper points.    (15.)

*Cast Iron Grating.*

Q. 1.

(72.).  2.  What are the essential qualities to be looked for in good sand, lime, and cement for the preparation of mortar ? In what proportions and in what manner should they be mixed ?    (12.)

(73.)  3.  What materials would you use, and in what proportions, for the concrete for (*a*) the foundations under heavy walls, and (*b*) a fire-proof floor in an upper story ?    (12.)

(74.)   4.   Draw to a scale of $\frac{1}{12}$ (1" to a foot) a section through the base of an 18" brick wall including three courses of the wall itself, with the usual footings and concrete. State how you arrive at the number of courses in the footings, and the width and thickness of the concrete.   (14.)

(75.)   5.   Draw to a scale of $\frac{1}{12}$ (1" to a foot) plans of two consecutive courses at the intersection at right angles of a 14" with an 18" brick wall built English bond.   (14.)

(76.)   6.   Draw full size a cross section of a 2" roll on 1" boarding for a lead flat showing the lead in position. State how far apart the rolls should be fixed, the quality of lead to be used and its thickness, how it should be dressed into place and the fall that should be given to the flat.   (12.)

(77.)   7.   Describe the nature and properties of only four of the following stones, and state for what purposes you would consider each suitable : Monks Park, Craigleith, Ancaster, Beer, Hopton Wood, Red Mansfield.   (14.)

(78.)   8.   Draw to a scale of $\frac{1}{48}$ ($\frac{1}{4}$" to a foot) a line diagram for a Kingpost roof truss of maximum span : figure the span and the scantlings of the various members.   Draw to a scale of $\frac{1}{8}$ (1$\frac{1}{2}$" to a foot) the junction of the principal rafter with the tie beam.   (14.)

(79.)   9.   On your squared paper draw neatly one-quarter full size the section of a 12" × 6" rolled steel joist and figure the thickness of the web. State the relation between the weight in lbs. per foot run and the sectional area in square inches.   (14.)

(80.)   10.   Draw to a scale of $\frac{1}{8}$ (1$\frac{1}{2}$" to a foot) a section through four courses of plain tiles laid on feather-edged boarding on 4" × 2" rafters, pitch of roof 45 degrees. Mark the dimensions for length, gauge and lap of the tiles, and describe how they are kept in position.   (12.)

(81.)   11.   Draw neatly on your squared paper a sketch of a jack plane : figure its real length and state for what purposes it is chiefly used. Show the metal portions on a separate sketch and describe how they are adjusted.   (14.)

(82.)   12.   Draw to scale of $\frac{1}{8}$ (1$\frac{1}{2}$" to a foot) a section through the jamb of an internal doorway in a 9" wall rendered on both sides, with framed jamb lining and the usual constructive details for a 2" door.   (15.)

*MAY*, 1906.

ELEMENTARY STAGE.

----

INSTRUCTIONS.

You must not attempt more than *seven* questions in all, and of these No. 1 must be one; that is to say, you are allowed to take not more than six questions in addition to No. 1.

(83.) *1. Make a neat tracing in ink of the drawing given, with the writing: the lines should be firm and solid and should finish accurately at the proper points. (15.)

(84.) 2. Describe fully what you know of blue lias lime, its origin, manufacture, preparation and the precautions to be taken in its use. (12.)

(85.) 3. What are the essential properties of a good brick? Distinguish between the following bricks, and state for what purpose they are chiefly used: Fletton, gault, red rubber, blue Staffordshire. (12.)

(86.) 4. Show by a sketch on your squared paper how a pole should be slung by a rope for lifting vertically. (12.)

(87.) 5. What is the size of a countess slate? Describe clearly and fully what is meant by "lap," and "gauge" in a slated roof, and illustrate your answer by sketches. (14.)

(88.) 6. Draw to a scale $\frac{1}{12}$ (1″ to a foot) the plans of two consecutive courses of a square three-and-a-half brick pier in Flemish bond: the joints may be shown by single lines. (14.)

(89.) 7. You have the choice of the following stones in building a mansion with stables attached: state in what parts you would use them, giving your reasons: Granite, Whinstone, Hard York, Craigleith, Portland whitbed, Box ground Hopton Wood, Derbyshire marble. (14.)

(90.) 8. Sketch full size on your squared paper a vertical section through the junction of two 3-inch round cast iron rain-water pipes, and describe the method of jointing. (14.)

# THEATRE OF MARCELLUS.

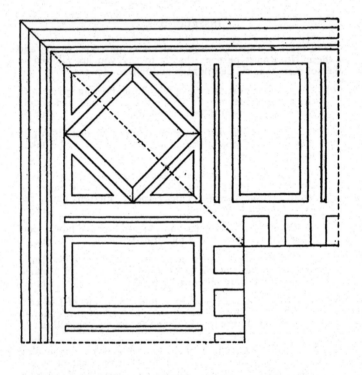

*Soffit of Cornice.*

(91.) 9. A York stone sill is described as "7" × 4½" rubbed,
weathered, and throated." Draw to a scale of ⅛ (1½" to a
foot) a cross section of this sill, and describe in their proper
order the operations of the mason in preparing it.      (14.)

(92.) 10. If lead weighs 710 lbs. per cubic foot what is the
thickness of 6 lb. sheet lead? What weight lead should be
used for flats, dormer cheeks, flashings, hips and valleys,
soil pipes?                                               (12.)

(93.) 11. A window opening 3 feet wide is spanned by a wooden
lintel 4" deep with 6" bearing at each end. Draw to a scale
of 1/12 (1" to a foot) an elevation of the opening and the lintel
with a segmental discharging arch over it in two half-brick
rings.                                                   (12.)

(94.) 12. A compound girder is composed of a 12" × 5" rolled
steel joist with a 9" × ½" steel plate top and bottom.
Sketch on your squared paper one quarter full size (3" to a
foot) a section through this girder: the rivets need not be
shown.                                                   (15.)

# INDEX.

| | PAGE |
|---|---|
| Abacus ... ... ... ... | 60 |
| Abutments ... ... ... | 59 |
| Advantages and Disadvantages in Slating ... ... | 386 |
| Angle Joints, crossing the Grain ... ... ... | 320 |
| Apex Stone ... ... ... | 90 |
| Appendix ... ... ... | 414 |
| Apron Pieces ... ... ... | 362 |
| Arcade ... ... ... ... | 61 |
| Arch Built with Orders ... | 73 |
| Arches ... ... ... ... | 58 |
| ,, Classification of ... | 62 |
| ,, Coloured Treatment of | 72 |
| ,, Discharging ... ... | 72 |
| ,, Drop ... ... ... | 67 |
| ,, Dutch ... ... ... | 62 |
| ,, Elliptical ... ... | 69 |
| ,, Equilateral ... ... | 69 |
| ,, Flat or Camber ... | 70 |
| ,, Florentine ... ... | 67 |
| ,, Four-centred ... ... | 67 |
| ,, Gauged ... ... | 65 |
| ,, Inverted ... ... | 62 |
| ,, Lancet... ... ... | 67 |
| ,, Pointed ... ... | 71 |
| ,, Rere ... ... ... | 71 |
| ,, Ring Courses of ... | 61 |
| ,, Rough Axed ... ... | 62 |
| ,, Rough Relieving ... | 62 |
| ,, Segmental ... ... | 66 |
| ,, Semi-circular ... ... | 67 |
| ,, Stilted ... ... ... | 67 |
| ,, Stone ... ... ... | 72 |
| ,, Three-centred... ... | 76 |
| ,, Trimmer ... ... | 65 |
| ,, Venetian ... ... | 67 |
| Architraves ... ... ... | 90 |
| Archivolt ... ... ... | 92 |
| Arris ... ... ... ... | 304 |
| Ashlar ... ... ... ... | 120 |
| Asphalte Paving ... ... | 84 |
| Astragal or Bead ... .. | 99 |
| Baluster ... ... ... | 92 |
| Base ... ... ... ... | 92 |
| Basement Floors ... ... | 204 |
| Bats ... ... ... ... | 19 |
| Batten and Button ... ... | 324 |
| Bead or Astragal ... ... | 99 |
| Beams, Built up ... ... | 183 |
| ,, Lengthening of ... | 175 |
| ,, Trussed Wood ... | 156 |
| Bearing Joints ... ... | 174 |
| ,, of Joists ... ... | 210 |
| ,, Surface ... ... | 158 |
| Bed Joints ... ... ... | 19 |
| ,, of Slates ... ... | 383 |
| ,, Surface ... ... | 92 |
| Beginners, Instructions for ... | 1 |
| Bending ... ... ... | 305 |
| Binders, Wrought Iron and Steel ... ... ... | 222 |
| Bird's Beak ... ... ... | 100 |
| Black Line Reproduction ... | 15 |
| Block-in Course ... ... | 119 |
| Blocking ... ... ... | 308 |
| Blocking Course ... ... | 92 |
| Block Joint ... ... ... | 381 |
| Blown Joint ... ... ... | 381 |
| Board Drawing ... ... | 2 |
| Board of Education, Syllabus | 414 |
| ,, ,, Examination Papers ... | 419 |
| Boards, Snow ... ... ... | 377 |
| Boasted Work ... ... ... | 109 |

PAGE

Bolts ... ... ... 172, 198
,, Anchor ... ... ... 127
,, Rag ... ... ... 127
Bond ... ... ... 29
,, Chimney, or Stretching.. 33
,, Definition of ... ... 29
,, Diagonal... ... ... 37
,, Double Flemish... ... 30
,, Dutch ... ... ... 33
,, English ... ... ... 30
,, English Cross ... 33
,, Facing ... ... ... 35
,, Heading ... ... 35
,, Herring Bone ... ... 37
,, Hoop Iron ... ... 38
,, in Footings ... ... 35
,, Raking ... ... 35
,, Single Flemish... ... 33
Bonders... ... ... ... 92
Boning ... ... ... ... 22
Boss ... ... ... ... 92
Bossing... ... ... ... 361
Bottom Rail ... ... ... 325
Box Girders ... ... ... 164
,, Gutters ... ... ... 377
Braces ... ... ... ... 241
Brads ... ... ... ... 194
Brandering ... ... ... 221
Breeze Bricks ... ... ... 78
,, Lintels ... ... ... 76
Bressummer, or Breastsummer 152
Bricklayer's Memoranda ... 397
,, Quantities ... 396
,, Tools ... ... 87
Bricklaying, Definition of ... 17
Bricknogged Partitions ... 234
Brick Partitions ... ... 228
,, Plan of ... ... ... 18
Bricks, Bedding of ... ... 21
,, Breeze... ... ... 78
,, Burning ... ... 17
,, Dimensions of ... 17
,, Moulding ... ... 17
,, Wetting of ... ... 21
,, Wood ... ... ... 78
Brickwork, Classification of ... 17
,, Definition of ... 17
,, Exercises ... ... 412
,, Stability of ... 23
Bridging Joist Joints ... ... 213
Bridle Joint ... ... ... 191

PAGE

Broach ... ... ... ... 92
Building Memoranda ... ... 398
,, Quantities ... 394
Building up ... ... ... 179
Built-up Girders ... ... 165
Burning in ... ... ... 379
Butt Hinges ... ... ... 351

Camber... ... ... ... 158
,, of Tie Rods ... ... 293
Cantilever ... ... ... 164
Capital .. ... ... ... 59
Carpenter's Memoranda ... 401
,, Quantities ... 402
,, Tools ... 349, 350
Carpentry, Joints in ... ... 173
Cased Frames ... ... ... 347
Casement Sashes and Frames . 342
Cast Iron ... ... ... 276
,, Cantilever ... ... 164
,, Girders ... ... 159
,, Nails ... ... 194
,, Shoes ... ... 270
,, Struts ... ... 272
Cast Lead ... ... ... 358
Cavetto ... ... ... ... 99
Ceiling Joist ... ... ... 219
Cement Fillets... ... ... 362
,, Joggles ... ... 123
,, Mortar ... ... 20
Centre Lines ... ... ... 5
Chamfered Panels ... ... 331
Chamfering ... ... ... 304
Characteristics of Partitions ... 223
Chase Mortices ... ... 186
Chimney Bond... ... ... 33
Chisel Draughted Margin ... 108
Circular Work ... ... ... 111
,, Sunk Work ... ... 111
Classification of Joints ... 174
Clear Span ... ... ... 158
Cleats ... ... ... ... 263
Closers ... ... ... ... 20
Coach Bolts ... ... ... 197
Coffer ... ... ... ... 92
Cogging .. ... ... ... 184
Coke Breeze Partitions ... 228
Collar Roofs ... ... ... 249
Colouring ... ... ... 10, 12

| | PAGE |
|---|---|
| Colours, Conventional ... | 10 |
| Column... ... ... ... | 93 |
| Combed Work... ... | 108 |
| Common Partitions ... ... | 233 |
| ,, Rafters ... ... | 263 |
| Compasses ... ... ... | 1 |
| Composite Roofs ... ... | 267 |
| ,, Trusses ... | 267, 275 |
| Composition Nails ... ... | 391 |
| Concrete Bricks ... ... | 78 |
| ,, Lintels ... ... | 76 |
| ,, Paving ... ... | 83 |
| Console... ... ... ... | 93 |
| Consoles ... ... ... | 134 |
| Construction of Trusses ... | 255 |
| ,, Scale ... ... | 5 |
| Copings... ... ... | 38, 93 |
| Copper Nails ... ... | 299, 391 |
| ,, Tacks ... ... ... | 387 |
| Corbel ... ... ... ... | 93 |
| ,, Table ... ... ... | 93 |
| Corbelling ... ... ... | 45 |
| Corbie Step Gable ... ... | 93 |
| Cornices, Brick ... | 47, 94 |
| Counter Cramp ... ... | 322 |
| Couple Close Roofs ... ... | 248 |
| ,, Roofs ... ... | 248 |
| Course, Label ... ... ... | 61 |
| ,, Projecting ... ... | 43 |
| ,, Ring ... ... ... | 61 |
| ,, String ... ... ... | 49 |
| Cooper Jointing ... ... | 308 |
| Cover Flashings ... ... | 361 |
| Coverings, Fixings for ... | 296 |
| ,, of Roofs ... | 358 |
| ,, ,, Iron Roofs ... | 296 |
| ,, Ridge ... ... | 365 |
| Cramps, Slate ... ... | 125, 127 |
| Crocket... ... ... ... | 94 |
| Cross Grooving ... ... | 302 |
| Crown ... ... ... ... | 61 |
| Cupola ... ... ... ... | 94 |
| Cyma Recta ... ... ... | 100 |
| ,, Reversa ... ... | 100 |
| | |
| | |
| Damp-Proof Course ... ... | 49 |
| ,, Stretching ... ... | 4 |
| Datum Lines ... ... ... | 5 |
| Designing Joints ... ... | 174 |

| | PAGE |
|---|---|
| Determining Resistance Wood Beams ... ... ... | 203 |
| Diagonal Bond ... ... | 37 |
| Diaper ... ... ... ... | 94 |
| Difference between Valleys and Hips ... ... ... | 367 |
| Dimensions of Roof Members | 264 |
| Discharging Arches ... ... | 72 |
| Dividers ... ... ... | 1 |
| Dogs ... ... ... ... | 198 |
| Dome ... ... ... ... | 94 |
| Door Frames, Solid ... ... | 332 |
| ,, Head Joint ... ... | 240 |
| ,, Studs ... ... ... | 240 |
| Doors, Classification of ... | 327 |
| ,, Five-Panel ... ... | 331 |
| ,, Four-Panel ... ... | 331 |
| ,, Framed and Braced ... | 327 |
| ,, Framed and Panelled | 331 |
| ,, Ledged ... ... | 327 |
| ,, Ledged and Braced ... | 327 |
| ,, Sash ... ... ... | 331 |
| Dormer... ... ... ... | 94 |
| Dots, Lead ... ... ... | 380 |
| ,, Soldered .. ... | 379 |
| Double Flemish Bond... ... | 30 |
| ,, Floors ... | 200, 221 |
| ,, ,, Sections ... | 222 |
| ,, Framed Floors ... | 225 |
| ,, Hung Sashes ... ... | 335 |
| ,, Nutted Screws... ... | 197 |
| ,, Tenons... ... ... | 325 |
| Doubles ... ... ... | 382 |
| Doubling Eaves Course ... | 387 |
| Dovetailed Keys ... ... | 323 |
| ,, Tenon ... ... | 188 |
| Dovetailing ... ... ... | 186 |
| Dovetail Halving ... ... | 184 |
| Dowel, Lead ... ... ... | 380 |
| ,, Pins ... ... ... | 193 |
| Dowelled Joint ... ... | 318 |
| Dowels... ... ... ... | 123 |
| Drafted Margins ... ... | 108 |
| Dragged Work ... ... | 108 |
| Drawing Boards ... ... | 2 |
| ,, Instruments... ... | 1 |
| ,, Paper ... ... | 3 |
| ,, Pencils ... ... | 2 |
| ,, Perspective ... ... | 14 |
| ,, Pins ... ... ... | 2 |
| Dressings ... ... ... | 94 |

| | PAGE |
|---|---|
| Drips ... ... ... ... | 364 |
| Drip Stone ... ... ... | 94 |
| Droved Work ... ... ... | 109 |
| Dutch Arches ... ... ... | 62 |
| „ Bond ... ... ... | 33 |

| | PAGE |
|---|---|
| Eaves Gutters ... ... ... | 259 |
| Effective Depth ... ... | 158 |
| „ Load... ... ... | 158 |
| „ Span... ... ... | 158 |
| Elevation ... ... ... | 14 |
| „ Sectional ... ... | 14 |
| Elliptical Arch ... ... | 69 |
| English Bond ... ... ... | 30 |
| „ Cross Bond ... ... | 33 |
| Entablature ... ... ... | 95 |
| Entasis ... ... ... ... | 95 |
| Erasure of Lines ... ... | 9 |
| Espagnolette Bolt ... ... | 356 |
| Examination Papers of the Board of Education ... | 419 |
| Excavator's Quantities ... | 394 |
| „ Tools ... ... | 85 |
| „ Work ... ... | 394 |
| Exercises ... ... ... | 412 |
| Expansion Joints ... ... | 294 |
| Expansion of Metals ... ... | 359 |
| External Mitres ... ... | 112 |
| Extrados or Back ... ... | 59 |

| | PAGE |
|---|---|
| Fanlights ... ... ... | 343 |
| Fascia ... ... ... ... | 258 |
| Fastenings and Locks... | 192, 355 |
| Feathers or Tongues ... | 158 |
| Feather-edged Copings ... | 93 |
| Feet of King or Queen Bolts... | 275 |
| „ Principal Rafters | 257, 269 |
| „ Struts ... ... | 275 |
| Fillet or Band ... ... | 99 |
| „ Cement ... ... | 362 |
| „ Tilting ... ... | 362, 388 |
| Finial ... ... ... ... | 95 |
| Fishing ... ... ... | 175 |
| Five-Panel Doors ... ... | 331 |
| Fixing for Plaster ... ... | 242 |
| „ Joinery to Brickwork... | 77 |
| „ Partitions ... ... | 241 |

| | PAGE |
|---|---|
| Fixings for Flashings ... ... | 361 |
| Flanges... ... ... ... | 164 |
| Flashings ... ... ... | 372 |
| „ Cover ... ... | 361 |
| „ for Chimney Stack... | 372 |
| „ Raking ... ... | 369 |
| „ Stepped ... ... | 369 |
| Flat Panels ... ... ... | 313 |
| „ Roofs ... ... ... | 248 |
| Flemish Bond ... ... ... | 30 |
| Flint Walls ... ... ... | 115 |
| Flitch Girders ... ... ... | 154 |
| Flooring Joints ... ... | 215 |
| Floors ... ... ... ... | 200 |
| „ Double ... ... ... | 200 |
| „ Exercises ... ... | 412 |
| „ Framed ... ... | 200, 225 |
| „ Material... ... ... | 221 |
| „ Single ... ... ... | 200 |
| „ Ventilation of ... ... | 226 |
| Florentine Arch ... ... | 67 |
| Flush Panels ... ... ... | 314 |
| Footings ... ... ... | 45, 95 |
| „ Brick ... ... ... | 45 |
| „ Stone ... ... ... | 114 |
| Forked Headings .. ... | 218 |
| Forms of Wooden Roofs ... | 245 |
| Four-Panel Doors ... ... | 331 |
| Foxtail Wedging ... | 191, 325 |
| Framed and Braced Doors ... | 327 |
| „ „ Panelled Doors... | 331 |
| Frames, Glueing-up ... ... | 311 |
| „ Solid ... ... | 343 |
| Framings ... ... ... | 311 |
| Frieze ... ... ... ... | 95 |
| Frog of Brick ... ... ... | 18 |
| Frost, Effect of, on Building... | 21 |
| Furrowed Work ... ... | 114 |

| | PAGE |
|---|---|
| Gable ... ... ... ... | 95 |
| Gablet ... ... ... ... | 95 |
| Galleting ... ... ... | 96 |
| Garden Wall Bond ... ... | 35 |
| Gargoyle ... ... ... | 96 |
| Gauge for fixing Slates near Middle ... ... | 384 |
| „ for Nailing Slates near Head ... ... | 384 |

|  |  | PAGE |
|---|---|---|
| Gauged Arches | ... ... | 65 |
| Geometrical Explanations | ... | 13 |
| German Truss ... | ... | 269 |
| Gibs and Cotters | ... | 260 |
| Girders ... ... | ... | 152 |
| ,, Box ... | ... | 164 |
| ,, Built up | .. | 165 |
| ,, Cast Iron | ... | 159 |
| ,, Exercises | ... | 412 |
| ,, Flitch ... | ... | 154 |
| ,, Plate ... | ... | 163 |
| ,, Steel ... | ... | 162 |
| ,, Warren and Lattice | ... | 170 |
| ,, Wood | ... | 151 |
| ,, Wrought Iron | ... | 152 |
| Glazier's Quantities | ... | 409 |
| ,, Sprigs... | ... | 196 |
| Grooved and Tongued Joints... | | 320 |
| Grounds ... | ... | 333 |
| Grout ... ... | ... | 96 |
| Guards, Snow ... | ... | 379 |
| Gutter Bearers | .. | 259 |
| Gutters ... ... | ... | 376 |
| ,, Eaves ... | ... | 259 |
| ,, Parallel Parapet | .. | 376 |
| ,, Parapet | ... | 376 |
| ,, Secret ... | ... | 372 |
| ,, Side . | ... | 372 |
| ,, Tapering Parapet | .. | 376 |
| ,, ,, Valley | ... | 377 |
| | | |
| Half-sawing ... | ... | 107, 397 |
| Hand Sawing ... | ... | 150 |
| Halving ... | ... | 184 |
| Hammer Dressing | ... | 107 |
| Hatching Sections | ... | 9 |
| Haunch... ... | ... | 61 |
| Haunchion ... | ... | 324 |
| Headers ... | ... | 19, 96 |
| Heading Bond... | ... | 35 |
| ,, Forked | ... | 218 |
| ,, Joints | ... | 217 |
| ,, Secret | ... | 217 |
| ,, Splayed | ... | 217 |
| Heads ... ... | . . | 240 |
| ,, of Principal Rafter | 261, 275 |
| Herring-Bone Strutting | ... | 214 |
| Hinges ... ... | ... | 351 |
| Hipped Roofs ... | ... | 292 |

|  |  | PAGE |
|---|---|---|
| Hips and Valleys | ... | 367 |
| Hoisting Apparatus | ... | 150 |
| Holdfasts ... | ... | 197 |
| Hollow Brick Partition | ... | 228 |
| ,, Rolls ... | ... | 364 |
| Hoop-Iron Bond | ... | 38 |
| Housed Joints ... | ... | 322 |
| Housing ... | 186, 302 |
| ,, Dovetailed | ... | 322 |
| | | |
| Impost .. | ... ... | 60 |
| Indian Ink ... | ... | 8 |
| Inking in ... | ... | 8 |
| Instruments, Drawing | ... | 1 |
| Internal Mitres | ... | 112 |
| Intertie ... ... | .. | 240 |
| Intrados or Soffit | ... | 59 |
| Inverted Arches | ... | 62 |
| Inband ... ... | ... | 99 |
| Iron, Bar ... | ... | 324 |
| ,, Cast .. | ... | 276 |
| ,, Flitch Girder | ... | 154 |
| ,, Roofs ... | ... | 276 |
| ,, ,, Details of | ... | 279 |
| ,, Roof Truss for 20' Span . | 278 |
| ,, ,, ,, Coverings of | 246 |
| ,, ,, ,, Details for | |
| 25' Span ... | ... | 282 |
| ,, Roof Truss, Details for | |
| 30' Span ... | 284–290 |
| Iron Roof Truss, Details for | |
| 35' Span ... | ... | 286 |
| ,, Roof Truss, Details for | |
| 40' Span ... | 287–289 |
| ,, Roof Truss, Details for | |
| 50' Span ... | ... | 280 |
| ,, Roof Trusses, Scantlings | |
| for ... ... | ... | 264 |
| Ironmonger's Quantities | ... | 404 |
| Isometric Projection ... | ... | 14 |
| | | |
| Jambs ... ..: | ... | 40, 96 |
| Joggles ... ... | ... | 122 |
| ,, Cement | ... | 123 |
| ,, Stub or Stump Tenon... | 188 |
| Joiner's Quantities | ... | 403 |
| ,, Tools ... | .. | 349, 350 |

|  | PAGE |
|---|---|
| Joinery ... ... ... ... | 299 |
| ,, Design of ... ... | 299 |
| ,, Exercises ... .. | 413 |
| Joints, Angle ... ... ... | 320 |
| ,, across the Flow ... | 362 |
| ,, across the Grain ... | 320 |
| ,, Bearing ... ... ... | 174 |
| ,, Block ... ... ... | 381 |
| ,, Classification ... 174, | 315 |
| ,, Cross .. ... ... | 322 |
| ,, Designing ... ... | 293 |
| ,, for Door Head ... ... | 241 |
| ,, for Lengthening Beams | 175 |
| ,, for Struts and Beams ... | 191 |
| ,, for Struts and Ties ... | 191 |
| ,, for Studs ... ... | 234 |
| ,, Forked Heading ... | 218 |
| ,, Framed ... ... ... | 324 |
| ,, Grooved and Tongued... | 215 |
| ,, Heading · ... ... | 217 |
| ,, Hook ... ... ... | 343 |
| ,, in Carpentry ... ... | 192 |
| ,, Joinery ... ... ... | 314 |
| ,, Lap ... ... ... | 362 |
| ,, Lead ... ... ... | 362 |
| ,, Masonry ... . | 121 |
| ,, Oblique and Shouldered | 174 |
| ,, of Flooring ... .. | 215 |
| ,, Parallel with Flow ... | 362 |
| ,, Plain or Square ... | 215 |
| ,, Ploughed and Tongued | 215 |
| ,, Preventing Warping ... | 323 |
| ,, Rebated 128, 217, | 320 |
| ,, ,, and Filleted ... | 217 |
| ,, Saddled or Water ... | 128 |
| ,, Secret ... ... ... | 217 |
| ,, Splayed Heading ... | 217 |
| ,, Square Heading ... | 217 |
| ,, Straight... .. ... | 218 |
| ,, Tenoned and Housed ... | 188 |
| ,, Weathered ... ... | 127 |
| ,, Wiped ... ... ... | 380 |
| ,, with the Grain... ... | 321 |
| ,, Wood ... ... ... | 78 |
| Junction of Iron Roof Members | 295 |
| ,, of Walls ... ... | 40 |
| Key ... ... ... ... | 59 |
| ,, for Plastering ... ... | 230 |
| Keyed Joint ... ... 80, | 318 |
| ,, Mitre ... ... ... | 320 |
| Keys, Hard Wood ... ... | 198 |

|  | PAGE |
|---|---|
| King Closers ... ... ... | 20 |
| ,, Post and Struts Joint ... | 261 |
| ,, ,, Tie-beam Joint | 261 |
| ,, or Queen Bolts, Feet of. | 275 |
| ,, Post Roofs ... ... | 264 |
| ,, ,, Joint ... ... | 261 |
| Kneeler... ... ... ... | 98 |
| Label .. ... ... ... | 98 |
| Label Course ... ... | 61 |
| Labours... ... ... ... | 107 |
| Lacing Courses ... 61, | 98 |
| Laminated Beams ... ... | 154 |
| Lap ... ... ... ... | 19 |
| ,, Joints ... ... ... | 362 |
| Lapping ... ... ... | 175 |
| Laps or Passings ... ... | 361 |
| Lath Nails ... ... ... | 196 |
| Lathing ... ... ... | 219 |
| Lattice Girders ... ... | 170 |
| Laying Sheet Lead ... ... | 358 |
| Lead Cast ... ... ... | 358 |
| ,, Coverings ... ... | 361 |
| ,, ,, Dowel or Dot ... | 380 |
| ,, ,, Milled ... | 358 |
| ,, ,, Plugs ... ... | 125 |
| ,, ,, Sheet Laying ... | 358 |
| ,, ,, Tacks ... ... | 361 |
| Lean-to Roofs ... ... ... | 248 |
| Ledged and Braced Doors ... | 327 |
| Lengthening of Beams ... | 175 |
| ,, Joints ... | 174 |
| Lettering ... ... ... | 12 |
| Levelling ... ... ... | 22 |
| Lights, Hospital ... ... | 340 |
| Lime Mortar ... ... . | 20 |
| Lines, Centre and Datum ... | 5 |
| ,, Dimension ... ... | 12 |
| Lintel, Joggled ... ... | 76 |
| Lintels, Arched, Breeze or Concrete ... ... | 76 |
| ,, Iron ... ... | 77 |
| ,, Wood... ... ... | 77 |
| Listel ... ... ... ... | 99 |
| Load, Effective ... ... | 158 |
| Loads upon Floors ... ... | 201 |
| Lock Rail ... ... ... | 325 |
| Long and Short Work ... | 112 |

| | PAGE |
|---|---|
| Mack Plaster Slabs ... ... | 229 |
| Manifolding of Writing ... | 16 |
| Margin, Chisel Draughted ... | 108 |
| Masonry Definition ... ... | 90 |
| ,, Exercises ... ... | 412 |
| Mason's Quantities ... ... | 397 |
| ,, Tools ... ... ... | 147 |
| Matched Joint ... ... ... | 318 |
| Metope ... ... ... ... | 98 |
| Milled Lead ... ... ... | 358 |
| Mitres, External ... ... | 112 |
| ,, Internal ... ... | 112 |
| Mitre and Tongue ... ... | 321 |
| ,, Butt ... ... | 321 |
| ,, Rebate ... ... | 321 |
| Modillion ... ... ... | 98 |
| Mortar ... ... ... ... | 20 |
| Mortice and Tenon ... ... | 186 |
| ,, Chase ... ... | 186 |
| Mosaic Paving ... ... ... | 84 |
| Moulded Work ... ... | 110 |
| Moulding ... ... 101, | 304 |
| Mouldings, Base ... ... | 99 |
| ,, Classic ... ... | 98 |
| ,, Connecting ... | 99 |
| ,, Crowning ... ... | 101 |
| ,, Gothic ... ... | 98 |
| ,, Joinery ... ... | 314 |
| ,, Supporting ... | 99 |
| Muntin ... ... ... ... | 330 |
| | |
| N Truss ... ... ... | 275 |
| Nails ... ... ... 193. | 390 |
| ,, Composition ... ... | 391 |
| ,, Slating ... ... | 196 |
| Needle Points ... ... | 194 |
| Niches ... ... ... ... | 144 |
| Nogging Pieces ... ... | 235 |
| Nosings ... ... ... ... | 365 |
| Notching ... ... ... | 184 |
| | |
| Objection to Composite Trusses | 221 |
| Open Slating ... ... | 386 |
| Order ... ... ... ... | 101 |
| Ordinary Mortice and Tenon 186, | 324 |
| Oriel ... ... ... ... | 103 |
| Out Band ... ... ... | 96 |
| Outline of Arches ... ... | 67 |
| Ovolo Moulding ... ... | 100 |

| | PAGE |
|---|---|
| Painter's Quantities ... ... | 409 |
| Panelling ... ... ... | 313 |
| ,, and Moulding ... | 313 |
| Panels, Chamfered ... ... | 331 |
| ,, Flat ... ... ... | 313 |
| ,, Flush ... ... | 314 |
| Paper, Drawing ... ... | 3 |
| Paperhanger's Quantities ... | 410 |
| Parallel Parapet Gutters ... | 376 |
| ,, Valley or Box Gutters . | 377 |
| Parapet ... ... ... 103, | 376 |
| ,, Gutters ... ... | 376 |
| Partitions ... ... ... | 227 |
| ,, Brick ... ... | 228 |
| ,, Bricknogged ... | 234 |
| ,, Common ... ... | 233 |
| ,, Dovetail Corrugated Steel ... | 231 |
| ,, Exercises ... ... | 413 |
| ,, Fixing of ... ... | 241 |
| ,, Hollow Brick ... | 228 |
| ,, Kulm Slabs ... ... | 231 |
| ,, Mack Plaster Slabs | 229 |
| ,, Timber ... ... | 233 |
| ,, Weight of ... ... | 241 |
| Passings ... ... ... | 362 |
| Patera ... ... ... ... | 103 |
| Paving ... ... ... ... | 81 |
| Pebbles ... ... ... ... | 123 |
| Pedestal ... ... ... | 103 |
| Pediment .. ... ... | 103 |
| Pencils, Drawing ... ... | 2 |
| ,, Sharpening ... ... | 3 |
| Pendant ... ... ... | 103 |
| Pens, Drawing ... ... ... | 2 |
| Perpends ... ... ... | 19 |
| Perspective ... ... ... | 14 |
| Piers ... ... ... 38, | 59 |
| Pilaster ... ... ... ... | 103 |
| Pillar ... ... ... ... | 103 |
| Pinnacle .. .. ... | 103 |
| Pins, Drawing ... ... ... | 2 |
| ,, Dowel ... ... | 193 |
| ,, Wood and Iron ... | 193 |
| Pipe Hooks ... ... ... | 196 |
| ,, Nails ... .. ... | 196 |
| Pitch of Roof ... ... | 245 |
| Pitches of Iron Roofs ... ... | 293 |
| Plain Work ... ... ... | 108 |
| Plain Joint-laid Folding ... | 215 |
| Plan ... ... ... ... | 13 |

| | PAGE |
|---|---|
| Planing... ... ... ... | 300 |
| Plasterer's Quantities ... ... | 408 |
| Plastering, Fixing for ... ... | 242 |
| ,, Key for ... ... | 230 |
| Plate Girders ... ... ... | 164 |
| Plates, Wall ... ... ... | 259 |
| Plinth ... ... ... 49, | 103 |
| Plough Grooving ... ... | 302 |
| Ploughed and Tongued Joint... | 215 |
| Plugs, Lead ... ... ... | 125 |
| ,, Wood ... ... ... | 78 |
| Plumber's Quantities ... ... | 406 |
| Plumbing ... ... ... | 358 |
| ,, Exercises ... ... | 413 |
| Plus and Minus Threads ... | 293 |
| Pockets in Walls ... ... | 226 |
| Pointed Arch with Rere-arch . | 71 |
| ,, Work ... ... ... | 110 |
| Pole Plates ... ... ... | 259 |
| Polishing ... ... ... | 111 |
| Porch ... ... ... ... | 103 |
| Prevention of Damp ... ... | 55 |
| ,, of Sound ... ... | 218 |
| Principal Rafters ... ... | 263 |
| ,, Feet of ... | 294 |
| ,, Heads of ... | 295 |
| Purlins ... ... ... ... | 261 |
| ,, Roofs ... ... ... | 250 |
| | |
| Quantities, Bricklayer's ... | 395 |
| ,, Carpenter's ... | 402 |
| ,, Excavator's ... | 394 |
| ,, Gas Fitter's, Electrician's... ... | 406 |
| ,, Glazier's ... ... | 409 |
| ,, Ironmonger's ... | 404 |
| ,, Joiner's ... ... | 403 |
| ,, Mason's ... ... | 397 |
| ,, Notes on ... ... | 411 |
| ,, Painter's ... ... | 409 |
| ,, Paperhanger's ... | 410 |
| ,, Plasterer's ... | 407 |
| ,, Plumber's... ... | 406 |
| ,, Smith's and Founder's ... | 405 |
| ,, Tiler's and Slater's | 399 |
| Queen Closers ... ... ... | 19 |
| Queen-post Roof Truss ... | 263 |

| | PAGE |
|---|---|
| Quoins ... ... ... 19, | 103 |
| ,, Squint ... ... ... | 43 |
| | |
| Rafters, Common ... ... | 263 |
| Racking ... ... ... | 23 |
| Rafters, Principal junctions of | 263 |
| Rag Bolts ... ... ... | 127 |
| Raglets ... ... ... ... | 361 |
| Raking Bond ... ... ... | 35 |
| Random Rubble ... ... | 115 |
| ,, Built in Courses | 117 |
| ,, Set Dry ... | 115 |
| ,, Uncoursed Set in Mortar ... | 117 |
| Rebate ... ... ... ... | 302 |
| ,, and Bead ... ... | 320 |
| ,, Butt and Staff Bead ... | 320 |
| Rebated and Filleted Joint ... | 217 |
| ,, Joints ... ... | 128 |
| Rebating ... ... ... | 300 |
| Regular Coursed Rubble ... | 119 |
| Regulations for Partitions ... | 242 |
| Rendering ... ... ... | 55 |
| Reproduction Work ... 9, | 15 |
| Rere-arch ... ... ... | 71 |
| Responds ... ... 61, | 104 |
| Returned Mitred and Stopped Moulding ... ... ... | 112 |
| Reveals... ... ... ... | 42 |
| Revetment Walls ... ... | 26 |
| Ridge Boards ... ... ... | 261 |
| ,, Coverings ... ... | 365 |
| Ring Courses of Arches ... | 61 |
| Ripper ... ... ... ... | 392 |
| Rise ... ... ... ... | 61 |
| Riveting and Bolting ... ... | 291 |
| Rivets ... ... ... ... | 171 |
| Rolls ... ... ... ... | 364 |
| ,, Hollow ... ... ... | 364 |
| Roof Coverings ... ... | 358 |
| Roofs ... 144, 245, 267, | 276 |
| ,, Composite ... ... | 267 |
| ,, Exercises ... ... | 413 |
| ,, Iron ... ... ... | 276 |
| ,, Steel ... ... ... | 276 |
| ,, Stone ... ... ... | 144 |
| ,, Wood ... ... ... | 245 |
| ,, ,, Classification of ... | 246 |
| Rough Axed Arches ... .. | 62 |
| ,, Relieving or Discharging Arches ... ... | 62 |

|  | PAGE |
|---|---|
| Rounded Corner | 322 |
| Rubbed Work | 11( |
| Rubble | 116 |
| Ruling Pen | 1 |
| Saddle Back Board | 331 |
| „ Stone | 90 |
| Saddled or Water Joint | 128 |
| Safe Working Load | 202 |
| Sash Doors | 331 |
| „ Fixing of | 347 |
| Sashes, Casements and Frames | 333, 342 |
| „ Description of | 342 |
| „ Double Hung, and Cased Frames | 335 |
| „ Hung on Centres | 340 |
| „ Opening Inwards | 343 |
| „ „ Outwards | 343 |
| „ Sliding | 344 |
| Sawing | 300 |
| Scabbling or Scappling | 107 |
| Scales | 5 |
| „ Diagonal | 7 |
| „ Vermer | 7 |
| Scantling, for Roofs | 264 |
| „ Calculation of | 201 |
| Scarfing | 178 |
| Scarf to Resist Compression | 182 |
| Science and Art Syllabus, &c., See Board of Education. | |
| Scontions | 104 |
| Scotia | 100 |
| Screws | 197 |
| „ Slot and Batten | 323 |
| Scribing | 304 |
| Seams or Welts | 365 |
| Secret Gutters | 372 |
| „ Headings | 217 |
| „ Joints | 217 |
| „ Tacks | 379 |
| Sectional Plans or Elevations | 14 |
| Sections | 14 |
| „ of Iron Girders | 159 |
| „ Hatching | 9 |
| „ Iron, to resist Tension | 292 |
| „ „ „ Compression | 292 |
| „ under Transverse Stress | 293 |
| Self Faced | 107 |
| Set Squares | 1 |

|  | PAGE |
|---|---|
| Shivers | 104 |
| Shooting | 300 |
| Short Work, Long and | 112 |
| Shouldered or Tusk Tenon | 188 |
| Shutters | 347 |
| Side Gutters | 372 |
| Sills | 104, 240 |
| Single Flemish Bond | 33 |
| Single Floors | 200 |
| Skewbacks | 59 |
| Skew Corbel | 93 |
| Slate Hips and Ridges | 388 |
| Slater's Memoranda | 399 |
| „ Tools | 392 |
| Slates, Dimensions for | 382 |
| Slating | 399 |
| „ Exercises | 413 |
| „ Open | 386 |
| Sleeper | 204 |
| Smith and Founder's Quantities | 405 |
| Snecked Rubble | 117 |
| Snow Boards | 377 |
| „ Guards | 379 |
| Soakers | 372 |
| Sockets | 199 |
| Soffit or Intrados | 347 |
| Soldered Dots | 379 |
| Solid Door Frame | 332 |
| „ Strutting | 214 |
| „ Window Frames | 218 |
| Sound, Prevention of | 343 |
| Spalls | 104 |
| Span | 61 |
| Spandril | 61 |
| Specification for Constructional Cast Iron | 159 |
| Specification of Slates | 391 |
| Spikes | 194 |
| Spire | 104 |
| Splayed Heading Joints | 217 |
| Spring Bows | 2 |
| Springers | 59 |
| Springing Points | 61 |
| Sprocket Pieces | 263 |
| Square Heading Joints | 217 |
| „ Joint Laid Folding | 215 |
| „ Rubble, Built in Courses | 119 |
| „ „ Uncoursed | 119 |
| Squares, Tee | 2 |
| Squinch Arches | 104 |
| Stability of Brickwork | 23 |

| | PAGE |
|---|---|
| Stanchions Cast Iron ... ... | 161 |
|   ,,    Steel ... ... | 171 |
| Steel Girders ... ... ... | 162 |
|   ,,  Roofs ... ... | 276 |
| Stepped Flashings ... ... | 369 |
| Stilted Arch ... ... ... | 67 |
| Stone Arches ... ... ... | 139 |
|   ,,  or Concrete Slabs ... | 82 |
|   ,,  Lintels ... ... 77, | 128 |
|   ,,  Slab Coverings ... ... | 388 |
|   ,,  Walling ... ... ... | 398 |
| Stones, Through ... ... | 106 |
| Straight Joints ... ... ... | 218 |
| Strains and Stress 173, 183, | 276 |
| Straps ... ... ... ... | 199 |
| Stresses on Trusses ... ... | 276 |
| Stretchers ... ... ... | 19 |
| Stretching or Chimney Bond ... | 33 |
|   ,,   Damp ... ... | 4 |
| String Course ... ... 49, | 106 |
| Strut and Beam Joints ... | 191 |
| Struts, Feet of ... ... ... | 275 |
|   ,,  and King-post Joint ... | 261 |
|   ,,  and Principal Rafters, | |
|       Junction of ... | 263 |
| Strutting, Herring-bone ... | 214 |
|   ,,   Solid ... ... | 214 |
| Stub or Stump Tenon ... ... | 188 |
| Studs ... ... ... ... | 240 |
| Sunk Work ... ... ... | 110 |
| Surface, Bed ... ... ... | 92 |
|   ,,  Bearing ... ... | 158 |
| Syllabus, Board of Education's | 414 |
| | |
| Tabling ... ... ... ... | 122 |
|   ,,  Joints ... ... ... | 122 |
| Tacks ... ... ... ... | 196 |
|   ,,  Copper ... ... ... | 387 |
| Tailing Iron ... ... ... | 106 |
| Tails of Slates ... ... ... | 388 |
| Taft, Joint ... ... ... | 381 |
| Tapering Parapet Gutters ... | 376 |
|   ,,  Valley Gutters ... | 369 |
| Tar Paving ... ... ... | 83 |
| Technical Terms ... 19, 59, 90 | |
| Tee Squares ... ... ... | 2 |
| Tenia ... ... ... ... | 106 |
| Templates ... ... ... | 106 |
| Tenon and Mortice ... ... | 186 |

| | PAGE |
|---|---|
| Tenoned and Housed Joints ... | 188 |
| Threads, Plus and Minus ... | 293 |
| Throating ... ... ... | 106 |
| Through Stones ... ... | 106 |
| Tie-beams ... ... ... | 259 |
|   ,,  and Principal Rafters ... | 261 |
| Tie-rods, Camber of ... ... | 293 |
| Tightening-up Joints ... ... | 293 |
| Tile Paving ... ... ... | 84 |
| Tiler's and Slater's Quantities | 399 |
| Tilting Fillets ... ... 362, | 388 |
| Timber Strut, Principal and | |
|     Queen Bolt ... ... | 275 |
| Tingles ... ... ... ... | 361 |
| Tongued, Ploughed and ... | 318 |
| Tongues or Feathers ... ... | 304 |
| Tooled Work ... ... ... | 109 |
| Tools ... ... 85, 147, | 149 |
|   ,,  Brick Cutting ... ... | 88 |
|   ,,   ,,  Laying and Pointing | 89 |
|   ,,  Carpenter's and Joiner's | 348 |
|   ,,  Excavator's ... ... | 85 |
|   ,,  Mason's ... ... 147-150 | |
| Toothing ... ... ... | 23 |
| Torus ... ... ... ... | 100 |
| Tracing ... ... ... ... | 15 |
| Transverse Stress ... ... | 183 |
| Trenails ... ... ... ... | 193 |
| Trimmer Arches ... ... | 65 |
| Trimming ... ... ... | 213 |
| Trussed Partitions ... 236. | 241 |
|   ,,  Wood Beams ... | 156 |
| Trusses, Construction of ... | 255 |
|   ,,  Mansard ... ... | 264 |
|   ,,  Open Steel ... ... | 281 |
|   ,,  Stresses on ... | 276 |
| Trussing ... ... ... | 234 |
| Turning Piece ... ... ... | 77 |
| Tusk Tenon ... ... ... | 188 |
| Tympanum ... ... ... | 106 |
| Typical Trusses ... 268, | 281 |
| | |
| Upper Floors ... ... ... | 204 |
| | |
| Valley Coverings ... ... | 369 |
|   ,,  Gutters ... ... ... | 377 |
|   ,,   ,,  Parallel ... | 377 |

| | PAGE |
|---|---|
| Valleys, Difference between Hips and ... ... ... | 367 |
| Vaulting ... ... ... | 107 |
| Veneering ... ... ... | 309 |
| Venetian Pointed Arch ... | 67 |
| Ventilation of Floors ... ... | 226 |
| Vermiculated Work ... ... | 112 |
| Vernier Scale ... ... ... | 7 |
| Voussoirs ... ... ... | 59 |
| | |
| Wall Hooks ... ... ... | 196 |
| ,, Plates ... ... ... | 259 |
| Walling Stone ... ... ... | 117 |
| Walls, Brick ... ... ... | 22 |
| ,, Classification ... ... | 25 |
| ,, Hollow . .. .. | 57 |
| ,, Pockets in ... ... | 226 |
| Warren Girders ... ... | 170 |
| Water-bath Reproduction ... | 15 |
| Water Joint ... ... ... | 128 |
| Weathering ... ... ... | 107 |
| Wedges ... ... ... ... | 198 |
| Wedging, Foxtail ... ... | 191 |
| Weights of Partitions and Floors ... ... ... | 241 |
| Welt ... ... ... ... | 365 |
| Wide Surfaces prevented from Warping .. ... ... | 323 |
| Wind Filling ... ... ... | 263 |

| | PAGE |
|---|---|
| Window and Door Jambs | 40, 96 |
| ,, Finishings ... | .. 347 |
| Windows, Three Light | ... 338 |
| Wiped Joint ... ... | ... 380 |
| Wire Nails ... ... | ... 196 |
| Wood Bricks ... ... | ... 78 |
| ,, Girders ... ... | ... 154 |
| ,, Joints ... ... | ... 78 |
| ,, Keys ... ... | ... 198 |
| ,, Lintels ... ... | .. 77 |
| ,, Pillars ... ... | ... 156 |
| ,, Plugs ... .. | ... 78 |
| ,, Roofs ... ... | ... 245 |
| Work, Axed ... . | 65, 109 |
| ,, Boasted or Droved | ... 109 |
| ,, Circular sunk ... | ... 111 |
| ,, Combed or Dragged | ... 108 |
| ,, Furrowed .. | ... 114 |
| ,, Moulded ... | ... 110 |
| ,, Plain ... ... | ... 108 |
| ,, Rubbed ... | ... 111 |
| ,, Tooled ... ... | ... 109 |
| Wrought Iron Nails ... | ... 194 |
| ,, ,, Girders | ... 152 |
| | |
| Zax ... ... ... | ... 392 |
| Zinc Nails ... .. | .. 391 |
| Zoccolo ... ... | ... 107 |

Bradbury, Agnew & Co., Ld., Printers, London and Tonbridge.